D1737908

AAA AND NETWORK SECURITY FOR MOBILE ACCESS

AAA AND NETWORK SECURITY FOR MOBILE ACCESS

RADIUS, DIAMETER, EAP, PKI AND IP MOBILITY

Madjid Nakhjiri
Motorola Labs, USA

and

Mahsa Nakhjiri
Motorola Personal Devices, USA

John Wiley & Sons, Ltd

Copyright © 2005   John Wiley & Sons Ltd, The Atrium, Southern Gate, Chichester,
West Sussex PO19 8SQ, England

Telephone   (+44) 1243 779777

Email (for orders and customer service enquiries): cs-books@wiley.co.uk
Visit our Home Page on www.wiley.com

Reprinted December 2006

All Rights Reserved. No part of this publication may be reproduced, stored in a retrieval system or transmitted in any form or by any means, electronic, mechanical, photocopying, recording, scanning or otherwise, except under the terms of the Copyright, Designs and Patents Act 1988 or under the terms of a licence issued by the Copyright Licensing Agency Ltd, 90 Tottenham Court Road, London W1T 4LP, UK, without the permission in writing of the Publisher. Requests to the Publisher should be addressed to the Permissions Department, John Wiley & Sons Ltd, The Atrium, Southern Gate, Chichester, West Sussex PO19 8SQ, England, or emailed to permreq@wiley.co.uk, or faxed to (+44) 1243 770571.

This publication is designed to provide accurate and authoritative information in regard to the subject matter covered. It is sold on the understanding that the Publisher is not engaged in rendering professional services. If professional advice or other expert assistance is required, the services of a competent professional should be sought.

Other Wiley Editorial Offices

John Wiley & Sons Inc., 111 River Street, Hoboken, NJ 07030, USA

Jossey-Bass, 989 Market Street, San Francisco, CA 94103-1741, USA

Wiley-VCH Verlag GmbH, Boschstr. 12, D-69469 Weinheim, Germany

John Wiley & Sons Australia Ltd, 42 McDougall Street, Milton, Queensland 4064, Australia

John Wiley & Sons (Asia) Pte Ltd, 2 Clementi Loop #02-01, Jin Xing Distripark, Singapore 129809

John Wiley & Sons Canada Ltd, 22 Worcester Road, Etobicoke, Ontario, Canada M9W 1L1

Library of Congress Cataloging in Publication Data

AAA and network security for mobile access : radius, diameter, EAP, PKI,
   and IP mobility / Madjid Nakhjiri, Mahsa Nakhjiri.
     p. cm.
   Includes bibliographical references and index.
   ISBN 0-470-01194-7 (cloth : alk.paper)
   1. Wireless Internet—Security measures.   2. Mobile computing—Security measures.
   I. Nakhjiri, Mahsa. II. Title
   TK5103.4885.N35 2005
   005.8—dc22
                                                                            2005016320

British Library Cataloguing in Publication Data

A catalogue record for this book is available from the British Library

ISBN-13 978-0-470-01194-2 (H/B)

Typeset in 10/12pt Times by Integra Software Services Pvt. Ltd, Pondicherry, India
Printed and bound in Great Britain by Antony Rowe Ltd, Chippenham, Wiltshire
This book is printed on acid-free paper responsibly manufactured from sustainable forestry in which at least two trees are planted for each one used for paper production.

To our parents for their love and patience,
To our daughter, Camellia, for giving us joy

Contents

Foreword   xv

Preface   xvii

About the Author   xxi

Chapter 1 The 3 "A"s: Authentication, Authorization, Accounting   1
1.1 Authentication Concepts   1
   1.1.1 Client Authentication   2
   1.1.2 Message Authentication   4
   1.1.3 Mutual Authentication   5
   1.1.4 Models for Authentication Messaging   6
      1.1.4.1 Two-Party Authentication Model   6
      1.1.4.2 Three-Party Authentication Model   6
   1.1.5 AAA Protocols for Authentication Messaging   7
      1.1.5.1 User–AAA Server   7
      1.1.5.2 NAS–AAA Server Communications   7
      1.1.5.3 Supplicant (User)–NAS Communications   8
1.2 Authorization   8
   1.2.1 How is it Different from Authentication?   8
   1.2.2 Administration Domain and Relationships with the User   9
   1.2.3 Standardization of Authorization Procedures   10
      1.2.3.1 Authorization Messaging   12
      1.2.3.2 Policy Framework and Authorization   12
1.3 Accounting   13
   1.3.1 Accounting Management Architecture   13
      1.3.1.1 Accounting Across Administrative Domains   14
   1.3.2 Models for Collection of Accounting Data   15
      1.3.2.1 Polling Models for Accounting   15
      1.3.2.2 Event-Driven Models for Accounting   15
   1.3.3 Accounting Security   17
   1.3.4 Accounting Reliability   17
      1.3.4.1 Interim Accounting   18
      1.3.4.2 Transport Protocols   18
      1.3.4.3 Fail-Over Mechanisms   18
   1.3.5 Prepaid Service: Authorization and Accounting in Harmony   19
1.4 Generic AAA Architecture   19
   1.4.1 Requirements on AAA Protocols Running on NAS   21

1.5 Conclusions and Further Resources	23
1.6 References	23

Chapter 2 Authentication — 25

2.1 Examples of Authentication Mechanisms	25
2.1.1 User Authentication Mechanisms	26
2.1.1.1 Basic PPP User Authentication Mechanisms	27
2.1.1.2 Shortcoming of PPP Authentication Methods	29
2.1.1.3 Extensible Authentication Protocol (EAP) as Extension to PPP	30
2.1.1.4 SIM-Based Authentication	30
2.1.2 Example of Device Authentication Mechanisms	31
2.1.2.1 Public Key Certificate-Based Authentication	32
2.1.2.2 Basics of Certificate-Based Authentication	32
2.1.3 Examples of Message Authentication Mechanisms	33
2.1.3.1 HMAC-MD5	34
2.2 Classes of Authentication Mechanisms	36
2.2.1 Generic Authentication Mechanisms	41
2.2.1.1 Extensible Authentication Protocol (EAP)	41
2.2.1.2 EAP Messaging	42
2.3 Further Resources	44
2.4 References	45

Chapter 3 Key Management Methods — 47

3.1 Key Management Taxonomy	47
3.1.1 Key Management Terminology	47
3.1.2 Types of Cryptographic Algorithms	49
3.1.3 Key Management Functions	50
3.1.4 Key Establishment Methods	51
3.1.4.1 Key Transport	51
3.1.4.2 Key Agreement	52
3.1.4.3 Manual Key Establishment	53
3.2 Management of Symmetric Keys	54
3.2.1 EAP Key Management Methods	54
3.2.2 Diffie–Hellman Key Agreement for Symmetric Key Generation	58
3.2.2.1 Problems with Diffie–Hellman	60
3.2.3 Internet Key Exchange for Symmetric Key Agreement	61
3.2.4 Kerberos and Single Sign On	62
3.2.4.1 Kerberos Issues	65
3.2.5 Kerberized Internet Negotiation of Keys (KINK)	66
3.3 Management of Public Keys and PKIs	67
3.4 Further Resources	68
3.5 References	69

Chapter 4 Internet Security and Key Exchange Basics — 71

4.1 Introduction: Issues with Link Layer-Only Security	71
4.2 Internet Protocol Security	73
4.2.1 Authentication Header	74
4.2.2 Encapsulating Security Payload	74
4.2.3 IPsec Modes	75
4.2.3.1 Transport Mode	76
4.2.3.2 Tunnel Mode	76

	4.2.4 Security Associations and Policies	77
	4.2.5 IPsec Databases	78
	4.2.6 IPsec Processing	78
	4.2.6.1 Outbound Processing	78
	4.2.6.2 Inbound Processing	79
4.3	Internet Key Exchange for IPsec	79
	4.3.1 IKE Specifications	79
	4.3.2 IKE Conversations	81
	4.3.2.1 IKE Phase 1	81
	4.3.2.2 IKE Phase 2	82
	4.3.2.3 Round Trip Optimizations	82
	4.3.3 ISAKMP: The Backstage Protocol for IKE	83
	4.3.3.1 ISAKMP Message Format	83
	4.3.3.2 ISAKMP Payloads in IKE Conversations	86
	4.3.4 The Gory Details of IKE	86
	4.3.4.1 Derivation of ISAKMP Short-Term Keys	86
	4.3.4.2 IKE Authentication Alternatives	88
	4.3.4.3 IKE Deployment Issues	90
4.4	Transport Layer Security	91
	4.4.1 TLS Handshake for Key Exchange	93
	4.4.2 TLS Record Protocol	95
	4.4.2.1 TLS Alert Protocol	95
	4.4.3 Issues with TLS	96
	4.4.4 Wireless Transport Layer Security	96
4.5	Further Resources	96
4.6	References	97

Chapter 5 Introduction on Internet Mobility Protocols — 99

5.1	Mobile IP	99
	5.1.1 Mobile IP Functional Overview	102
	5.1.1.1 Mobile IP Registration	103
	5.1.1.2 Mobile IP Reverse Tunneling	106
	5.1.2 Mobile IP Messaging Security	107
	5.1.2.1 Caveat: Key Establishment	109
5.2	Shortcomings of Mobile IP Base Specification	109
	5.2.1 Mobile IP Bootstrapping Issues	110
	5.2.1.1 Dynamic Home Address Assignment	111
	5.2.1.2 Dynamic Home Agent Assignment	111
	5.2.1.3 Dynamic Key Establishment	113
	5.2.2 Mobile IP Handovers and Their Shortcomings	113
	5.2.2.1 Layer-2 Triggers and Fast Handovers	114
	5.2.2.2 Candidate Router Discovery Issues	115
	5.2.2.3 Delay and Disruption Tolerance by Applications	116
	5.2.2.4 Establishment of Network Services	116
5.3	Seamless Mobility Procedures	117
	5.3.1 Candidate Access Router Discovery	118
	5.3.2 Context Transfer	120
	5.3.2.1 Design Considerations	122
	5.3.2.2 Messaging Overview	124
5.4	Further Resources	125
5.5	References	126

Chapter 6 Remote Access Dial-In User Service (RADIUS) — 127
6.1 RADIUS Basics — 127
6.2 RADIUS Messaging — 128
   6.2.1 Message Format — 129
   6.2.2 RADIUS Extensibility — 130
   6.2.3 Transport Reliability for RADIUS — 130
   6.2.4 RADIUS and Security — 131
      6.2.4.1 RADIUS Message Integrity Protection — 131
      6.2.4.2 Attribute Hiding — 132
      6.2.4.3 Security Vulnerabilities of RADIUS — 134
      6.2.4.4 RADIUS over IPsec — 135
6.3 RADIUS Operation Examples — 135
   6.3.1 RADIUS Support for PAP — 135
   6.3.2 RADIUS Support for CHAP — 136
   6.3.3 RADIUS Interaction with EAP — 138
   6.3.4 RADIUS Accounting — 139
      6.3.4.1 Basic Operation — 139
      6.3.4.2 Security and Reliability of RADIUS Accounting — 140
6.4 RADIUS Support for Roaming and Mobility — 141
   6.4.1 RADIUS Support for Proxy Chaining — 142
      6.4.1.1 Roaming Concepts — 142
      6.4.1.2 Proxy Chaining Operation — 143
      6.4.1.3 Issues with Proxy Chaining — 143
6.5 RADIUS Issues — 143
6.6 Further Resources — 144
   6.6.1 Commercial RADIUS Resources — 144
   6.6.2 Free Open Source Material — 145
6.7 References — 145

Chapter 7 Diameter: Twice the RADIUS? — 147
7.1 Election for the Next AAA Protocol — 147
   7.1.1 The Web of Diameter Specifications — 148
      7.1.1.1 Diameter Base Specification — 148
      7.1.1.2 Security Specifications — 149
      7.1.1.3 Diameter Transport Profile — 150
      7.1.1.4 Diameter NAS Application — 150
   7.1.2 Diameter Applications — 151
   7.1.3 Diameter Node Types and their Roles — 152
7.2 Diameter Protocol — 153
   7.2.1 Diameter Messages — 153
      7.2.1.1 Diameter Message Format — 154
      7.2.1.2 Diameter Command Code (Message Types) — 154
      7.2.1.3 Attribute-Value Pair (AVP) Format — 155
      7.2.1.4 Examples of Diameter Base Specification AVPs — 156
   7.2.2 Diameter Transport and Routing Concepts — 157
      7.2.2.1 Diameter Transport Concepts — 157
      7.2.2.2 Diameter Routing Concepts — 158
      7.2.2.3 Diameter Message Routing and Forwarding — 159
   7.2.3 Capability Negotiations — 159

7.2.4 Diameter Security Requirements	160
7.2.4.1 Use of IPsec or TLS for Diameter	161
7.2.4.2 Path Authorization: Impact of Security on Authorization and Accounting	161
7.3 Details of Diameter Applications	162
7.3.1 Accounting Message Exchange Example	162
7.3.2 Diameter-Based Authentication, NASREQ	163
7.3.2.1 Commands Introduced by NASREQ	164
7.3.2.2 NASREQ AVPs	164
7.3.2.3 Diameter NAS Messaging	165
7.3.3 Diameter Mobile IP Application	167
7.3.4 Diameter EAP Support	167
7.4 Diameter Versus RADIUS: A Factor 2?	168
7.4.1 Advantages of Diameter over RADIUS	168
7.4.1.1 Fail-Over	168
7.4.1.2 Server-initiated Messages	169
7.4.1.3 Reliable Transport	169
7.4.1.4 Capability Negotiation	169
7.4.1.5 Security and Audibility Issues	169
7.4.1.6 Diameter Support for Agents and Inter-Domain Roaming	170
7.4.1.7 Peer Discovery and Configuration	170
7.4.1.8 Backward Compatibility with RADIUS	170
7.4.2 Issues with Use of Diameter	170
7.4.3 Diameter-RADIUS Interactions (Translation Agents)	171
7.5 Further Resources	172
7.6 References	172
Chapter 8 AAA and Security for Mobile IP	**175**
8.1 Architecture and Trust Model	177
8.1.1 Timing Characteristics of Security Associations	178
8.1.1.1 Pre-established SAs (PSA)	178
8.1.1.2 Mobility Security Associations (MSA)	179
8.1.1.3 AAASA	179
8.1.1.4 Lifetimes	180
8.1.1.5 Security Parameter Index (SPI)	180
8.1.2 Key Delivery Mechanisms	181
8.1.3 Overview of Use of Mobile IP-AAA in Key Generation	182
8.2 Mobile IPv4 Extensions for Interaction with AAA	184
8.2.1 MN-AAA Authentication Extension	184
8.2.2 Key Generation Extensions (IETF work in progress)	186
8.2.3 Keys to Mobile IP Agents?	187
8.3 AAA Extensions for Interaction with Mobile IP	187
8.3.1 Diameter Mobile IPv4 Application	188
8.3.1.1 Diameter Model for Mobile IP Support	188
8.3.1.2 New Diameter AVPs for Mobile IP Support	190
8.3.1.3 Diameter Mobile IP Messaging Overview	193
8.3.2 Radius and Mobile IP Interaction: A CDMA2000 Example	196
8.3.2.1 Mobile IP Support Within CDMA2000	196
8.3.2.2 RADIUS Support, or Not!	197
8.3.2.3 CDMA2000 Messaging Procedure	199

8.4	Conclusion and Further Resources	200
8.5	References	201

Chapter 9 PKI: Public Key Infrastructure: Fundamentals and Support for IPsec and Mobility — 203

- 9.1 Public Key Infrastructures: Concepts and Elements — 204
  - 9.1.1 Certificates — 204
  - 9.1.2 Certificate Management Concepts — 205
  - 9.1.3 PKI Elements — 209
  - 9.1.4 PKI Management Basic Functions — 210
    - 9.1.4.1 Basic PKI Transactions — 211
    - 9.1.4.2 Enrollment and Authentication — 211
  - 9.1.5 Comparison of Existing PKI Management Protocols — 212
    - 9.1.5.1 PKCS #10 — 213
    - 9.1.5.2 SSL Protection for PKCS #10 — 214
    - 9.1.5.3 PKCS #7 Protection for PKCS #10 — 215
    - 9.1.5.4 IETF Certificate Management Protocol (CMP) — 219
    - 9.1.5.5 Certificate Management Using CMS (CMC) — 221
    - 9.1.5.6 Simple Certificate Enrollment Protocol (SCEP) — 221
  - 9.1.6 PKI Operation Protocols — 221
    - 9.1.6.1 PKI Certificate Discovery and Validation Protocols — 222
- 9.2 PKI for Mobility Support — 222
  - 9.2.1 Identity Management for Mobile Clients: No IP Addresses! — 222
    - 9.2.1.1 Certificate Subjects for Mobile Devices — 223
    - 9.2.1.2 Certificate Subjects for Human Users — 224
  - 9.2.2 Certification and Distribution Issues — 225
    - 9.2.2.1 Validity Checking and CRL Distribution — 225
    - 9.2.2.2 Roaming and Certification — 226
    - 9.2.2.3 Device Certificates — 226
    - 9.2.2.4 User Certificates — 226
- 9.3 Using Certificates in IKE — 227
  - 9.3.1 Exchange of Certificates within IKE — 229
    - 9.3.1.1 Certificate Data Type Profiling for ISAKMP — 229
    - 9.3.1.2 In-Band Versus Out-of-Band Exchanges — 230
    - 9.3.1.3 Certificate Authority and Certificate Chains — 230
  - 9.3.2 Identity Management for ISAKMP: No IP Address, Please! — 231
- 9.4 Further Resources — 232
- 9.5 References — 232
- 9.6 Appendix A PKCS Documents — 233

Chapter 10 Latest Authentication Mechanisms, EAP Flavors — 235

- 10.1 Introduction — 235
  - 10.1.1 EAP Transport Mechanisms — 237
  - 10.1.2 EAP over LAN (EAPOL) — 237
  - 10.1.3 EAP over AAA Protocols — 238
- 10.2 Protocol Overview — 239
- 10.3 EAP-XXX — 242
  - 10.3.1 EAP-TLS (TLS over EAP) — 244
    - 10.3.1.1 EAP-TLS Architecture and Message Format — 244
    - 10.3.1.2 Protocol Overview — 246
    - 10.3.1.3 Drawbacks with EAP-TLS — 248

        10.3.2 EAP-TTLS	248
            10.3.2.1 EAP-TTLS Functional Elements	250
            10.3.2.2 Messaging Overview	252
            10.3.2.3 Protocol Overview	253
            10.3.2.4 Session Resumption: EAP-TTLS Support for Mobility	254
            10.3.2.5 Example: CHAP Over EAP-TTLS	255
        10.3.3 EAP-SIM	257
    10.4 Use of EAP in 802 Networks	259
        10.4.1 802.1X Port-Based Authentication	259
            10.4.1.1 EAPOL in 802.1X and Interaction with RADIUS	260
            10.4.1.2 Security Flaws of 802.1X, WPA/RSN and 802.1aa	260
        10.4.2 Lightweight Extensible Authentication Protocol (LEAP)	260
        10.4.3 PEAP	262
    10.5 Further Resources	262
    10.6 References	263

**Chapter 11 AAA and Identity Management for Mobile Access: The World of Operator Co-Existence	265**
    11.1 Operator Co-existence and Agreements	265
        11.1.1 Implications for the User	266
        11.1.2 Implications for the Operators	267
        11.1.3 Bilateral Billing and Trust Agreements and AAA Issues	269
            11.1.3.1 Identity Management and Security Issues	271
        11.1.4 Brokered Billing and Trust Agreements	272
        11.1.5 Billing and Trust Management through an Alliance	274
    11.2 A Practical Example: Liberty Alliance	275
        11.2.1 Building the Trust Network: Identity Federation	276
            11.2.1.1 Identity Services	276
            11.2.1.2 Circle of Trust	278
            11.2.1.3 Building the Circle of Trust	278
        11.2.2 Support for Authentication/Sign On/Sign Off	279
            11.2.2.1 Enabling Protocols	281
        11.2.3 Advantages and Limitations of the Liberty Alliance	282
    11.3 IETF Procedures	283
    11.4 Further Resources	285
    11.5 References	285

**Index	287**

Foreword

The market for mobile computers and commmunication devices continues to grow, which means that every year there are more and more of them. This is creating numerous opportunities for network providers and operators of all sorts, because many of these devices derive their usefulness from their ability to get access to the Internet. Recently, within the IETF, there has been a surge of interest in creating new protocols and protocol interfaces to better enable operators to take advantage of these opportunities. These new protocols, taken as a whole, bring about a new kind of operator operation known as "AAA services", thus the title of the book. Madjid, one of the two authors of this book, is known to me as a regular in several IETF working groups, and his work is well represented within this book.

There is no doubt that AAA services are already of tremendous importance in today's Internet, given that much of the access control is mediated already by RADIUS servers and associated protocols. Even so, I think that the true value of AAA services is still in the process of emerging, as we transition from laptop computing to wireless mobile communications in the future. As we begin to store more of our credentials on our wireless gadgets, and as the needs for user authentication continue to expand, it seems very natural that today's AAA practice will adapt to the needs of the new wireless technologies. These needs include higher performance, improved roaming facilities, and interface to a multiplicity of security technologies. Already, my experience is that I have to carry around a bag of strange connectors, security cards, credit cards, and telephone numbers in order to be mobile. It seems that when traveling, leaving any of these behind is much worse than forgetting to pack a toothbrush, soap, or even shirts or socks. After all, I can usually find a place to buy those latter items.

Within the book, we can see the first glimmerings of how this new wireless mobile world will look to the user desiring to make use of local Internet connectivity. Several recent specifications have been finally approved and are dutifully described in this book. In particular, the ideas of seamless mobility and context transfer provide great hope for the desired user productivity and the experience of well-engineered convenience. Clearly, there is a big gap separating the barebones specification and widespread deployment. It is to fill just these gaps that books such as this one are needed. But filling known gaps is only the beginning. Once the basic hurdles are cleared, I am confident that many new applications will soon be imagined and built to use the simplified access models provided by the new AAA services.

Charlie Perkins

Preface

In today's world, where computer viruses and security threats are common themes in anything from Hollywood movies and TV advertisements to political discussions, it seems unthinkable to ignore security considerations in the design and implementation of any network. However, it is only in the past 4–5 years that talkative security experts have been invited to the design table from the start. The common thinking only 5 years ago was either: this is somebody else's problem or let us design the major functionalities first, then bring in a cryptographer to secure it! This treatment of security as an add-on feature typically led either to design delays, overheads and extra costs when the "feature" had to be included, or to ignored security provisioning when the "feature" was not a must. The problem, of course, stemmed from the fact that security "features" have rarely been revenue-makers. As we all know, many political, social and economic events in the last half decade have forced the designers, regulators and businessmen to adjust their attitudes towards security considerations. People realized that although security measures are not revenue-makers, their lack is indeed a deal breaker, to say the least, or has catastrophic aftermaths, at worst.

The Internet Engineering Task Force (IETF) has also played an important role in establishing the aforementioned trend by making a few bold moves. The rejection of some very high profile specifications due to the lack of proper security considerations was a message to the industry that security is not to be taken lightly. This was done in a dot.com era where the Internet and its applications seemed to have no boundaries and security provisioning seemed to be only a barrier rather than an enabler.

As a result of this trend, the field of network security gained a lot of attention. A profession that seemed to belong only to a few mathematically blessed brains opened up to a community of practitioners dealing with a variety of networking and computing applications. Many standards, such as 802.1X, IPsec and TLS, were developed to apply cryptographic concepts and algorithms to networking problems. Many books were written on the topics of security and cryptography, bringing the dark and difficult secrets of fields such as public key cryptography to a public that typically was far less mathematically savvy than the original inventors. Many protocols and procedures were designed to realize infrastructures such as PKIs to bring these difficult concepts to life. Still, cryptographic algorithms or security protocols such as IPsec are not enough alone to operate a network that needs to generate services and revenues or to protect its constituency. Access to the network needs to be controlled. Users and devices need to be authorized for a variety of services and functions and often must pay for their usage. This is where the AAA protocols came in. In its simpler form a AAA protocol such as a base RADIUS protocol only provides authentication-based access control. A few service types are also included in the authorization signaling. RADIUS was later augmented with accounting procedures. Diameter as a newer protocol was only standardized less than 2 years ago. Both RADIUS and Diameter are still evolving at the time of writing. This evolution is to enable AAA mechanisms and protocols to provide powerful functions to manage many complicated tasks ranging from what is described above to managing resources and mobility functions based on a variety of policies. In the near future the networks need to allow the user through a variety of interfaces, devices and technologies to gain access to the network. The user will require to be mobile and yet connected. The provision of the connection may at times have to be aided by third parties. The interaction between AAA and security procedures with entities providing mobility and roaming capabilities is a very complicated one and is still not completely understood. Despite this complexity, there seem to be very few books on the market that discuss more than a single topic (either security, or mobility or wireless technology). The topic of AAA is largely untouched. Very little text in the way of published literature is available on AAA protocols, let alone describing the interaction of these protocols with security, mobility and key management protocols.

The idea for writing this book started from an innocent joke by the IETF operation and management area director during an IETF lunch break a few years ago. When we asked about the relations between the use of EAP for authentication and Mobile IP-AAA signaling, the answer was "Maybe you should write a book about the subject". Even though this was considered a joke at a time, as we started to work on deploying AAA infrastructure for Mobile IP and EAP support, the need for easy-to-understand overview material was felt so strongly that the joke now sounded like black humor. We had to write a book on AAA as a community service!

The book is geared towards people who have a basic understanding of Internet Protocol (IP) and TCP/IP stack layering concepts. Except for the above, most of the other IP-related concepts are explained in the text. Thus, the book is suitable for managers, engineers, researchers and students who are interested in the topic of network security and AAA but do not possess in-depth IP routing and security knowledge. We aimed at providing an overview of IP mobility (Mobile IP) and security (IPsec) to help the reader who is not familiar with these concepts so that the rest of the material in the book can be understood. However, the reader may feel that the material quickly jumps from a simple overview of Mobile IP or IPsec to sophisticated topics such as bootstrapping for IP mobility or key exchange for IP security. Our reasoning here was that we felt that there are a number of excellently written books on the topics of Mobile IP and IPsec, to which the reader may refer, so it would not be fair to fill this book with redundant information. Instead, the book provides just enough material on those topics to quickly guide the reader into the topics that are more relevant to the rest of the material in this book. The book may also serve as a reference or introduction depending on the reader's need and background, but it is not intended as a complete implementation reference book. The tables listing the protocol attributes are intentionally not exhaustive to avoid distractions. Most of the time, only subsets that pertain to the discussions within the related text are provided to enable the reader to understand the principles behind the design of these attributes. At the same time, references to full standards specifications are provided for readers interested in implementation of the complete feature sets.

Chapter 1 of this book provides an overview of what AAA is and stands for. It provides thorough descriptions of both authorization and accounting mechanisms. Unfortunately the field and standardization on authorization mechanisms is in the infancy stage at this point and accounting, compared to authentication, has received far less attention in the research and standards community due to its operator-specific nature. Due to the enormous amount of research done on authentication, we devote Chapter 2 entirely to authentication concepts and mechanisms and also provide a rather unique classification (from IAB) of authentication mechanisms in that chapter. We will come back to the topic of authentication and describe more sophisticated EAP-based authentications in Chapter 10, but after Chapter 2, we go through the concepts of key management in Chapter 3 to lay the groundwork for most of the security and key management discussions in Chapter 4 and the rest of the book. Chapter 4 discusses IPsec and TLS briefly, but provides a thorough discussion on IKE as an important example of a key management and security association negotiation protocol. As mentioned earlier, the aim of that chapter is not to describe IPsec or TLS thoroughly. Both these protocols are provided for completeness and to provide the background for the later discussion of security topics. Chapter 5 discusses mobility protocols for IP networks. It describes basic Mobile IP procedures and quickly goes through the latest complementary work in IETF, such as bootstrapping. This chapter also describes two IETF seamless mobility protocols, context transfer and candidate access router discovery, which may be required to achieve seamless handovers. This chapter also describes the security procedures for Mobile IPv4 and lays the groundwork for Mobile IP-AAA discussions in Chapter 8. Chapters 6 and 7 describe the two most important AAA protocols, namely RADIUS and Diameter and their applications for authentication and accounting. Many of the specifications that are considered work in progress in IETF are covered here.

Chapter 8 finally covers the topic discussed in the IETF joke we mentioned earlier: Mobile IP-AAA signaling to provide authentication and key management for Mobile IP signaling.

Chapter 9 goes on to provide a description of public key infrastructures (PKI) and the issues and concerns with management of PKIs, certificates and their revocation.

Chapter 10 describes the EAP authentication framework, EAP signaling transport and the structure for a generic EAP-XXX mechanism. It also provides overviews of a variety of EAP authentication methods, such as EAP-TLS, EAP-TTLS, EAP-SIM, and so on.

Finally, Chapter 11 makes a humble attempt at describing the overall problem of AAA and identity management in a multi-operator environment and discusses various architectural models to tackle the problem. This chapter also provides an overview of the Liberty Alliance.

We wish the readers a joyful read.

Acknowledgements

Finally, it is the time to give acknowledgement to the people who have provided help, encouragement and support. First, we would like to thank Mike Needham of Motorola Labs for showing enormous enthusiasm and full confidence when we broached the idea of writing a book at a time when we were not fully confident ourselves that this was a task we could tackle. We would like to specially thank Dorsa Mirazandjani for acting as our test audience, reading and providing comments and corrections on many chapters of this book, despite her busy work and graduate school schedule. We would also like to thank Jeff Kraus for taking the time and reading through Chapter 8 and providing technical and editorial feedback. A special thanks you goes to Mana Mirazanjani for the first draft of the beautiful cover design. Another very special thank you goes to Charlie Perkins who despite his very busy schedule took the time and wrote a generous foreword for this book. We would like to thank the IETF for providing open standards and specifications, without which the material for this book would have been very hard to find. We would also like to thank the Liberty Alliance for accommodations they made in the process of writing Chapter 11.

Finally, we want to thank the John Wiley publishing team, especially Birgit Gruber and Joanna Tootill for their kindness, patience, encouragement and support throughout the project.

About the Author

Madjid Nakhjiri is currently a researcher and network architect with Motorola Labs. He has been involved in the wireless communications industry since 1994. Over the years, Madjid has participated in the development of many cellular and public safety mission-critical projects, ranging from cellular location detection receiver design and voice modeling simulations to the design of architecture and protocols for QoS-based admission, call control, mobile VPN access and AAA procedures for emergency response networks. Madjid has been active in the standardization of mobility and security procedures in IETF, 3G and IEEE since 2000 and is a coauthor of a few IETF RFCs. Madjid has also authored many IEEE papers, chaired several IEEE conference sessions and has many patent applications in process.

Mahsa Nakhjiri is currently a systems engineer with Motorola Personal Devices and is involved in future cellular technology planning. Mahsa holds degrees in Mathematics and Electrical Engineering and has specialized in mathematical signal processing for antenna arrays. She has been involved in research on cellular capacity planning and modeling, design and simulation of radio and link layer protocols and their interaction with transport protocols in wireless environments. Mahsa has also worked with cellular operators on mobility and AAA issues from an operator perspective.

# 1

The 3 "A"s: Authentication, Authorization, Accounting

For the road travelers in the United States, especially the parents who take their children in the family car on the long road trips, the letters AAA stand for a peace of mind. They feel that any time their car breaks down, they can call the number for the American Automobile Association and ask for roadside assistance. Even though this book is not about that sort of AAA, the 3 "A"s that we talk about here, when designed properly, can bring the same peace of mind to the network operator and its customers. Authentication, authorization, and accounting are three important blocks used in the construction of a network architecture that helps protect the network operator and its customers from fraud, attacks, inappropriate resource management, and loss of revenue.

In this chapter, we describe each of the "A"s in the AAA first as a separate topic, and then as a piece that interacts with the other "A"s in an effort to justify why all the 3 "A"s should be treated by the same framework and servers. At the end of the chapter, we provide a model for a generic AAA architecture.

1.1 Authentication Concepts

According to the dictionary, the word "authentic" refers to something that is not false, or a fake imitation, but is worthy of acceptance as a truth or a fact. From the times of early civilizations, where people have run 26 miles only to deliver a message and then fall over and die, to today, when information can travel across the globe in fractions of a minute with a mouse click, proof of authenticity is the first thing the receiver of a message checks.

Authentication consists of two acts: first, the act of providing proof of authenticity for the information that is being delivered or stored, and second, the act of verifying the proof of authenticity for the information that is being received or retrieved. In the early ages, an emperor would use his personal seal on his letters to provide assurance for the authenticity of the letter. The letter could then be carried by any messenger, whose identity was not important. The local lord would recognize the emperor seal and trust authenticity of the letter. He would break the seal, read the letter, start an attack or collect taxes accordingly. In the days of digital information delivery, delivering proof of authenticity is equally important but poses its own challenges, as we will see.

The message delivery example above presents one type of authentication problem where authenticity of the information is important, while the identity of the messenger is not. However, in most of the cases, the identity of a person we are dealing with is an important factor in how we handle that interaction. When we go to a bank or through customs into a new country, we have to show identification to prove our identity. At first, the problem of identification does not seem to be related to the authentication. However, when one thinks about the possibility of a person lying about her identity or privileges, verification of authenticity of the provided identity becomes an authentication problem as well. Stating a name is typically not enough for identification, while showing a sort of identification issued by a trusted authority typically is. The acts of providing proof and verifying the authenticity of the identification presented are again the two acts of authentication.

Today, the two mentioned forms of authentications, i.e. providing information integrity and identity verification, are among the most fundamental security mechanisms required for providing access to network users and clients. In this introductory section, we provide a relatively short overview of various authentication concepts to allow the reader to understand the distinction between the constantly confused types of authentication. In Chapter 2, we will delve into more details of various authentication procedures.

1.1.1 Client Authentication

Client authentication means that a client wishing to gain access and connect to the network presents its identity along with a set of credentials. As proof of authencity for the presented identity. The credentials are then used by the network to verify that the identity actually belongs to the client.

We intentionally used the term client, since it can be interpreted both as a device as well as a human user, who is a consumer of a network service. For that reason, the client authentication needs to be further refined into two categories: user authentication and device authentication. Until recently, very few network security designs made a visible distinction between user authentication and device authentication. In the following we will explain the reason. Traditional architectures dealing with network access control could be divided into two categories:

- Architectures accommodating users that arrive at a fixed location, such as a local area network with fixed devices, such as data terminals, already connected to an infrastructure. The user needs to use its personal credentials to log into the network through a device (terminal), which itself typically resides in a computer room and is trusted through its wired connections. A good old world college campus terminal room scenario! The student simply trusts the network set up by the campus, as long as the college is an accredited one and the terminal is not asking for credit card numbers as login credentials! In this basic scenario, the distinction between device and user authentication although very clear for a human, is not important. The device is not authenticated at all. The user credential with a central server is the main criterion for allowing network access to the user.
- Architectures accommodating mobile users carrying their personal devices to gain access to a wide area network. A perfect example is cellular phone systems. The user registers

with a cellular service provider and purchases a cellular phone that works with operator's network. The operator creates a set of credentials specific for the user. The cellular phone, that the user carries, is then programmed with such credentials to access the network. Many times, the user is unaware of these credentials and the actual process of authentication during connection establishment. The device is the entity that interacts with the network and presents the credentials needed to perform authentication. The idea is: since the user always carries the same cellular phone (as long as she is loyal to her service provider) no distinction between the user and device credentials has to be made. The downside is that if the device was lost, stolen, or even cloned, the rouge or unintended user could use the device to gain access to the network without having her real identity exposed, and this could go on until the legitimate user would report a lost or stolen phone or illegitimate entries in her monthly statement from the service provider.

With the proliferation of public local network providers, such as wireless hot spots serving passing customers, many sorts of vulnerabilities will appear at various corners of the architecture:

- The long-term customer–operator business and legal relationship no longer exists, which means the network operator and the user cannot trust each other as.
- The one-to-one mapping of user–device does not exist. Even if the operator could trust a user, the operator would not know what device the user may use every time they try to access the network. In other cases, such as in service organizations, government agencies, or police department, users that belong to a team can share their devices with each other.

Such refinements require more precise definition of the network usage and security policies that in turn means the architecture must be designed more carefully. The network operator may need to make sure that both device and user are authentic, possibly using separate processes. In some cases, the device may have to even be manufactured and configured with credentials for access to the network. (For instance, cable modems for cable Internet Service Provider (ISP) networks are produced this way.) In such cases, proper protection must be in place, so that the credentials on the device can not be tampered with. The network operator also needs to make sure the user presents accurate identity and proper credentials that can identify her at the time of use, so that the various users can be distinguished even when they are using a shared device.

Device authentication credentials can be certificates or cryptographic keys that are loaded in the devices either by the manufacturer in the factory or by the network operator at the time of service initiation. When designed properly and in a modular manner, the device authentication process should be transparent to the user. In fact, the user should not even be allowed to access device authentication credentials. On the other hand, user authentication credentials are personalized, typically given to the user in an out-of-band method. Examples are over a phone call by the user stating some secrets about her identity or through a face-to-face meeting after presentation of a driving license, company badge, and so on. Upon identification of the user, the system operator issues the authentication credentials for the user. The credentials must be carried by the user at all times either memorized (password) or in the form of a token such as a secure ID card, a certificate on some sort of cryptographic module. The user applies her authentication credentials on the device provided for access to the network to connect to the network.

It should be noted that the security architecture may require both device authentication as well as user authentication in various steps of a network access process. An example would be the case of IP networks: in order to communicate to the IP network, the device needs to acquire an IP address. The IP address is not only an important resource for the network, but also allows the user to gain access to many other network services. Furthermore, the IP address can be used as a backdoor to launch active or passive attacks. Hence, IP address acquisition should be tightly controlled. From a security standpoint, it means a device needs to first authenticate itself to the network before being able to gain an IP address from the network. Once the device is authenticated, has gained IP address from the network, and is registered with various agents, it allows the user to present her credentials to the network and gain access to the services to which she is entitled. The latter brings another point and that is the user credentials may be used to determine authorization levels for the users based on their pre-configured service profile. We will discuss the topic of authorization later on.

1.1.2 Message Authentication

Device or user authentications deal with ensuring that the end points of the communications are legitimate and who they claim they are. Message authentication, on the other hand, ensures and verifies the integrity of the data at hand (remember the example of sealed letter from the emperor). When message authentication is performed, the receiver of the message can be sure that the information included in the message has been produced by a legitimate source and not been altered by other parties in transit. This is why message authentication is usually considered as a data integrity protection mechanism. Unlike device or user authentications that are typically performed at the beginning of a session and require their own messaging mechanism, message authentication may have to be done quite often during the session and for a variety of traffic, such as control messaging, important data packets, and sometimes even data session.

Note that the goal of data integrity protection is to prevent malicious and intended corruption of data by the so-called men in the middle (MITM), trying to tamper with the message contents. This is different from the information-theoretical codes and cyclic redundancy checks designed to mitigate the random and natural data corruptions caused by physical communications media imperfections. Aside from the cause of corruption being different, calculations required for message authentication for security protection is also different from its signal processing counter parts; an attacker can always alter the data and re-calculate the information-theoretical checks to lure the receiver, while the attacker cannot re-calculate the message authentication data added to the message, since message authentication is typically performed on the basis of knowledge of a shared secret.

More details are provided on message authentication in Chapter 2, so we will not go into any details in this introductory chapter. In short, this is how message authentication works: the sender provides proof for data integrity by running a so-called secret hash algorithm over the contents of the message and adds the results of the algorithm (called digest or hash) to the end of the message. Hash algorithms are mathematical one-way functions. In other words, while it may be straightforward to calculate the output of a hash function (digest) from an input data packet, it is extremely hard to determine what input packet has been used to create the output digest. However, hash algorithms are rather well known, and hence if no secrets are used, it is possible for an attacker to change the packet in flight and re-calculate a new digest and then to replace both the original packet and digest with the altered packet and the new digest. In order to prevent the MITM from running the same algorithm over an altered version of the message, the sender uses a secret when creating the hash. The secret is only known to the sender and the receiver. This process is often referred to as running a secret hash algorithm, even though the algorithm itself is usually known. Once the receiver receives the message, it runs the same hash algorithm over the same portion of the data and compares the result of its calculation with the hash provided by the sender and if there is a match, the receiver considers the data authentic, otherwise the receiver needs to discard the data.

The secret hash algorithms are examples of symmetric key authentication methods. Another method of providing message authentication is to create the so-called digital signatures. Digital signatures are typically based on public key cryptography. We will go through the details of these procedures in Chapters 2 and 3.

1.1.3 Mutual Authentication

We finalize the description of various authentication concepts with a short note on mutual authentication. Client authentication that was explained earlier is a unilateral authentication, where only one end of a communications channel proves that the identity it has presented to the other end is authentic. It would make sense that when establishing communications both ends authenticate to each other. However, the reason this type of unilateral authentication is so widespread is that in many cases, the client simply trusted the network. So the network does not have to prove to the client that it is also authentic. In the earlier cellular phone example, due to the large expense involved with setting up a cellular network, it would be hard to imagine that attackers will go through the trouble of setting up a whole network and its high towers simply to hijack a cellular phone subscription. For those reasons until very recently, neither the device nor the user required any authentication from the network.

With a proliferation of large number of access providers competing for customers, a user may be confronted with a number of competing networks with which she does not have any business or trust relationships. Take a mom and pop coffee shop for example: a customer entering the coffee shop does not know whether the mom and pop that own a coffee shop actually know how to operate a wireless local access network (WLAN) access point (AP) or how to protect their access points from being loaded with viruses and Trojan horses. For all a user knows, another customer may have found the WLAN AP behind the coffee grinder machine, disconnected it and simply installed another AP to re-route all the traffic in the shop to some place else, or simply passively copy traffic to a server. Imagine, if the customer was planning to do some Internet shopping while sipping coffee and was entering her credit card number at an e-commerce web site not using proper encryption methods. In this case, it makes sense for the coffee shop AP to authenticate itself to the user's device as well. As we can see the unilateral client authentication is not sufficient and has to be upgraded to a mutual authentication where the server also authenticates to the client.

The mutual authentication between the client and the server is a special case of a more generic case of mutual authentication, where the two parties are simply peers as opposed to client and server. In that case, each peer authenticates to the other, either sequentially, or in parallel.

1.1.4 Models for Authentication Messaging

1.1.4.1 Two-Party Authentication Model

This model is used when the two peers interact with each other through a direct line of communication without the involvement of any middle nodes such as gateways or proxies. In such cases, the two entities directly authenticate to each other. The most prominent example is a direct client–server interaction, during which the client has to authenticate unilaterally to the server to gain access to service. However, mutual authentication can also be performed in direct two-party manner. As we will see later on, many key exchange mechanisms, such as Internet Key Exchange (IKE), require a direct two-party authentication exchange.

1.1.4.2 Three-Party Authentication Model

With the increase in size of networks and number of users wishing to access the network and its services, the networks have started to deploy specific points of presence (POP), which are low-cost unsophisticated devices without large processing power or database capabilities. The POPs typically interact directly with the users, but refer to central internal servers for many tasks and decisions regarding interactions with users. One such task is authentication. The POPs are typically incapable of authentication processing. Therefore, the two-party authentication model has been expanded to include three parties:

1. *The supplicant*, the user trying to gain access. In case of dialup networks, the user is configured with a phone number for the ISP modem pool and is given a password. The user dials the number of the modem pool and reaches one of the modems.
2. *Authenticator*, at edge interacting with user. In case of dialup, the authenticator is at the modem pool. The authenticator has no authority by itself, and is like the security guard at the door of a corporate building. When a visitor, who has no badge, walks in, the guard needs to call the authorities to ask whether it should let the visitor in or not. As we will see later on, in the AAA model, this entity is called the network access server (NAS), which acts as a AAA client.
3. *Authentication server*, who has the real authority and the necessary information database (e.g. a list of who is authorized, user names, and passwords) to make decisions regarding granting access to the user. In the security guard analogy, this is the boss upstairs that the guard calls to and the one who makes the ultimate decision. The AAA server in the AAA model is shown in Figure 1.1.

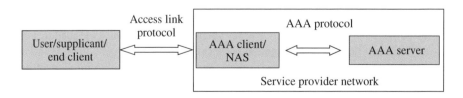

Figure 1.1 The three-party authentication model deploying an AAA infrastructure

1.1.5 AAA Protocols for Authentication Messaging

In small networks, an authentication server could be configured by the system administrator at the authenticator, so the authenticator and authentication server are co-located. However, as we mentioned earlier, in large networks or multi-domain networks, this is not practical, and many network points of presence, acting as NASs are deployed and the authentication must happen according to the three-party model. Naturally, the end-to-end authentication exchange needs to happen between the supplicant and the AAA server, but the NAS is also involved in the exchange of authentication-related messaging. Since the NAS controls communications into and out of the private network, it needs to intercept messaging between the supplicant and AAA server. Furthermore, the NAS typically acts as a protocol dividing point, since the communications to the AAA server side of the NAS typically happens over a private and typically wired and trusted part of the network, while the communications on supplicant-side of the NAS is over an untrusted and many times wireless medium. In order to provide interoperability between various network equipments, protocols for various segments of the three-party model have been standardized. The protocols for communications between the NAS and the supplicant typically depend on the type of access technology that the network provider offers and many times are at a lower layer (link layer). However, the NAS communicates with the central authentication server over standard UDP/IP or TCP/IP protocols and can use a standard protocol for carrying authentication messages on behalf of the supplicant and the server. As we will see later on, experience has shown that the authentication server can be co-located with entities performing authorizations and accounting as well (a AAA server). For that reason, the protocol between NAS and the authentication server is now a AAA protocol as follows.

1.1.5.1 User–AAA Server

As mentioned earlier, there is no direct communication between user and AAA server. The most prominent AAA protocol today, RADIUS (stands for remote access dial in user service), was designed to allow a NAS forward a user's request and its credentials to the server, and then carry the server's response back to the user. The Access-Request, Access-Challenge message structure in RADIUS shows that it was designed to accommodate password-based authentication methods, so that NAS can forward an authentication request message from the user to the authentication server and issue eventual challenges created by the network and present to the user.

1.1.5.2 NAS–AAA Server Communications

As mentioned earlier, this communication is typically on a private part of network. A AAA protocol, such as RADIUS protocol, is used for this purpose. Original assumption was that there is only one hop between the NAS and the AAA server. However, multiple hops deploying AAA proxies may be required. We will discuss AAA proxies in Chapters 6 and 7. It is important to note that when proxies are involved, the NAS–AAA server communication may no longer be over a private network. This means the information carried between the NAS and AAA server over the AAA protocol may need special security protection.

1.1.5.3 Supplicant (User)–NAS Communications

This communication is typically over a single-hop link called access link, since it is part of an access network. The access link provides a physical channel and a link layer protocol. The physical channel only provides the coding and modulation of information bits into electrical signals. Neither the earlier phone line dialup connections nor the later cellular systems provided layer-2 services, such as packet formatting, framing, and multiple-access mechanisms, and required specific layer-2 protocols such as point-to-point protocol (described in Chapter 2) for this purpose. The new wireless access technologies such as 802.11 WLAN have their own framing mechanisms and do not need additional protocols such as PPP for carrying layer-2 level messaging. However, the main point is that as we mentioned before, the IP-level service should be established after the initial authentication and hence, the user–NAS exchange of authentication messages need to run directly over an access technology-specific layer-2 protocol.

1.2 Authorization

Authorization is defined as the act of determining whether a particular privilege can be granted to the presenter of a particular credential. The privilege can be right of access to a resource, such as a communications link, an information database, a computing machine, or many other things owned by a network or service provider. The presenter of the credential can be either a device or a user.

1.2.1 How is it Different from Authentication?

The problem of authorization and its distinctions from the problem of authentication can be easily explained with the following example. Let us assume that a person, holding a personal handheld device such as a wireless-link enabled PDA, has subscribed with a high-priced network operator. This person requests to see some movie clips on his personal device. She uses her personal device to connect to a wireless network, and based on the credentials that the network provider has given to her at the time of subscription, she can authenticate through her PDA and connect to the network. In many of the networks today, this authentication would be enough for her to access the movie clips from a server located inside the operator's network. However, imagine if the network provider would charge different prices for different movies or download speeds. A lower paying user is allowed to download the clips at a much slower speed. The user may request a higher quality of service (QoS) by agreeing to pay a one-time premium. In this case, a mere verification of user identity would not be enough for granting the service requested by the user. The network must somehow access the user profile, consult entities controlling the amount of available bandwidth, and then make a decision on whether or not to authorize the user to access the service.

Another commercial example where authorization is important is networks that provide service to users that have purchased pre-paid cards. A cellular phone user buys a pre-paid card that is supposed to allow her to make phone calls for three hours. Every time such user requests to make a phone call, the network must check to see whether there is any credit left on the user's card before allowing the user to connect.

In commercial applications, the problem of authorization usually either translates to protection of revenue or to exercising the right to a service. However, in public safety, military, and security applications, the consequences may be more severe. A low ranking officer should not be authorized to join a conference call held between army generals. Public safety responders may have special units dealing with specific emergencies, such as high-rise building fires. Dispatch calls or conversations within the unit need to remain private within the group, both for privacy issues and to save bandwidth. A fire fighter who is not part of a special unit should not be allowed to join the channel reserved for that special unit. In such a case, once the fire fighter and his device is authenticated to the network, the network must pursue with the act of authorization, i.e. check his profile against what is required to authorize an agent to join specific calls on specific channels.

Historically, a private enterprise owning the computer or radio networks simply authorized its employees or affiliates for use of its resource as long as they could prove their affiliation to the enterprise. The credentials for proof of affiliation were the authentication credentials that were given to the user at the time of initiation with the enterprise (as described in section 1.1 before). Once the user was authenticated, she was also authorized for service, in other words not only the authorization was equated with authentication, but also the credentials presented for authentication were also used for authorization. This model still exists in many enterprise networks: the second A (authorization) is coupled with the first A (authentication) in AAA. In commercial networks, authorization is based on acquisition of revenue, i.e. if the network provider knows it can collect payment from the user of the service, the user will be authorized for service. Here the second A is coupled with the third A (accounting). The user subscribes for a service and agrees to pay a fee for the service and the network provider agrees to provide the user with the service. However, even though we said that authorization is coupled with accounting and billing, this is only in theory (business agreement paper work). In practice, the network operator implements this agreement by giving the user the authentication credentials and again bases authorization on authentication.

Now that we have established that there is a difference between authentication and authorization, let us discuss another extreme example from real life, where authorization is actually more important than authentication. When we want to go see a movie at a movie theatre, we will pay the cashier and buy a ticket. Let us assume we pay cash and we are adults that are not bound by the viewing rating; once we receive the ticket, that ticket alone is all that we need to go to a hall and see the movie. The ticket is the credential needed to authorize us to see the movie and hence is a perfect example of authorization credentials. The movie ticket cannot be used instead of a passport to get onto an international flight. In other words, it is completely worthless as authentication credential (to verify our identity).

The networking counterpart of the movie ticket is the pre-paid calling card that is becoming popular. The pre-paid card allows the card holder to make a phone call through a network provider's infrastructure and the provider does not care who the user is as long as there is still credit left on the card.

1.2.2 Administration Domain and Relationships with the User

At this point, it is useful to also describe the concept of administrative domain: According to RFC 1136 [ADMRT1136], an administrative domain is "a collection of end systems, intermediate systems and sub-networks operated by a single organization or administrative authority". In AAA language, this typically means that the domain is served by the same AAA server (or pool of synchronized AAA servers, if failure recovery is important) and is ruled by the same network policies. When a user affiliates with a private network or subscribes with a commercial network provider, it is said to belong to the administrative domain. The AAA server within the domain is said to act as the user's home AAA server (AAAH or HAAA). The user's information and credentials are then stored at that AAA server and the user is said to have a trust relationship (and business relationship for commercial networks) with the administrative domain and its AAA server.

In a more general case, a user may roam to a network that is not part of its home administrative domain. The user may request services from this visited domain, which does not have any trust relationship with the user. In this general case, the service equipment, providing the service to the user, is part of a different administrative domain than the user's home domain and the visited network cannot grant any services to the user unless it has some relationship with the user's home domain. Two kinds of relationships exist between the user and the network (or between networks):

1. *Contractual relationships*: The user must have a contractual relationship with the service domain it is trying to connect to. The contractual relationship of the user is not with a specific network entity, but the entire domain. The authorization to obtain bandwidth or QoS is always processed on the basis of existing contractual relationship.
2. *Trust relationship*: The trust relationship is required for authentication or securing the communications between any two entities. In contrast to contractual relationship, the trust relationship is between the user and a specific component of the network not just with the entire domain. The trust relationship is usually instantiated in the form of a security association, which defined a set of pre-defined security algorithms and related keys to use to provide communications security.

Note that a contractual relationship is independent of trust relationship.

1.2.3 Standardization of Authorization Procedures

One can safely say that the second "A" (authorization) of the three "A"s within "AAA" is the one to which the least attention is paid in the standards developing organizations, such as Internet Engineering Task Force (IETF). Lack of attention may be something that most middle siblings in a large family of children complain about. Authorization has received its share of hand-me-downs from its bigger sibling, authentication: as we mentioned before, most of the time authorization is performed on the basis of same credentials as those used for authentication.

Realizing that distinction of authentication and authorization deserves a focused research and standardization effort, the IETF AAA working group formed a separate subgroup called authorization subgroup to start looking at procedures and standardization for authorization. However, eventually this activity gave way to more hard pressing issues in the AAA working group and was moved into a research group in Internet Research Task Force (IRTF), called AAA architecture group (AAAArch). The AAAArch group worked on several informational documents describing framework [AUTHFR2904], requirement [AUTHREQ2906], and application examples [AUTHAPP2905] for authorization. In this section, we give a few highlights from those specifications. However, the fact that these specifications come from the research branch of IETF (IRTF) rather than its engineering branch attests to the fact that these specifications have not caught on in the real world yet. Realizing that equating authentication and authorization and their credentials is becoming an issue, the IETF steering group (IESG) has recently urged network and AAA procedure designer within the engineering teams to start decoupling the authentication and authorization procedures through using a variety of tools such as separating authentication and authorization credentials. We have yet to see any significant progress in this area, but feel that it is important to recite the more important results of the IRTF documentations in the following.

A simplified architecture for systems implementing authorization mechanisms is shown in Figure 1.2. Upon receiving the request for a service or a resource, the service provider network first consults an authorization server that holds user profiles and can then make a determination on whether the user is authorized to use the service it has requested or not. The IETF authorization model reasonably assumes that authorization requests are only processed for authenticated users. In other words, we do not need to perform authorization in case we have not been able to successfully verify the end party's identity with certainty. Still, it would make sense that the same server performs both authentication and authorization, even though this server may fetch authentication credentials and user profile information from different databases. As we will see later on, since an accounting server also collects user's resource usage information, accounting is typically also performed by the same server, namely the AAA server: authentication, authorization, accounting server. The model also includes an entity that provides the actual service and is called the service equipment. The service can be something that the user considers as a service or something that provides some functionality within the network. Example of the first type of service is a music stream or a conference call, supported by an application server. In this case, the service equipment is the application server. Network functionality types of service are usually transparent to the user. For instance, a mobility agent is a service equipment that provides mobility services for the user's traffic.

The model shown here assumes that service equipment is administered by the same service provider network as the one that the user is subscribed to. In a simple case, typically called single administrative domain case, the service equipment is also part of the home administrative domain and can directly interact with the home AAA server of the user. In more general cases, the service equipment and the service provider network may be different from user's home domain.

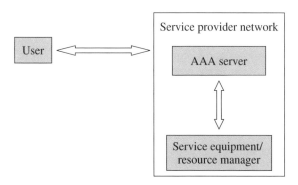

Figure 1.2 Service authorization model

1.2.3.1 Authorization Messaging

Authorization of the user for service is accomplished on the basis of interaction between the user, the AAA server, and the service equipment. Three major scenarios are recognized for this interaction, each of which leads to a different sequence for the order in which the operations are performed:

- *Agent sequence*: This sequence is used for scenarios where the user contacts an entity in the AAA infrastructure first. The contact point is typically an edge entity that is also an AAA client. The user sends its service request, which can be seen as authorization request, towards the AAA server. The AAA server authorizes the user based on the information it has on the user or after consulting other entities such as a policy server or a resource manager, such as a bandwidth manager. After the authorization is successful, the AAA server sends the authorization possibly along with other configuration information to the service equipment. The service equipment prepares itself (e.g. sets up states and so on) for providing the service and possibly acknowledges to the AAA server that it has completed the configuration procedure. The AAA server replies to the user that authorization is complete and service is set up. Here, the AAA server acts as an agent for the user, and hence the sequence of events is called agent sequence.
- *Pull sequence*: In this scenario, the user sends the request directly to the service equipment. One example is, as we will see later on, when a user requests to use the services of a Mobile IP agent for support of her mobility. The service equipment forwards the request to domain's AAA server. Note that, in case, the user belongs to a different administrative domain, this AAA server is a local AAA server that must contact the user's home AAA server. The home AAA server evaluates the request and returns a response that eventually gets back to the service equipment. The service equipment accepts or rejects the service based on the AAA server response and notifies the user accordingly.
- *Push sequence*: This model is the one most similar to the movie theatre ticket. The user gets a ticket or certificate from the service provider AAA server. Anytime the user requests a service from the service equipment, the user presents the ticket to the service equipment as a way to show that it has been authorized by the AAA server to access that service.

1.2.3.2 Policy Framework and Authorization

When we were discussing the differences between authorization and authentication, we mentioned that after verifying the identity of a user requesting a service, the network needs to check the user profile and make an authorization decision based on that profile. In order to keep the authorization process consistent and scalable, the decision is often made with the help of a pre-set policy. Since many types of policies, such as security policy, group affiliation policy, and roaming policy exist, having a policy framework in place is important. The policy framework defines various architecture elements such as a policy repository, policy decision points (PDP), and so on. Policy repository typically includes the following information: (1) available services, (2) resources about which authorization decisions can be made, (3) policy rules to make authorization decisions, and (4) authorization event log for cases when authorization may be conditioned on the log of some previous events that must have happened.

The policy framework also defines the processes for managing and sharing the policy information with other entities in the network. We do not go into the details of the policy framework and refer the reader to the work done by the Policy Framework working group in IETF. However, it is important to know that the AAA server may need to interact with entities, that we simply call policy servers in order to make appropriate authorization decisions. This interaction must allow the AAA server to retrieve the policy and enforce it during the authorization process.

1.3 Accounting

The final "A" in "AAA" is for Accounting. Even though a majority of engineers believe the terms accounting and billing have the same meaning, accounting involves more than tracking a user's total number of phone call minutes or data packets. A variety of applications are defined for accounting:

- *Auditing*: The act of verifying the correctness of an invoice submitted by a service provider, or the conformance to usage policy, security guidelines, and so on.
- *Cost allocation*: With the convergence of telephony and data communications, there is increasing interest in understanding the cost structure associated to each of the telephony and data portions.
- *Trend analysis*: Typically used in forecasting future usage for the purpose of capacity planning.

Each of the applications above may be processed at a different logic management entity. But in general, accounting is concerned with collection of information on resource consumption at all or specific parts of the network. This information is generally referred to as accounting data or accounting metrics. Typically, the network device providing services to a user collects information about user's resource consumption according to the accounting application's needs.

The accounting data collected by the network device is then carried by the accounting protocols over to the management entities responsible for each accounting application. As will be explained later on, each of these different accounting applications may have varying security and reliability requirements from accounting protocols; thus it is difficult to devise a single accounting protocol that meets the needs of every application. The goal of accounting management is to provide a set of tools that can be used to meet the requirements of each application. In the following sections, we go through the details of accounting management architecture.

1.3.1 Accounting Management Architecture

The accounting management architecture specifies interactions between network devices and accounting servers and any possible billing servers. It also defines procedures for collecting usage data. In the following, we define each of the concepts and components of accounting management architecture.

- *Accounting metrics*: The data collected by the network device and transferred to the accounting server via the accounting protocol.
- *Charging*: Charging derives non-monetary costs for accounting data based on service and customer-specific tariff parameters. Different cost metrics may be applied to the same accounting records.
- *Accounting server*: The entity that is responsible for processing the accounting data received from the network device. This processing may include summarization of fractions of accounting information (called interim accounting), elimination of duplicate data, or generation of session records. In order to reduce the volume of accounting data and the bandwidth required to accomplish the transfer, the processed accounting data can be submitted to a billing server. Such session records may be batched and compressed by the accounting server prior to submission to the billing server.
- *Billing*: Billing translates costs calculated by the charging into monetary units and generates a final bill for the customer. Billing policies define among others the type of bill, e.g. invoice or credit card charge, the form of the bill, e.g. itemized or not, and the times for the bills, e.g. weekly, monthly.
- *Billing server*: The entity that handles rating and invoice generation, but may also carry out auditing, cost allocation, trend analysis, or capacity planning functions.
- *Accounting proxy*: The entity that acts as both a client and a server. When a request is received from a client, the proxy acts as an AAA server. When the same request needs to be forwarded to another AAA entity, the proxy acts as an AAA client.
- *Local proxy*: A AAA server that satisfies the definition of a proxy and exists within the same administrative domain as the network device (e.g., NAS) that issued the AAA request. Typically, a local proxy will enforce local policies prior to forwarding responses to the network devices, and are generally used to multiplex AAA message from a large number of network devices.

1.3.1.1 Accounting Across Administrative Domains

In networks with a large amount of users, large geographical footprint or in cases where users roam between networks owned by different service providers, the administration of each network can be managed by a separate accounting and billing server.

- *Intra-domain accounting* involves the collection of information on resource usage and it's processing within one administrative domain. In intra-domain accounting, accounting packets and session records typically do not cross administrative boundaries.
- *Inter-domain accounting* involves the collection of information on resource usage within an administrative domain, for use within another administrative domain. In inter-domain accounting, accounting packets and session records will typically cross administrative boundaries.

The architecture shown in Figure 1.3 depicts interaction between various entities within a multi-domain accounting model. The accounting server needs to distinguish between inter- and intra-domain accounting events and route them appropriately. A specific type of entity identifier, called network access identifier (NAI) is generally used for this purpose. The NAI consists of a user identity followed by a realm or domain identity in the form of userID@realmID. If the session record contains the NAI for the end node, the involved server can by examining the domain (realm) portion of the NAI identify the domain to which

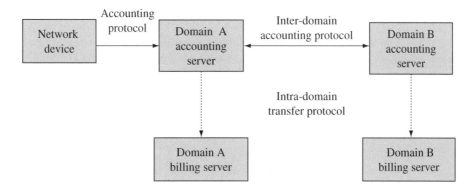

Figure 1.3 An illustration of accounting management architecture

the data need to be routed to. If the domain portion is absent or corresponds to the local domain, then the session record is treated as an intra-domain accounting event. Intra-domain accounting events are typically routed to the local billing server, while inter-domain accounting events will be routed to accounting servers operating within other administrative domains.

1.3.2 Models for Collection of Accounting Data

Several accounting data collection methods are implemented in the industry. Examples are the polling model and the event-driven model. In the following sections we provide a short description for a few of these models.

1.3.2.1 Polling Models for Accounting

In the polling model, an accounting manager will poll devices for accounting information at regular intervals. In order to ensure against loss of data, the polling interval will need to be shorter than the maximum time that accounting data can be stored on the polled device.

Without non-volatile storage, the polling model may result in loss of accounting data in case the device reboots. The polling model performs poorly in implementation of roaming applications, i.e. in scenarios where the accounting data may be collected at multiple network devices due to user's roaming. For example, if a roaming user is receiving services in a network other than its home domain, to allow issuance of a single bill to the customer, the foreign domain needs to have roaming agreement with the user's home domain and send the accounting data collected for the user to the user's home domain for processing. In order to retrieve accounting data for the user within a given domain, the accounting manager would need to periodically poll all devices in all domains, which would add a processing delay to the billing process, not to mention the routing problems that can arise.

1.3.2.2 Event-Driven Models for Accounting

In an event-driven model, a device will contact the accounting server when it is ready to transfer accounting data. Most event-driven accounting systems, such as RADIUS-based accounting systems, do not perform batching and hence transfer only one accounting event per packet. This model is called even-driven model without batching and is rather inefficient. An event-driven model typically stores accounting events that have not yet been delivered only until a timeout interval expires. Once the timeout interval has expired, the accounting event is lost, even if the device has sufficient buffer space to continue to store it. As a result, the event-driven model has the smallest memory requirement, and is the least reliable, since accounting data loss will occur due to device reboots, sustained packet loss, or network failures of duration greater than the timeout interval. The event-driven model is frequently used in networks with roaming applications, since this model sends data to the recipient domains without requiring them to poll a large number of devices.

Since the most basic event-driven model does not support batching, it permits accounting records to be sent with low processing delay, enabling application of fraud-prevention techniques. However, because roaming accounting events are frequently of high value, the poor reliability of this model is an issue. As a result, an event-driven polling model may be more appropriate.

In the event-driven model with batching, again, a device will contact the accounting server or manager when it is ready to transfer accounting data. However, the device contacts the server when a batch of a given size has been gathered or when data of a certain type are available or after a minimum time period have elapsed. Since, such systems transfer more than one accounting event per packet, they are more efficient. An event-driven system with batching will store accounting events that have not yet been delivered up to the limits of memory. As a result, accounting data loss will occur due to device reboots, but not due to packet loss or network failures of short duration. Note that while transfer efficiency will increase with batch size, without non-volatile storage the potential data loss from a device reboot will also increase.

Through implementation of a scheduling algorithm, event-driven systems with batching can deliver appropriate service to accounting events that require low processing delays. For example, high-value inter-domain accounting events could be sent immediately, thus enabling use of fraud-prevention techniques, while all other events would be batched. As a result, this approach can have good scalability and flexibility characteristics.

In the event-driven polling model, an accounting manager will poll the device for accounting data only when it receives an event. Events are typically generated by accounting clients. Examples of events include: when a batch of a given size has been gathered, when data of a certain type are available or lapse of a minimum time period. Without non-volatile storage, an event-driven polling model will lose data due to device reboots, but is resistant to packet loss, or network partitions of short duration due to its buffering nature. Unless a minimum delivery interval is set, event-driven polling systems are not useful in monitoring of device health.

The event-driven polling model can be suitable for use in roaming, since it permits accounting data to be sent to the roaming partners with low processing delay. At the same time, non-roaming accounting can be handled via more efficient polling techniques, thereby providing the best of both worlds.

Where batching can be implemented, the state required in event-driven polling can be reduced to scale with the number of active devices. Note that processing delay in this approach is higher than in event-driven accounting with batching since at least two round-trips are required to deliver data: one for the event notification, and one for the resulting poll.

1.3.3 Accounting Security

In an accounting framework, two types of data communications are required: the exchange of accounting policies and the collection of accounting records. Both communications introduce potential security hazards. However, accounting records provide the basis for billing. Therefore, the motives and potential for fraud is extremely high in the collection of accounting data. Thus, different accounting applications may impose different requirements on security of accounting protocol. In this section, we describe security requirements for accounting protocols:

- *Secrecy of accounting policies and accounting data*: Unauthorized entities should not be able to read or modify accounting policies or accounting records.
- *Authentication of accounting data and accounting policy sources*: One should ensure that the data are originated from the original source. Source-authentication can be achieved by using digital signatures.
- *Protection of the integrity of accounting policies and records*: It should be ensured that the data was not modified on the way from sender to receiver. Data-authentication can also be achieved with digital signatures.
- *Verification of generated accounting data for correctness*: Correctness of accounting data generated by a service provider must be ensured. A provider may generate incorrect accounting records either deliberately or unintentionally through faulty configuration. These incorrect accounting records probably have the consequence of incorrect bills. Customers can verify the correctness of the accounting data through their measurements and/or through data collected by a trusted third party. A trusted third party can be an independent accounting service provider or a more general entity providing an auditing service.

1.3.4 Accounting Reliability

In this section, we describe reliability requirements for accounting protocols. Typically, accounting faults include packet loss, accounting server failures, network failures, and server reboots, and it is important that accounting management systems be scalable and reliable. However, different accounting applications may impose different requirements on accounting protocol. In applications such as usage-sensitive billing, cost allocation, and auditing, loss of accounting data can translate to revenue loss and there is an incentive to engineer a high degree of fault resilience. Some of the application-specific reliability requirements are listed below.

- *Billing*: When accounting data are used for billing purposes, the requirements depend on whether the billing process is usage-sensitive or not. Non-usage-sensitive billing does not require usage information; in theory, all accounting data can be lost without affecting the billing process. This would, however, affect other tasks such as trend analysis, auditing, and data on wholesale information. Usage-sensitive billing processes depend on usage information; therefore, packet loss may translate directly to revenue loss. As a result, the billing process may need to conform to financial reporting and legal requirements, and therefore an archival accounting approach may be needed.

- *Trend analysis and capacity planning*: In trend analysis and capacity planning, the forecasts are inherently imperfect, high reliability is typically not required, and moderate packet loss can be tolerated.

1.3.4.1 Interim Accounting

Billing applications typically require the NAS to send messages to the AAA server, indicating the start and termination of sessions for which accounting is to be processed. However, the bulk of information, necessary for the billing process, such as session time, number of bytes transferred, etc. is only available in the messages that are sent at the end of session (such as accounting stop message). This means, if for some reasons, such as power failure, reboot, network problem, the NAS becomes unavailable for some time, it becomes impossible for a service provider to bill for the sessions initiated on that NAS.

A procedure called interim accounting provides a remedy to this problem by having the NAS sending periodic updates with information pertaining to the sessions to the AAA server. This way, interim accounting provides checkpoint information that can be used to reconstruct the session record in the event that the session summary information is lost. To accomplish this, a new accounting message called interim accounting message was introduced in [RADINTDR]. Interim accounting message is sent periodically from NAS to the AAA server. It should be noted that the applicability of interim accounting is limited. For instance, in case of packet loss due to network congestion, sending interim accounting data over the wire can make the problem worse by increasing bandwidth usage. Therefore, the use of on-the-wire interim accounting is best restricted to only long-lived sessions, where accounting data are of exceptionally high value. This is accomplished by simply choosing the interim-reporting interval to a value larger than the average session duration. This ensures that most sessions will not result in generation of interim accounting events, and the additional bandwidth as well as memory space consumed by interim accounting will be limited.

1.3.4.2 Transport Protocols

Accounting protocols dealing with session data need to be resilient against packet loss especially in inter-domain accounting, where packets often pass through network access points. Resilience against packet loss can be accomplished via implementation of a retry mechanism on top of UDP, or use of TCP or SCTP. UDP-based transport is frequently used in accounting applications. When implementing UDP retransmission, there are a number of issues to keep in mind: one of them is retransmission behavior, which is important that the retry timers relate to the round-trip time, so that retransmissions will not typically occur within the period in which acknowledgments may be expected to arrive.

1.3.4.3 Fail-Over Mechanisms

Congestion control is another issue where without non-volatile storage, the combination of congestive back off and buffer exhaustion can result in loss of accounting data. Accounting server fail-over is a powerful tool for providing resiliency against failures. In the event, a primary accounting server fails, it is desirable for the device to fail-over to a secondary server, which can take over the responsibilities of the primary device in a seamless manner. Providing one or more secondary servers can remove much of the risk of accounting server failure. Secondary fail-over servers have become commonplace.

1.3.5 Prepaid Service: Authorization and Accounting in Harmony

A new trend that has been emerging in the cellular telephony during the past few years is that the users instead of binding themselves to a single cellular operator through a long-term contract, simply buy a calling card for specified number of minutes. The pre-paid card allows the user to place phone calls up for a number of minutes up to the face value of the card through the network of any operator honoring the card. In the networking jargon, this is called pre-paid service. We consider pre-paid service as a perfect example where authorization and accounting go hand in hand. The user is authorized to make a call as long as there is credit available on her call. As the call proceeds, the accounting mechanism collects the usage information and informs a specific server, typically called pre-paid server to make sure that the amount of the credit on the card is depleted according to the usage information.

1.4 Generic AAA Architecture

We explained the three-party authentication model deploying a AAA server, and throughout this chapter attempted to show why authentication, authorization, and accounting procedures should be performed by the same server: the AAA server. In the remainder of this chapter, we complete this discussion by providing a brief overview of the generic AAA architecture defined by a former subgroup of the AAA working group in IETF. These activities later moved into the AAAArch group within IRTF.

The AAAArch team defined a generic AAA architecture described in the form of the experimental RFC 2903 [GENAAA2903], which specifies how the AAA architecture can interact with other network management entities. Each of these management entities provides a specific service or function, but since it receives assistance from the AAA infrastructure it is seen as an application for the AAA infrastructure (called AAA application). Examples of such services or applications are bandwidth management, quality of service, and mobility services. The fact that RFC 2903 is an experimental RFC indicates that this specification provides guidance (rather than a standard) for how such architectures can be designed in the future. As we will see in later chapters, early AAA protocols such as RADIUS were not designed on the basis of architecture in mind. However, newer protocols such as Diameter are designed this way. Diameter has separate specifications for each application, and this is why we feel paying some attention to the generic AAA architecture defined by RFC 2903 is worth the time. RFC 2903 defines a new concept called Application-Specific Module (ASM) to abstract the functionality of entity that the AAA server interacts with. The ASM is the entity that manages an application. Examples of ASMs are QoS managers, bandwidth brokers, and Mobile IP agents. RFC 2905 [AUTHAPP2905] provides more details on each application and its interactions with the AAA servers. The RFC 2903, however, by defining the concept of ASM, can define a generic interface between AAA servers and any type of management entity without getting into details of each application.

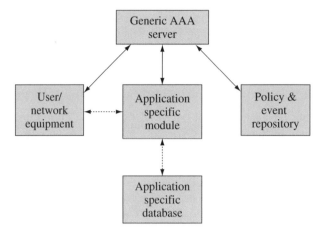

Figure 1.4 Generic model for interaction of AAA server with various management entities

Figure 1.4 shows the interaction between AAA server and other network entities according to the generic AAA architecture in RFC 2903. The authorization messaging process that we described earlier is an example of how such AAA architecture is deployed for authorization processes. As we saw, the details of the interaction between the user, AAA server, and application module depend on the specific deployment scenario and hence must be designed accordingly. The architecture, however, provides a framework for this interaction. The picture also shows that, besides the application-specific module, the AAA server may also need to interact with a policy repository to fetch the policy rules that apply to the user and the application at hand. The protocols used for interaction between the AAA server and the policy repository may be different from those used for interaction between the AAA server and the ASM. The same applies for interaction between ASM and its database (if this database is not co-located with the management entity itself).

It is important to note that each service has application-specific information (ASI), such as amount of bandwidth, that only the application-specific module (bandwidth broker in this case) understands. The AAA server that is supposed to authorize the request may not understand the details of each piece of information. Furthermore, the information may have a unique structure as well. Therefore, in general the AAA server is not required to have application-specific knowledge and needs to refer to the ASM, which was called service equipment during the discussion of authorization process earlier in this chapter. In order to divide the tasks between AAA server and ASM, the ASI needs to be separated from the logic needed for authorization, so the network equipments and AAA servers can treat the ASI in an opaque manner and send it back and forth as needed. All the AAA server needs to know is the location (ASM) to which it needs to send the application-specific data. As we will see later on, for most AAA protocols, specific data units called attributes are designed for each type of ASI. As long as the AAA can recognize the attribute type, it can simply tell what application it relates to. Typically, some sort of identifier for the ASM is included in the request to the AAA server, so that it can route the packet to the proper destination. For instance, an NAI may be used to help the AAA server identify the ASM or to let the AAA server know that the ASM may belong to a completely different administrative domain.

The following provides an example on the chain of actions that should take place in the generic AAA architecture.

- The user or the network equipment that acts as an application agent for the user (such as an NAS) sends a combined authentication and service request to AAA server. For this request to be granted, the least required from the user is to present proper authentication and authorization credentials.
- AAA server verifies the authentication credentials for the user after consulting its user databases and possibly the policy repository. When authentication is successful, the AAA server inspects the contents of the authorization request and determines what the authorization is requested for.
- AAA server retrieves policy rules from the repository (if needed) and performs an authorization decision on each component (attribute) of the request according to one the following alternatives:
  - When the component is an ASI that must be processed by the ASM, the component is sent to the ASM to be evaluated.
  - Query the policy repository or an event log (if implemented) for an answer.
  - Forward the component to another AAA server, if other administrative domains are involved.
- The AAA server informs the application-specific network equipment (service equipment) about the authorization and possibly provides necessary information for setting up the service at the point it is being delivered to the user (typically network point of presence), and if needed orders that POP to report accounting data back to the AAA server.

Two notes are in order: First, if any of the information components are to be maintained private and/or secure, the AAA server must make sure the components are encrypted and/or authenticated while in transit. Depending on the type of protocol used, message routing and security may be performed hop by hop or end to end as will be discussed in Chapters 6 and 7. Second, the decision of authorization is still made by the AAA server, even if the services, resources, or policy decisions and management are performed at other entities. For instance, the AAA server decides whether the user is authorized to receive a high bandwidth data stream or a new temporary IP address, while the bandwidth broker or the mobile IP agent manage the admission control and bandwidth allocation or routing configurations, respectively.

1.4.1 Requirements on AAA Protocols Running on NAS

We discussed the role of network access server (NAS) in a three-party authentication model. However, since the NAS functionality is typically implemented inside network edge devices (network point of presence), the role of NAS has now changed from simply acting as a pass-through for authentication to a device that performs many access control and set up functions. Not only is the NAS a full blown AAA client that has a role in all the As in AAA (rather than just authentication), but also may support many different types of protocols such as access protocols, device management protocols, and routing protocols. So the general assumption is that the NAS *must* be able to support many of not all these protocols, and this makes the general perception of an NAS a rather non-standard one, since different network designers may expect different functionalities from the network edge devices. Hence, from an AAA standpoint, it is useful to define requirements on an NAS when it comes to support for AAA protocols. RFC 3169 [NASCRIT3169] has done just that. Both authentication, authorization, and accounting requirements are covered in this specification. In this final section of the chapter, we provide a few examples from this specification. It should be noted that these requirements are not specific for support of a specific AAA protocol; as a matter fact as we will see in Chapter 6, RADIUS clients (NAS) actually do not meet many of these requirements.

- AAA protocols are considered application level protocols and hence must be carried by other communications protocols, including transport layer protocols. Transport reliability requirement can be different for authentication messages and accounting messages. However, it is stated that AAA protocol *must* be able to detect a failure of the transport protocol to deliver a message within a known and controllable time period.
- In case of a AAA server failure, the AAA protocol *must* allow any sessions between an NAS and a given AAA server to fail-over to a secondary server without loss of state information.
- AAA protocol at NAS must support at least the RADIUS attribute set.
- The AAA protocol *must* provide mutual authentication between AAA server and NAS. At minimum, the AAA protocol should support use of shared secrets between NAS and AAA servers and avoid using the same shared secret for all NASs. More details on this will be provided in Chapter 6. The AAA protocol *must* also support public key infrastructure (PKI)-based methods for mutual verification of identity (authentication). We will discuss PKIs in Chapter 9.
- Some attributes (application specific information) may need to be encrypted. The AAA protocol *must* support selective encryption of attributes on an attribute-by-attribute basis.
- The NAS and AAA protocol *must* support transport (between NAS and AAA server) of credentials needed for authentication of users to servers and servers to users. Examples of credentials are password, response to challenges, X.509 certificates, and Kerberos tickets. Chapter 2 provides more detail on various authentication methods. The AAA *must* also provide means to hide these credentials from non-AAA entities during the transport.
- The NAS must be able to support multiphase authentication, such as prompting a text from NAS to user, engaging in challenge–response exchanges, and so on.
- The NAS and the AAA protocol must be able to support common types of identity, such as NAI and EAP identity request packets (see Chapter 2), and may be other network-specific identifiers.
- The AAA protocol *must* be able to provide an unambiguous location code that reflects the geographic location of NAS. It *must* also be able to provide unambiguous time stamp needed for authentication and auditing purposes.
- The AAA *must* provide means of assigning IP addresses to an incoming terminal as needed. The AAA protocol *must* provide a means of applying IP protocol filters to user sessions.
- For proper network operation management, it is important that the network load and usage is measured with accuracy, and for this purpose real-time accounting is important to monitor the status of sessions served by the NAS in real time. Therefore, the AAA protocol and NAS *must* support delivery of accounting information triggered by a number of events such as: start and end of sessions, expiration of certain pre-determined intervals, re-authorizations.

1.5 Conclusions and Further Resources

This chapter attempted to present the basic problems as well procedures regarding to authentication, authorization, and accounting. Since the topic of authentication is a rather complicated one that has gone under tremendous amount of progress, we will devote another chapter on that topic. As mentioned earlier, authorization is less standardized. However, we expect that with the increase in user mobility and service offering, the field of authorization will go through much further progress. For now, most of the work related to authorization is categorized under policy frameworks. The reader is referred to the Policy Framework working group in IETF for further information in the policy area: http://ietf.org/html.charters/policy-charter.html.

Also more information about authorization can be found in the AAA applications examples [AUTHAPP2905].

IETF documentations such as [ACCMGM2975] and [POLACC3334] can provide more information on accounting.

In the area of billing and accounting, many service operators may deploy their own specific technologies that follow their own policies. More information can be found from the companies providing AAA servers to these service operators. Examples are

- Bridgewater systems: http://www.bridgewatersystems.com/document_library/
- Interlink networks: http://www.interlinknetworks.com/resource/wp5-1.htm
- Funk software: http://www.funk.com/

1.6 References

1. [GENAAA2903], Laat, C. *et al.*, "Generic AAA Architecture", IETF, RFC 2903, August 2000.
2. [AUTHFR2904], Vollbrecht, J. *et al.*, "AAA Authorization Framework", IETF, RFC 2904, August 2000.
3. [AUTHREQ2906], Farrell, S. *et al.*, "AAA Authorization Requirements", IETF, RFC 2906, August 2000.
4. [AUTHAPP2905], Vollbrecht, J. *et al.*, "AAA Authorization Application Examples", IETF, RFC 2905, August 2000.
5. [ADMRT1136], Hares, S. and Katz, D., "Administrative Domains and Routing Domains, A Model for Routing in the Internet", IETF, RFC 1136, December 1989.
6. [ACCMGM2975], Aboba, B. *et al.*, "Introduction to Accounting Management", RFC 2975, October 2000.
7. [RADINTDR], Calhoun, P. *et al.*, "RADIUS Accounting Interim Accounting Record Extension", draft, January 1998.
8. [POLACC3334], Zseby, T. *et al.*, "Policy Based Accounting", RFC 3334, October 2002.
9. [NASCRIT3169], Beadles, M. and Mitton, D., "Criteria for Evaluating Network Access Server Protocols", IETF, RFC 3169, September 2001.

2

Authentication

Basic definitions and various types of authentication such as user authentication and message authentication were described in Chapter 1. As we saw, the basic philosophy of authentication is easy to understand. However, providing a complete solution typically requires understanding of all the dimensions of the problem typically through (thread analysis). This can be rather difficult as more and more sophisticated networks and services arise, and more and more sophisticated hackers deploy ever stronger computing devices against these networks everyday. Thus, it is no surprise that a large number of authentication mechanisms have been developed and standardized through a variety of standard bodies. The plethora of authentication mechanisms is so large that the Internet Architecture board (IAB) has decided to conduct a survey and a classification of various authentication mechanisms to aid the designers to better understand and track the process in the field [AUTHSRV].

In this chapter, we first delve into more details of authentication procedures and models and describe some of the most common authentication mechanisms. We also lay the groundwork for the latest and more advanced authentication mechanisms described in the final chapters of this book. Finally, at the end of this chapter, we describe the IAB classification of the authentication mechanisms, since we feel this overview enables the reader to simply place any new authentication mechanism he/she may come across in future into the proper category, and thereby quickly judge the security strengths and flaws of the mechanism.

2.1 Examples of Authentication Mechanisms

In this section, we describe the most widespread legacy methods used for authentication mechanisms of users and messages. Some of these mechanisms may be considered inadequate from a security standpoint these days. However, we feel this overview will help us pave the way for discussing more sophisticated extensible authentication protocol (EAP) based authentication mechanisms, especially since EAP was originally designed as extension to legacy authentication mechanisms. We described how device authentication mechanisms are now gaining distinction and significance of their own, reading to more sophisticated methods for authentication of devices compared with those traditionally utilized for authentication of users. A few of the device authentication methods are highlighted in this section for completeness. In later chapters, we will provide more details on these more sophisticated authentication methods.

2.1.1 User Authentication Mechanisms

For a long time, users that needed to remotely access the network through wired connections were using dial-up services to a modem pool provided by the network operator. A majority of the dial-up services used a protocol called point-to-point protocol (PPP) that was designed to provide a number of link layer functionalities. Several user authentication mechanisms were designed for controlling access to the dial-up services to users. PPP was therefore designed to support facilities for performing several user authentication mechanisms and carrying their related data. We describe these user authentication mechanisms, since they are considered as important legacy mechanisms. However, to understand how these authentication mechanisms are embedded inside PPP, we first provide some details on the PPP itself.

Overview of Point-to-Point Protocol (PPP)

The traditional dial-up networks and even some of the newer wireless access networks, such as CDMA2000 cellular networks, provide data service to mobile users relying on PPP for establishment of layer-2 links to network access node. From the standpoint of these networks, the main role of PPP is to provide layer-2 framing by encapsulating the data between the PPP header and the cyclic redundancy (CRC) checks (see Table 2.1). Note that the CRC provides protection against random channel errors, but not active attacks. It is also important to note that PPP provides a set of services beyond just framing services. As explained in the following, PPP includes three phases that are vital for setting up dial-up connections:

- *Link control protocol (LCP) phase*: It is the first phase of PPP during which PPP allows the user and the network to negotiate link parameters such as maximum frame size and link speed. The important point for this discussion is that LCP also allows the two parties to negotiate a mechanism for the authentication to be performed during the next phase.
- *Authentication phase*: The second phase of PPP is designed to handle authentication exchanges according to the authentication protocol that was negotiated during the LCP phase of PPP. The PPP end point within the network can either authenticate the supplicant directly, or act as a mediator and pass the authentication credentials to an authentication server (or an AAA server). PPP originally supported two authentication methods: password authentication protocol (PAP) and challenge handshake protocol (CHAP). As shown in Table 2.1, the protocol field in the PPP header can indicate which authentication protocol is being used.
- *Network control protocol (NCP) phase*: It is the phase during which network layer parameters such as header compression protocol and the network protocol (such as IP or IP control protocol, IPCP) are negotiated.

Table 2.1 shows the format for PPP frames. This format is used for both PPP control messaging and actual user data. In other words, as we mentioned earlier and will discuss in

Table 2.1 PPP frame format

Field name	Sub-field	Description
PPP header	Address	FF for multidrop links
	Control	To distinguish error and flow control messages (otherwise usually 03)
	Protocol	ID for protocol, whose data are encapsulated in PPP frame as PPP data; 0021 for IP, C023 for PAP, C223 for CHAP
PPP data1 (protocol header)	Message Code	01 for configure request, 02 and 03 for configure Ack and Nack, 04 for configure reject
	Message ID	Acts a sequence number to match request and responses.
	Message length	Message length
PPP data 2 (protocol data)		
CRC		Performed on original data

detail later on, PAP- and CHAP-related data during the authentication phase can be carried using this format. Once PPP control signaling is complete (all three PPP phases are complete) and the PPP link is established, this format is used to encapsulate user data, such as IP data packets inside the PPP frames. The data packets are wrapped within the PPP header and the CRC field of the PPP frame.

2.1.1.1 Basic PPP User Authentication Mechanisms

As mentioned earlier, the initial PPP design according to RFC 1661 [PPP1661] supports PAP and CHAP as authentication mechanisms. In this section, we describe how PAP and CHAP exchanges are performed during PPP link setup.

PAP is the most basic user authentication protocol used with PPP. A user (supplicant), conforming to a password-based authentication mechanism, submits its credentials (user name: joe, password: SecRet) to the PPP end point (authenticator). The authenticator creates a PPP authentication request by encapsulating the user's credentials inside a PPP frame as shown in Table 2.2. This attests to our previous statement that the PAP used by PPP is not secure, since the password information is carried in the clear without any protection over the PPP link. As we will see later on, AAA protocols such as RADIUS provide additional protection of such sensitive data (password hiding).

In order to avoid sending the sensitive password information over unprotected communication links, CHAP [CHAP1994] was developed. Instead of directly asking for the password, the server issues a randomly generated value (challenge) and expects the user to provide a response to the issued challenge based on her knowledge of a secret. The user takes the challenge and using the key it shares with the server (can be the password), and a hash protocol calculates the response. The server does the same calculation and compares the results of its own calculation with that received from the user and if there is a match, the user is authenticated. The hash protocol is usually a one-way message digest, such as MD5 [MD51321], which is negotiated during the LCP phase of PPP. To ensure anti-replay attacks, the server should use each challenge value for a short period of time. After that period has expired, the server should create a new challenge for the next challenge/response

Table 2.2 PPP frame carrying PAP authentication request

Field name	Sub-field	Description
PPP header	Address	FF
	Control	03
	Protocol	C023, which means authentication protocol is PAP
PAP header	PAP code	01 for authentication request (configure request)
	PAP ID	01 (first message)
	PAP authentication request length	00 0F
PAP data	Peer ID length	3 (3 octets for joe)
	Peer name	6A 6F 65 (letters j o e in ASCII)
	Password length	06 (6 octets for 6 letter for SecRet)
	Password	53 65 63 52 65 74 (ASCII for letters S e c R e t)
CRC		85 6B for CRC 16

handshake. Each challenge value is accompanied by a challenge ID to distinguish that challenge from others. To indicate that a new challenge has been issued, the server assigns a new challenge ID to each newly issued challenge. This way, a user may receive different challenges every time and when responding, the user includes the new challenge ID in the hash calculation.

The CHAP protocol defines challenge and response packets, each including a CHAP header, the CHAP data itself (value), and other related fields as shown in Tables 2.3 and 2.4. The CHAP header consists of a code, an identifier, and a length field. In a manner

Table 2.3 CHAP challenge packet (sent as PPP data)

Field name	Sub-field	Description
CHAP header	CHAP code	01 for challenge
	CHAP ID	01 (first message)
	CHAP length	00 23 (for 35 octets)
CHAP data	Value size	10 (for 16 octets or 128 bits challenge)
	Value	A randomly generated 128 challenge
	Server name	Name of dial-up server challenging the user

Table 2.4 CHAP response packet (sent as PPP data)

Field name	Sub-field	Description
CHAP header	CHAP code	02 for response
	CHAP ID	01 (copied from challenge ID)
	CHAP length	00 18 (for 24 octets)
CHAP data	Value size	10 (for 16 octets or 128 bits hash value included)
	Value	A 128 bit MD5 hash calculated over ID, shared secret (SecRet) and challenge value sent by server (see RFC 1321 for MD5)
	Peer name	Name of the challenged user

similar to PAP, when CHAP is used as the authentication mechanism over PPP link, the CHAP packets themselves carried inside PPP frames as PPP data.

The identifier field is one octet that *must* be changed each time a challenge is sent and serves to match the corresponding challenge and response packets. Therefore, the response identifier must be copied from the identifier field of the challenge packet, which caused the response. The challenge value is a variable stream of octets and hence must be changed each time a challenge is sent. The length of the challenge value is independent of the hash algorithm used and depends on the type of random generator.

As mentioned earlier, when presented by a challenge in a PPP frame carrying the challenge packet, the user responds to the PPP server's challenge by sending a response calculated based on the server's challenge and her key. The response value is the one-way hash calculated over a stream of octets (concatenation of) of the identifier, the "secret", and the challenge value. The secret is a key (could be a user password) that the user shares with the server. The length of the response value depends upon the hash algorithm used (16 octets for MD5). CHAP can use different hash algorithms and MD5 is one of them, but MD5 is the only one specified in the CHAP RFC.

As will be evident from our classification of authentication mechanisms, CHAP is not completely secure either. Although the user's password is not sent in clear, the password storage can still be compromised. PAP prevents this by storing hash values of passwords, but this is not possible in CHAP. The problem is that since both challenge and its hash value are sent along the same message, it is subject to dictionary attacks, i.e. the attacker records the message, and then runs a brute force algorithm on the challenge until it finds the hash value and that way finds the key.

2.1.1.2 Shortcoming of PPP Authentication Methods

As the name applies, PPP link only runs on a single hop between the supplicant and the PPP end node and therefore carries the PAP and CHAP data only over the access link. When a three-party authentication model is applied, i.e. the authenticator passes the PAP or CHAP credentials to an authentication server, another protocol (typically an AAA protocol) must be used to carry this data from the PPP end point to the authentication server in a secure manner. We will discuss methods for secure transport of sensitive data over AAA protocols in Chapters 6 and 7, so for this discussion we focus on the security of the access link provided by the PPP link.

Contrary to popular belief, PAP and CHAP over PPP are merely user authentication protocols and not encryption protocols, i.e. they do not provide data privacy protection for the PPP link. PPP link by itself does not provide data encryption service for protection of PAP or CHAP data either. Hence, the PAP or CHAP data carried by PPP essentially goes over unprotected media. The reason PPP was designed this way is because it was designed for phone lines and phone lines are physically difficult to hijack. Furthermore, PAP and CHAP themselves are not considered as cryptographically strong mechanisms. As we will see in the final parts of this chapter, PPP and CHAP are vulnerable to a variety of attacks. For those reasons PPP-based authentication phase is deemed unfit for wireless channels, such as CDMA2000 and 802.11 wireless local area networks (WLAN). These systems tend to deploy specific methods for authentication of data traffic as explained later on.

A final drawback with PPP authentication mechanisms is that, when PPP is used for link layer connections, the protocol and mechanism by which the user were to authenticate to the network was negotiated during LCP phase of PPP. The actual act of authentication is performed during the authentication phase of PPP. This means that the network point of presence with which the user engaged needs to understand the negotiations regarding the authentication mechanism. Adding new authentication mechanisms, algorithm, or even parameter changes requires upgrading the network points of presence.

2.1.1.3 Extensible Authentication Protocol (EAP) as Extension to PPP

Due to the inflexibility of PPP toward new authentication mechanisms, the IETF decided to extend the PPP protocol. IETF PPP extensions group designed the EAP specifications in the form of RFC 2284 [EAP2284] as an extension to PPP. EAP was designed to provide extensibility to PPP by providing support for generic and newer authentication mechanisms. EAP can be chosen as one of authentication mechanisms over a PPP link. However, the difference between using EAP versus PAP or CHAP over PPP links is that the two ends of the PPP negotiation negotiate the use of PAP or CHAP as well as the authentication parameters (such as hash function, etc.) during LCP link negotiation phase, while, EAP as an authentication method is negotiated during the authentication phase of PPP as opposed to during LCP phase. This way, when EAP is chosen as authentication protocol, the actual authentication is performed at a later stage. The EAP messages can be carried over PPP frames as shown in Table 2.5.

Since the days of PPP and original creation of EAP as a PPP extension, EAP has found applications in many different environments and scenarios, especially in access environments, where IP is not implemented over the access link. EAP is now considered as an authentication framework that can support multiple and even future authentication mechanisms. We will go through EAP in more detail later on in this chapter and will devote an entire chapter (10) to discussing a variety of applications of EAP in authentication and key management. We will also talk about the support for EAP by the AAA protocols in Chapters 6 and 7.

2.1.1.4 SIM-Based Authentication

A very popular authentication mechanism coming from the cellular world (GSM) is subscriber identity module (SIM) based authentication. SIM is a sort of smart card hardware that as the name suggest is configured for a specific subscriber identity and can be inserted into any GSM standard phone. Since the SIM card can potentially be moved between different phones, it is seen as a rather strong user authentication mechanism.

Table 2.5 PPP Frame Carrying EAP Messages

Field name	Sub-field	Description
PPP header	Address	FF
	Control	03
	Protocol	C227, which means authentication protocol is EAP
EAP header	As described in Table 2.6	
EAP data	As described in Table 2.6	
CRC		

SIM-based authentication is essentially a challenge/response mechanism, during which the SIM card is presented with a random challenge called RAND (128 bits) by the network authentication server. The cryptographic processor on the SIM card creates a response (SRES) based on an operator-specific algorithm and key Ki that is recorded in its memory. The authentication server also holds a copy of Ki and performs the same calculation on the RAND, and by comparing the results with SRES received from the SIM card on the user's phone can verify that the user does in fact hold the proper SIM card (and the key Ki).

Besides the SRES, the card also produces an encryption key (Kc) that is later on used for encryption of wireless link traffic.

$$[SRES, Kc] = F(RAND, Ki)$$

In the GSM world, the group (RAND, SRES, Kc) is sometimes referred to as the "triplets". It is important to note that this is a good example of an authentication mechanism that not only provides client authentication, but also helps establish encryption keys for traffic that follows after the initial authentication. A final note on SIM authentication is that although a SIM card is widely considered as a user authentication mechanism, since it is actually a piece of hardware that is rarely removed from the cellphone by the user, it is actually a device authentication method. Unless the user carries the card separately from the device and only inserts it in the phone prior to the phone call or protects the use of the SIM card by her personal password, it cannot be seen as a user authentication method. Even then, since the SIM card can be unlocked based on the password, its cryptographic strength as a user authentication mechanism is reduced to the strength of a password mechanism and all that effort behind its clever design goes to waste.

We will come back to this during our discussion on device and user certifications in Chapter 9, but want to provoke the reader's thoughts by saying that the SIM card should truly be identified as a device authentication mechanism, despite the popular belief.

2.1.2 Example of Device Authentication Mechanisms

We briefly went through the distinction between the user and device authentication in Chapter 1 to emphasize that it is important to allow devices to authenticate to the network in the absence of human users. To strengthen our argument against SIM cards as a user authentication mechanism, we will provide a perfect example on how a SIM card can be considered as device authentication. Assume that a cellular phone including a valid SIM card is lost or stolen, and the user never had time to configure the phone with her personal password. Until the owner reports the loss of theft, the lucky new owner can simply use the phone including the SIM card to make phone calls. The SIM card unknowing about its new owner will proceed with the challenge/response, using the key in its memory.

Instead of going through a variety of device authentication mechanisms, we like to quickly announce that we feel the most pronounced way of providing device authentication is through device certificates and provide a short overview of certificate based authentication. However, before doing that, we provide several pieces of advice for the reader to consider when dealing with device authentication mechanisms:

- Device authentication typically needs to precede user authentication, so that the device can proceed with the initial link and network configurations that should otherwise be transparent to the user. However, unless absolutely necessary for the device to be able to function in the absence of a user, device authentication should take the interaction with the network only to a certain point. The network needs to require the user to perform an authentication before granting full authorization to full utilization of network services. Otherwise, a device in the wrong hands can be the open gate to an otherwise well-guarded network.

- The credentials such as keys and certificates must be safeguarded properly on the device. Depending on the network policy, making the storage of the credentials tamper evident may not be enough and the device designer may have to provide proof that the device credentials are tamper resistant, i.e. they are destroyed if tampered with.

2.1.2.1 Public Key Certificate-Based Authentication

We will provide more details about the public key cryptography methods in Chapter 3 of this book. However, the concept of public key cryptography is by now intuitive for many people; so here we will only provide a brief introduction on how it can be used for mutual authentication between two end parties or for message authentication. The basic assumption is that each party possesses a private and a public key. Only the holder knows the private key, while the public key is known to other people as well. To understand the distinction, the following expression is rather useful: you can never keep the private key private enough, while you can never make the public key public enough. In the public key-based authentication, the authenticatee (sender) uses its own private key (that is secret to the rest of the world) to sign whatever information that is to be authenticated by the authenticator. The authenticator uses the sender's public key (the well-known key in the public key pair) to check the signature.

In contrast to authentication, when public keys are used for encryption, usage of the keys seems to be reversed: A sender (Alice) that needs to protect a message to a receiver (Bob) encrypts the messages using receiver's (Bob's) public key. Upon reception, Bob uses his own private key to decrypt the message from Alice. Since Bob is the only person that knows his private key, he is the only one that can decrypt the message.

Despite the seemingly different process, a common theme exists:

- The owner of the key always uses the private key either to sign a message, when sending one, or to decrypt the message when receiving one.
- The other party needs to know the public key either when verifying a signature by the owner of the key pair during an authentication process or when sending an encrypted message to the owner of the key pair.

In either case, the public key needs to be conveyed to the other party either prior to the communication or along with the information being passed. The problem of passing public keys to other parties is solved by public key certificates.

2.1.2.2 Basics of Certificate-Based Authentication

A certificate includes the entity's identity along with its/her public key, both signed by a trusted authority. We will describe the details of public key certificates and their design in Chapter 9 of this book. For now, we provide an overview of the usage of certificates for authentication. The basics of certificate-based mutual authentication procedure as described by NIST standard for entity authentication in Federal Information Processing Standard publication 196 ([FIPS196]). The message exchange defined by NIST is rather generic and does not include any specifics on the types of challenges, identifiers, or certificates used in the process, but provides the concept in a rather concise manner as depicted in Figure 2.1:

- In the first step, which is an optional step, the initiator (entity A) of the exchange sends an authentication request to the responder, with which the initiator wishes to engage in mutual authentication. The format of the request is not defined by NIST in order to allow a choice for a protocol that fits the exact deployment scenario.
- The responder (entity B) generates a random number (R_B) as a challenge and sends it to entity A. This challenge is called TokenIR1 and must be used by entity A in a manner that fits the challenge/response protocol, being used. The responder also includes an optional token identifier (TokenID) along with the token in its response to initiator.
- The initiator receives the R_B and creates a random challenge (R_A) of its own and takes the two random numbers along with the name of entity B, and optionally other useful data as part of its own token (TokenRI). The token also includes a signature of all that information with the private key of the initiator. The initiator creates a message and includes its public key certificate (Cert I) to the message, so that the responder can verify the signature.
- The responder receives the message, verifies the signature. If the verification passes, it means the initiator is authenticated. The responder then creates a new token (TokenIR2) including R_B, R_A, and name of initiator and possibly other useful data and a signature over these data using the responder's private key. The responder sends a message back to the initiator and includes its own public key certificate (Cert R) to help the initiator with the verification of the signature and thereby authentication of responder to the initiator.

2.1.3 Examples of Message Authentication Mechanisms

Methods that protect the information from tampering by illegitimate parties are referred to data integrity protection methods. When information is carried through messages over communications channels, the integrity protection is typically provided by message authentication mechanisms. In order to provide data integrity protection for the message, the sender needs to provide a proof of authenticity for the message. In real life, all the legal documents are signed by the involved parties. The signatures not only provide proof of authenticity since the signature for each person is unique, but also prevent the documents from forgeries.

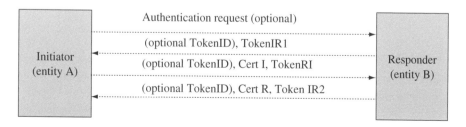

Figure 2.1 Basic exchange for two-challenge mutual authentication based on certificates

Of course, anybody who has asked a big sister to forge the parent signatures on a report card knows that there are artistic ways to get around the weakness of authentication methods. After that analogy, understanding message authentication in the world of digital communications becomes very simple: the sender of a message can provide proof of authenticity for the message by signing the message with a secret that is unknown to the outside world and adds the signature to the end of the message. However, it should be noted that in contrast to the personal signatures that always look the same and hence can be easily forged, the signatures in the digital world depend on the message content and take a different form (bit string) every time. It is as if your parents and the teacher had a secret agreement that your parent would sign your report card in a different way for every season. This way, if a man-in-the-middle (MITM) tries to change the message contents, without the knowledge of the secret, she cannot reproduce a signature that matches the content of the message. A typed legal document that has crossover and handwritings over it must again be signed by the involved parties, otherwise it has no legal bearings. To produce the digital signature, the sender needs to run the message through an algorithm that takes the secret (key) as a secondary input. However, since running these algorithms over entire messages are computationally expensive, the sender compresses the data using a so-called hash algorithm (H) and arrives at a digest value, which is typically called message authentication code (MAC). The MAC value is added by the sender to the end of the message sent to the receiver and is checked by the receiver.

Hash algorithms are based on hash functions that are mathematical one-way functions, meaning that it is close to impossible (depending on the dispensable amount of time and computing power, of course) to guess the input of the hash function, given the output of the hash function. This is a very important characteristic of the hash function to be used for message authentication. The attacker should not be able to guess the input of the hash function from its output. Now an observant reader might say, when the data is simply sent in the clear, the attacker can easily read the message as well as the MAC. If both message and MAC value are readable, then both input and output of the hash function are exposed and there is no use for the hash function. Well, not quite, the sender and receiver also share a secret that they use as input to the hash function while calculating the hash value (MAC). The hash functions that can accept secret keys are often referred to as keyed (or secure) hash functions. An attacker that does not know the secret cannot tamper with the message data without being exposed, since she cannot re-calculate the hash value based on the altered data. However, secret hash algorithms have been put to test by hackers as well as cryptoanalysts, who attempt to break the existing algorithms as part of their day jobs. Experience and science have shown that in majority of cases increasing the size of key, i.e. the number of bits in the key, tends to make the keyed hash function more resilient to attacks.

Over the years, many hash functions have been developed for message authentication. In the following section, we describe HMAC as a standardized mechanism for providing MACs.

2.1.3.1 HMAC-MD5

As mentioned earlier, the value of a hash function is based on the difficulty with which the input of the hash function can be guessed from its output; so an attacker cannot easily alter the message and re-calculate the hash to present the forged message to the unsuspecting receiver. Many hash functions were developed during the course of several decades of research on cryptography. However, as the processing power of the computer CPUs and cryptographic analysts grows, many previously developed hash functions are rendered less effective, and the key sizes that were previously deemed long enough seem too short to provide adequate protection against cryptographic attacks. The hash function design can be compared to clothing out of fashion industry: they are both cool when they come out and passé a few years later, with the difference that the hash functions tend to not make any comebacks. That could (only a guess) be due to a more vivid imagination of hackers and cryptographers in their own field even though they typically have an unusual sense of fashion.

With all that digression, it should now be obvious that when providing integrity protection, the need for replacing the less secure hash functions with the new and more secure ones existed frequently. For those reasons, providing a framework that defines the usage of a generic hash function was deemed useful. A framework developed by IETF, called HMAC [HMAC2104], enables the designers and implementers to deploy any generic keyed hash function for a generation of MACs. We provide an overview of the HMAC framework in the following. In order to prove that the message is authentic, the sender compresses the data to be protected using a hash algorithm (H). If the data is too long, it may be divided into blocks of B-octets long and then fed into the hash function, which produces an output value of length L ($L=16$ bytes by MD5 and $L=20$ for SHA1).

$$HMAC = H(K \otimes opad, H(K \otimes ipad, M))$$

where, the K is the secret key shared between the sender and the receiver of the message (M). To maintain a minimum strength for the security of the HMAC procedure, the HMAC specification recommends that the key length is at least as large as the length of the output of the hash function (L).

The ipad and opad simply refer to the inner and outer padding applied to the original message, respectively and are simply constant values that have the same length as the input data:

- ipad is the one-byte value of 0×36 ($\times$ stands for hexadecimal notation) repeated B times (to make B bytes).
- opad is the value of $0 \times 5C$ repeated B times.

Now that we have gone through the details of HMAC calculations, for the rest of this book we use the following simplified form to refer to an HMAC calculated over a message M, using a key K:

$$HMAC(K, M)$$

The sender simply appends the calculated HMAC value to the message M, before sending the message to the receiver:

$$M, HMAC(K, M)$$

As shown above, the HMAC specification abstracts the exact details of the hash function (or choice of hash function in case of those who are not cryptographers and cannot design hash functions the way RSA folks did) out. Any hash function that is deemed cryptographically suitable for any future computing platform and/or communications scenario can be inserted in the process of calculating the HMAC and hence the security the HMAC process is always is good as the hash function being used. At the time of generation of the RFC, there were many candidates for the hash function to use. Since then MD5 and recently secure hash algorithm (SHA1) [SHANIST] have gained the most popularity. When the HMAC framework is used with MD5 as the hash algorithm, it is typically referred to as HMAC-MD5. It should be noted that at time of this writing usage of HMAC-MD5 is being discouraged for applications with high security requirements. NIST is requiring the use of SHA1 with at least 128-bit long keys in conjunction with the HMAC protocol (HMAC-SHA1) for federal government applications.

Regardless of the hash algorithm being used with HMAC protocol, the HMAC protocol is a wide and powerful tool in security design to provide not only authentication of a message or an entity (the holder of the key), but also anti-replay protection. The following is a generic form of what one may come across in many signaling procedures:

HMAC(Key, information, nonce)

The purpose of the key is to provide assurance of integrity of the data, since the key is only known to the sender and the receiver. However, the nonce is not a secret value and is simply a value that is used only once to provide the so-called anti-replay protection. A replay attack is when an illegitimate party records a message between two other entities and replays the message at a later point acting as the original sender. Using each nonce value only one time ensures the receiver that the message is not a replayed version of older message.

2.2 Classes of Authentication Mechanisms

In earlier sections, we provided a few well-known authentication mechanisms. Since authentication is a very old problem, the list of authentication mechanisms out there can be very long, especially when we consider that people customized many of those of mechanisms for a variety of deployment scenarios and security requirements. For that reason, instead of going on and providing more authentication examples, we provide a classification of authentication mechanisms that is the result of a survey conducted by IAB[V]. The IAB classified the authentication mechanisms into seven different classes, which covers almost all the authentication models "out there". In this classification, the following three fundamental criteria are considered:

1. Authentication based on something the authenticating party has, such as a physical hardware token or a card.
2. Authentication based on something the authenticating party knows, such as a secret or a password.
3. Authentication based on something the authenticating party is, such as a physical characteristic of the link it is attached to.

The seven classes of authentication mechanisms are as follows:

1. *Passwords in the clear*: This is the oldest and simplest user authentication method, by which the user supplies a (user name, password) pair along with its authentication request or network access request to the network. The request is processed by a server, which

looks up the password in a password file using the user name as the lookup key. The password file is usually not in a clear, meaning that the server uses some sort of hash algorithm and a secret (known only to the server) to convert all the passwords into data that are unintelligible to outsiders. Once the server receives a request including a password, it runs the same hash algorithm on the password and compares the results with what is stored in the hashed password file. This authentication method is subject to many problems such as sniffing, online password guessing, offline dictionary attacks, and so on. Sniffing is a problem especially if the request is sent over wireless links not providing confidentiality mechanisms. The attacker can read and store the user name and password and later on launch a replay attack, i.e. replay these credentials in an attempt to gain access by pretending to be a legitimate user. Bad passwords can easily be guessed without any listening capabilities. The attacker simply tries a number of passwords, until it guesses the right password. Typically, the administrator can alleviate this problem by allowing the user only try a limited number of password retries, after which the account is locked. Unfortunately, this inherently paves the way for launching Denial of service (DoS) attacks through which the users are blocked from the accounts. Offline dictionary attacks are possible if the attacker has access to a large number of digested (hashed) passwords or the actual server password file. The attacker simply runs the hash algorithm on a large number of passwords, mostly words chosen from dictionaries (hence the name) until it finds some password matches or the actual secret for digesting the passwords. One remedy for offline dictionary attack is to use the so-called salting method. When calculating the hashed password for each user, the server injects a random value called salt in the hash process and stores [salt || hash (salt || password)] instead, where || stands for ordered concatenation. This way, even two users with the same password will have different hashed password values.

2. *One-time passwords*: A One-time password can, as it sounds, be only used once. This mechanism was invented to solve the problem of replay attacks and several other active attacks without requiring too many changes on systems that use legacy user name/password methods. The communications protocol is basically the same, while both client and server require minimal changes. The client is either equipped with a list of passwords, each of which can be used only once, or with a processor or card (SecureID card) that can calculate these passwords based on a predetermined algorithm. Since each password is used only once, it can be sent in the clear. One prominent example of one-time passwords in S/key [SKEY1760]. The passwords in S/key method are related to each other through some hash algorithm, i.e.

$$P[i] = hash(P[i-1])$$

The process is additionally protected by a secret that is used when calculating P[0]:

$$P[0] = hash(Starting\_seed \parallel secret)$$

The passwords are all calculated first (P[0], ..., P[N]) and then are sent in the reverse order, i.e. P[N] first, P[N−1] later and so on. This is to prevent the attackers from guessing the consequent passwords based on her knowledge of the previous password and hash algorithm. The strength is drawn from the one-way property of the hash function, which means, the input of the hash function can not be calculated based on the knowledge of the output. The server simply stores the last received password P[i], and when it receives another password

P[i − 1] from the same user next time, it simply runs the hash on the previously received and stored password (P[i − 1]) to confirm the validity of the currently received password. The S/key if designed properly allows the server to keep the password file in the clear, if such a file is needed. The SecureID cards created by RSA security are examples of one-time password mechanisms. The card generates new time-dependent keys (number sequences shown on the card display) every minute or so. The user has a memorable password and when logging to the network, the user combines her own password with the set of numbers, shown by the card, into the login application. Unfortunately, an attacker can capture the entire data and replay it before the next time-independent key is generated.

3. *Challenge/response*: Challenge/response authentication mechanism is based on the assumption that the user can use a password or a more sophisticated secret to calculate the response to a challenge issued by the network. The challenge is typically a randomly generated value, which is hopefully generated so that it does not repeat itself easily. The response is calculated based on a pre-negotiated hash algorithm and the knowledge of the secret, i.e. response = hash (challenge || key). Challenge/response methods were designed for two purposes: first to avoid sending the password over the wire or air and thereby avoiding the password-sniffing problem and second to prevent replay attacks, assuming that the challenge is generated with large entropy. Note that if not carefully designed, the security of challenge/response method can be worse than the password method. For instance, when simple hash algorithms with short-length keys are used, by sniffing a single (challenge, response) pair, the attacker can in all privacy conduct a dictionary search till she finds the key/password. Another example of flawed challenge/response design is when the server holds a local copy of the user password in the clear to verify the response to its challenge. Not holding the passwords in a hashed password file, as in password-based authentication methods, means that the attacker can simply attack the server and access the file (in the clear) and thereby access to the password for the user (or all users) or the password to other servers, if the user is using the same password for many servers. These two problems are referred to as weak password equivalence and strong password equivalence by IAB [AUTHSRV]. A solution to this problem is to use a two-stage hash process, in which the server calculates a hashed value of the password, using a salt and stores the hash, instead. The server then sends challenge and the salt to the user and expects a response from the user, which in turn performs the two-stage hash. The first hash leads to the password version stored at the server and the second hash leads to the response the challenge has if the stored password has been applied to the challenge. This way, the server can run a simple hash on the stored password, and when the challenge/response is sniffed and attacked, at best only the stored password will be compromised. The mathematical process is shown below.

$$stored\_password = hash(user\_password \parallel salt)$$
$$User\_response = hash(hash(user\_password \parallel salt) \parallel challenge)$$

Hypertext transfer protocol (HTTP) is a protocol used for requesting and distributing information over the World Wide Web. The original HTTP standards only required a password-based authentication (called HTTP basic authentication), which was considered extremely weak for the scenario it was being used for. The challenge/response mechanism is enhanced for HTTP requests to web servers, so that not only the user but also the HTTP request for receiving content from the server is authenticated. The new challenge/response-based procedure is called HTTP digest authentication, since a digest (hash) is calculated over both user password, the challenge and part of the actual HTTP request. However, the HTTP digest requires two round trips, during which the user first requests for some information by sending an HTTP request. The server rejects original request, indicating its requirement for digest authentication and sends a challenge toward the user. In the second request, the user includes the digest that is calculated based on the user's credentials. Obviously, this would lead to doubling the authenticated part of web traffic, so one solution that has been suggested is for the server to include (after the first authentication) a challenge in its response to the user, so the challenge can be used during the next authenticated request. For the reasons mentioned above, HTTP over secure socket layer (SSL) (explained below) has gained more popularity than HTTP digest.

4. *Anonymous key exchange*: If the communication channel carrying the authentication credentials between the two parties is protected by added encryption and integrity protection, then many of the basic authentication mechanisms such as password in the clear can be used over this secure channel. However, establishing a secure channel that provides data encryption requires that encryption keys are somehow shared between the two parties beforehand. Diffie and Hellman provided a very clever method, called righteously "the Diffie–Hellman method", that allows two parties, that have no previous relationships with each, establish a shared key and thereby a secure channel between themselves. We will describe the Diffie–Hellman method in Chapter 3, but for now, suffice to say that it requires the peers to send their public keys to each other. Based on the exchange public keys, the peers arrive at the shared secret. Note that the public keys and cryptographic methods are only required for initial key exchange and not for the actual authentication itself. The first disadvantage is that this method still requires the client to have the ability to perform public key cryptography. Furthermore, in a raw form of Diffie–Hellman, the peers do not provide a proof of their identity to each other; when the identities are not verified, the key exchange is considered to be anonymous. Since any active attacker can hijack the message, including a peer's public key and insert her own public key instead and pretend to be the legitimate peer, the anonymous key exchange is said to be prone to MITM attacks. The attacker can later on even get a hold of the passwords that the user is sending to the server over the channel she believes to be secure. Secure shell (SSH) uses the anonymous key exchange mechanism, during which the server sends a raw public key (without additional information) to the client. The client caches the public key to guard itself against future MITMs, where the attacker could replace the server key with her own key. The caching, however, does not help if the initial key exchange is under an MITM attack. Many methods have been suggested to alleviate the MITM problem. However, without a proper identity verification, the MITM threat persists and since the MITM is able to also sniff the channel, all the added complexity of key exchange does not buy any added security over the initial password mechanism. A way to circumvent this problem is to have the public key of each party signed by a trusted authority. The result is called a public key certificate that includes both the public key and the signature from the trusted authority. We will explain certificates in detail in Chapters 3 and 9.

5. *Zero-Knowledge password proofs*: These methods are designed very cleverly to avoid the problem of the authentication mechanisms requiring pre-shared keys, or password that can either be sniffed or attacked by an MITM. Zero-knowledge password proof methods enhance the Diffie–Hellman methods by added encryption using user-generated passwords. However, the methodology is heavily patented and hence has not gained any wide

acceptance. The reader is referred to literature on encrypted key exchange (EKE) for further information [EKE].

6. *Server certificates plus client authentication*: This mechanism is designed based on the premise that if the two peers can establish a secure channel based on a unidirectional authentication (authentication of only one peer to another), the authentication of the second peer to the first can then happen over the established secure channel in a far less sophisticated manner. Any of the password in the clear, one-time passwords, challenge/response and anonymous key exchange mechanisms can be used for authentication of the second peer to the first. However, with the existence of the secure channel that can provide both message authentication and encryption, the advantage of more sophisticated authentication methods over the less sophisticated ones is significantly reduced, since passwords can easily be carried over the secure channel. This mechanism is also based on the very practical assumptions that servers are typically much fewer in numbers and less mobile than their clients and therefore providing certificates only to a few servers is more practical and economic than providing certificates to a large number of clients. Hence, the server can authenticate to the client using its certificates, while authentication of client to the server is deferred to a later point after a secure channel is established based on server-to-client authentication. The idea of providing server side certificates gained popularity with the advent of SSL to promote shopping over the Internet. The client needed to know that the e-commerce website was legitimate before making any credit card payments for her purchase, so the server needed to prove its legitimacy by presenting a certificate. However, SSL methods used for e-commerce were not true mutual authentication mechanisms, since they did not require anything from the client besides her valid credit card number! True examples of server certificate plus client authentication are EAP-TTLS, which is explained in detail in a later chapter, and HTTP over SSL (HTTPS). HTTPS assumes that the server can support SSL and owns certificates that can be verified by the client. Once SSL is established after server authentication, the client performs a password authentication through either HTTP basic or digest methods. Another popular practice for user authentication is the use of application layer authentication mechanisms available through hypertext markup language (HTML) forms that prompt the user for her password.

7. *Mutual public key authentication*: The most robust and secure method for mutual authentication is to have both server and client to present public key certificates to each other for the purpose of authentication. By robustness, we mean that the authenticating peers do not rely on any previous arrangements for establishment of a contract, trust relationship or shared secrets prior to the first direct communication. By security, we mean there is no way a middleman can forge a certificate or a signature that one of the parties has provided for her authentication. However, managing certificates is a complex business requiring advanced expertise. We will devote an entire chapter on describing public key infrastructures (PKIs) for support and management of certificates. For now, we suffice to say that besides the PKI complexity issue, the main problem for this method is safeguarding the private keys; a lost private key leads to a compromised certificate and can thereby create large security problems depending on the authority associated with the certificate. The server private key may be hardened by use of specialized hardware accelerator processes and other tamper resistant (not just tamper evident) mechanisms. Protecting the client private key is typically more difficult, since the client device is less sophisticated and can be stolen and tampered with. So the main problem for this method is to keep the client private key safe. Two methods are suggested for this problem: (1) have

the client carry a secure, portable and tamper resistant token that carries a private key and (2) have the user memorize a password, from which the client device can derive the private key. The token usually has a USB jack or a smart card interface that is plugged into a computer, but the problem is that in order to access the private key on the token, the connecting computer needs to have knowledge of a personal identification number (PIN). Hence, any computer, on which the token has been used, can be potentially used to hijack the PIN and consequently jeopardize the private key. Furthermore, due to cost limitations, tokens are not completely made tamper resistant, so the primary problem with tokens is theft. The password method for deriving the private key works so that the password is used as a seed to a cryptographic process that arrives at the private key. The password-derived keys are vulnerable to dictionary attacks. The attacker can run many passwords through the dictionary till the password generates a private–public key pair that generates the same signature as that used during the certificate-based authentication process.

One of the important goals of the exercise of classification of authentication mechanisms was to help the designers and implementers to understand the strength of the authentication mechanism they were dealing with. The authentication mechanisms that belong to the same class provide the same level of strength.

2.2.1 Generic Authentication Mechanisms

Now that we have completed the description of all the seven classes of authentication mechanisms, it is useful to describe a new and widely used concept that the IAB calls generic authentication mechanism. The idea here is to provide a framework that can support many different authentication mechanisms and allow the peers to negotiate their authentication mechanism of choice rather than to be restricted to specific authentication mechanism. This idea is attractive to network designers, since it allows them to upgrade their access control and security procedures as newer and more secure authentication mechanisms pop up without the need changes to the network architect. Unfortunately, due to the provided facility for negotiation of a mechanism of choice, this method is prone to downgrade attacks. A downgrade attack is when an active attacker intercepts the negotiations and convinces the two parties to choose a weaker, less secure authentication mechanism than their strongest common mechanism. The attacker can then simply crack the weaker mechanism. A countermeasure for this attack is to have each peer provide a digest on the authentication proposal to the other peer. However, if the digestion mechanism being offered is weak, then this protection will be weak as well. For instance, if the peer plans to offer password authentication, providing a password-based digest would not provide a very strong protection.

2.2.1.1 Extensible Authentication Protocol (EAP)

EAP is a prime example of a generic authentication framework. EAP has gained popularity in the recent years. The popularity arises from the pressure imposed on the network designers and administrator to not only support the legacy authentication mechanisms required to run their existing platforms but also to support the newer and stronger authentication mechanisms. However, one of the important goals of the exercise of classification of authentication mechanisms was to help the designers and implementers to realize when the authentication mechanisms they provide belong to the same or different classes. The IAB recommends designers to resist the pressure of supporting multiple authentication mechanisms that essentially belong to the same class to the EAP framework. Also the IAB recommends designers to resist the pressure of supporting legacy authentication mechanisms due to increased risk of complexity and interoperability problems.

The main advantage of EAP is the flexibility it provides to the three-party authentication models. Whenever, new authentication methods are developed, rather than requiring the authenticator (NAS) to be updated for support of the new method, EAP permits the authenticator to simply pass the message exchange for a new method through to a backend server that in turn understands the new authentication method. This way, only the backend server (typically an AAA server) needs to be upgraded. This saves the administration the trouble of upgrading a large number of NASs (at the edge of network).

We will describe the use of various authentication mechanisms within EAP in Chapter 10 and for now suffice to provide a very short overview of the EAP messaging.

2.2.1.2 EAP Messaging

As mentioned earlier, EAP by itself does not perform the act of authentication; it merely provides a mean for the negotiation between the user and the authentication server (with NAS in the middle). In a way, EAP acts like a real estate agent; it introduces the seller and the buyer and sits back while they are exchanging information and negotiating and then finally helps closes the deal. So it makes sense that EAP includes start and end messages and a bunch of middle messages. There are four types of messages in EAP

- EAP request message (authenticator to user)
- EAP response message (user to authenticator)
- EAP success
- EAP failure

Table 2.6 shows the format for EAP packets.

Table 2.6 EAP packet format

Field name	Sub-field	Description
EAP header	EAP code	1 for EAP request, 2 for EAP response, 3 for EAP success, 4 for EAP failure
	EAP ID	Transaction ID: used for matching request and response messages (1 for first message)
	EAP message length	2 octets for the length of the message (including code, ID and length field itself)
EAP data	Type	Type field is only used for EAP request and response message and indicates the type of information being requested from the peer corresponded to the server. Examples are: 1 for Identity, 3 for NAK(response only), 4 for MD5 challenge, 5 for one-time password (OTP)
	Type data	

Request and response messages are used for all information exchanges and by using an EAP type field, further EAP request and response messages can be created for exchanging a variety of information types. For instance, EAP request identity is a message sent by the server, requesting the user to present her identity (name or any other type of identity such as NAI or one produced by a smart card). However most of type numbers are used to identify authentication methods such as EAP-TLS. EAP-TLS is an authentication method that has type 13 and is explained in a later chapter. The flexibility in use of type field for creating a variety of exchanges is the reason for EAP being called *extensible*: new authentication methods can be introduced very easily. The authentication types are assigned by IANA and that way the AAA servers can, based on the type value, determine whether they can authenticate a client supporting the specified type of authentication. If the authentication method is not supported, the user receives an EAP response with type 3 (NAK) as an indication for a negative response.

The role of the NAS (authenticator) in the signaling is very simple: in general, the NAS forwards the requests from the server to the user and responses from the user to the server without understanding much of the EAP messaging. However, if the NAS is configured to understand EAP success or failure messages, the NAS can simply wait until it receives an EAP success or failure and that way knows whether the authentication process is completed or not. This is why EAP success or failure messages serve a very useful purpose.

EAP can run directly on link layer without requiring a network layer protocol, such as IP, and is designed to include its own support for in-order delivery and retransmissions. Although EAP was originally designed to run over PPP, it can be carried over wired or wireless LAN segments, as well as dedicated circuit switched links. 802 based LANs do not require PPP framing services and as we will see later on EAP will run directly over the link layer for the LAN. This is referred to as EAP over LAN (EAPOL) as described later. However, the link layer provides a transport mechanism for EAP messages only over the link between the user and the NAS. As mentioned earlier, EAP is designed to also support three-party authentication models, when a NAS, which is typically located at the edge of the network forward the messaging to a backend authentication server (such as an AAA server). Figure 2.2 shows the model for transport of EAP messaging over the entire path from the user to the authentication server and vice versa.

As can be seen from Figure 2.2, the model comprises the following elements:

- *Peer*: It represents the end of layer-2 communication link attempting to connect to the network through the edge device. Typically, no distinction is made between the end user and end device when the term peer is discussed. We do the same unless the actual scenario imposes a restriction between the user and her device when it comes to EAP authentication. In IEEE 802.1x context, the peer is called the supplicant.

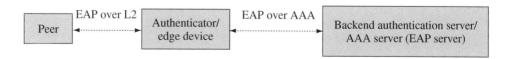

Figure 2.2 EAP key management for three-party authentication model

- *Authenticator*: The end of the layer-2 communication link, and the entity in charge of access control on behalf of the network. This is typically the edge device that either authenticates the peer by itself (two-party authentication model), or as is more often the case in modern networks outsources the authentication task to a central server using a protocol that carries the EAP messaging (three-party authentication model). In this chapter and most of the rest of this book, we assume this protocol is a standard AAA protocol such as RADIUS or Diameter.
- *Backend authentication server*: It is the central entity that accepts the authentication requests from the authenticators (on behalf of the peer) and perfoms the authentication. Since we assumed the authenticator and authentication server speak a AAA protocol, we can also assume that the backend authentication server is also a AAA server.
- *EAP server*: It is the entity that terminates the EAP authentication method and other EAP-related functions. In the two-party authentication model, the EAP ends at the authenticator, and hence the EAP server is co-located at the authenticator. But in this book, since we mostly deal with the three-party authentication model, we assume that the EAP server is also co-located with the AAA server, or even simpler, the AAA server also functions as the EAP server.

We will refer back to this model many times during the course of our discussions, in this book.

A final note is that, despite its popularity, EAP seems to not be very popular with IAB, since EAP creates multiple layers of negotiations. First, the EAP needs to be accepted during the authentication phase of the PPP link establishment. Later on during the EAP exchange, an authentication mechanism is chosen. One of the choices provided by EAP is transport layer security (EAP-TLS), which as we will see, itself provides methods for authentication mechanism negotiations. Three layers of negotiations can seem excessive, and IAB suggests using frameworks that allow the peers to choose mechanisms that themselves are frameworks instead of mechanisms. Despite all this downplay, as we see later on EAP-TLS is one of the popular authentication mechanisms today.

Threat Analysis

Finally, another concept that is useful to be familiar with, when studying the security related literature, is the concept of threat model or threat analysis. Threat analysis is a very important step in choosing authentication and security mechanisms. During the threat analysis, the designers brainstorm on what security problems can arise for their network architecture and its usage scenarios, what sort of attacks can be performed on the ongoing communications or on the network entities and how these attacks should be countered. An important factor to consider is also to assess how technically savvy or competent are the users. Practical examples of security measures to consider include: should the data being communicated protected against tampering or eavesdropping.

2.3 Further Resources

The field of authentication is a very well explored one and has seen many years of research and standardization. It seems however that with the advent of many new wireless and wired network technologies and many new services and application, this field is still going through a serious expansion as well as large number of customizations. For those reasons, it is very difficult for us to provide a comprehensive list of resources for further study; so we simply opt to provide links for the technologies that are covered in this chapter. Also the aim of this book is to provide a much better understanding of the various authentication issues for various scenarios throughout this book, so we are by no means done with the topic of authentication at this point.

Here are a few links to topics discussed in this chapter

- More information on EAP can be found on the IETF EAP working group website: http://www.ietf.org/html.charters/eap-charter.html.
- More information on PPP and its extensions can be found at the concluded IETF PPP working group website http://www.ietf.org/html.charters/OLD/ppp-charter.html and the PPP extension working group website http://www.ietf.org/html.charters/pppext-charter.html and book written by J. Carlson [PPPCARL].
- There is a new working group formed in IETF that works on the problem of allowing the clients to authenticate themselves to the access network using the IP protocol. The name of the group is "protocol for carrying authentication for network access" nicknamed PANA. More details about PANA's work can be found at the group's web site: http://www.ietf.org/html.charters/pana-charter.html.
- For a wealth of information on computer security and authentication mechanism, US government requirements and national standard one can go to the National Institute of Standards and Technology (NIST) website for security at http://csrc.nist.gov/.
- Finally, we refer the readers to the IAB survey of authentication mechanisms [AUTHSRV] for very useful insights on choosing authentication mechanisms for their scenarios at hand.

2.4 References

1. [AUTHSRV], Rescorla, E., "A Survey of Authentication Mechanisms", Internet Architecture Board (IAB), work in progress, draft-iab-auth-mech-02.txt, October 2003.
2. [SKEY1760], Haller, N., "The S/Key One-Time Password System", RFC 1760, February 1995.
3. [EKE], Bellovin, S. and Merritt, M., "Encrypted Key Exchange: Password-Based Protocols Secure Against Dictionary Attacks", Proceedings of IEEE Symposium on Research in Security and Privacy, May 1992.
4. [PPPCARL], Carlson, J., *"PPP Design, Implementation and Debugging"*, Addison-Wesley, 2000.
5. [PPP1661], Simpson, W., "The Point to Point Protocol", IETF, RFC 1661, July 1994.
6. [CHAP1994], Simpson, W., "PPP Challenge Handshake Authentication Protocol", IETF, RFC 1994, August 1996.
7. [MD51321], Rivest, R., "The MD5 Message Digest Algorithm", IETF, RFC 1321, April 1992.
8. [EAP2284], Blunk, L. and Volbrecht, J., "PPP Extensible Authentication Protocol", IETF, RFC 2284, March 1998.
9. [HMAC2104], Krawczyk, H. *et al.*, "HMAC, Keyed-Hashing for Message Authentication", IETF, RFC 2104, February 1997.
10. [SHANIST], NIST, "Secure Hash Standards", FIPS PUB 180–1, April 1995.
11. [FIPS196], US Department of Commerce, "Entity Authentication Using Public Key Cryptography", FIPS 196, Federal Information Processing Standards publication, February 1997, http://www.mirrors.wiretapped.net/security/info/reference/nist/fips/fips-196.pdf.

3

Key Management Methods

Well-designed security mechanisms intended to protect privileged data or resources typically rely on existence of some sort of secret that is known only to the intended users of that mechanism. In general, the secret is called the key, even though it can be as simple as a memorized password. Many times, the strength of a security mechanism greatly depends on how the key/s, used by that mechanism, are generated and handled. The fact that, the US National Institute of Standards and Technology (NIST) has held several key management workshops, emphasizes the importance of key management mechanisms. As a result of these workshops, NIST has produced a set of guidelines provided in documentations such as [KMGNIST1]. The NIST documentations are very important, especially since compliance to government standards is a necessary requirement for many security products being deployed for government and public safety agencies. Since this book is not a text on cryptography (and we do not claim that we can write one!), we suffice devoting this chapter to main key management concepts and mechanisms without going into the details of the mathematics involved. We will provide more detail on protocols deploying these concepts in Chapter 4 when describing security mechanisms for the Internet and in Chapter 9 when describing infrastructure for managing public key certificates.

3.1 Key Management Taxonomy

Before getting into the actual discussion of key management methodology, it is useful to go through the fundamental terminology for key management.

3.1.1 Key Management Terminology

The NIST key management guidelines provide a comprehensive glossary along with a list of 22 different key usage scenarios, each including a separate terminology. In the following, we provide a short excerpt of that glossary (in alphabetical order) in a way that serves our discussions within this book.

Cryptographic key: A parameter, used in conjunction with a cryptographic algorithm to perform any of the following, is considered a cryptographic key:

- Transformation of plaintext data into ciphertext and vice versa.
- Computation of a digital signature from data and verification of digital signature from data.
- Computation of an authentication code from data.
- Derivation of the keying material to be used by another cryptographic process.

Cryptoperiod: The period of time (typically a start and an end date) over which a specific key is valid for use within a given system or in conjunction with an application.

Ephemeral key: A short-lived cryptographic key that is unique to each execution of a key establishment process as opposed to long-term secrets that can be used for multiple executions or procedures.

Group keys: Keys shared by all the members of a trusted group. This is in contrast to pairwise keys (see following text), only used by the peers in a pair. A pairwise key hierarchy versus a group key hierarchy describes the relationships between the keys used between the devices in each case.

Initialization vector (IV): A vector used in defining the starting point of an encryption process within a cryptographic algorithm.

Key de-registration: A stage in the lifecycle of the keying material, defined by removal of all or some of the records of the keying material. Such records are typically registered at a registration authority. This is an item whose definition is still under flux, since the NIST guidelines are still subject to change. See the definition of key registration in the following text.

Key encrypting key: A cryptographic key that is used for encryption or decryption of other keys. The key encryption is used for protection of the key during transport or storage. The cryptographic key that has been encrypted is called the encrypted key, while the key that is used for that encryption is called key encryption key (KEK). Key encryption keys are typically used in conjunction with symmetric key encryption methods.

Key labeling: A label may be used to provide information for the use of a key. The label can contain a subset of the following information: identity of the key and key owner or sharing entity, key length, cryptoperiod, key type, the intended application (such as e-mail), and protection mechanisms or requirements (such as wrapping and integrity protection mechanisms) for the key itself.

Keying material: The data (e.g. keys and IVs) necessary to establish and maintain cryptographic keys or trust relationships are called keying material. Most of the times keying material may be referred to data that are needed to derive the keys rather than the keys themselves.

Key recovery: The mechanisms and processes that allow authorized entities to retrieve keying material from key backups or archives during the lifecycle of the keying material.

Key registration: It is a stage in the lifecycle of keying material, and is defined as the process of officially recording the keying material by a registration authority.

Key revocation: The process whereby the affected entities that are notified/ordered to remove the key material from operational use prior to the end of their established cryptoperiod (lifetime). This can be due to compromise of a key through anything from a lab technician writing the key on Post-It note and posting it on the monitor (there actually happen) to compromise or loss of the device holding the key. Another example is key revocation may also be due to intended causes such as change in user affiliation due to resignation of the user from her position with the company.

Key type: If used properly, the term key type refers to the cryptographic purpose for which the key is intended. Examples of key types are signing keys, encryption keys, and master keys. However, one comes across literature where the type of the key is tied to the type of service application (such as e-mail) for which the key is being used.

Key wrapping: Encrypting a symmetric key using another symmetric key. The latter key, which is used for wrapping the first key, is called key encrypting key as explained earlier.

Manual key transport: A "non-electronic" mean of transporting cryptographic keys. This refers to the old fashion and many times more secure way of transporting the keys by physically storing the keys on a CD ROM or hard copy and carrying them in person or by mail. This is in contrast to carrying keys using electronic communication protocols.

Master key: A key used to derive other keying material. The master key could potentially be either a symmetric key or part of a key pair.

Nonce: A nonce is a non-repeating value (typically a pseudorandom number) that is used only once for a particular purpose, such as preventing replay attacks.

Pairwise key: A key that is used only for communications between a pair of entities and is unique for only those entities. This means that a device that needs to communicate with multiple devices needs to have a set of pairwise keys for each of its peers, as shown in Figure 3.1. A device dealing with N other devices needs N different pairwise keys with each of those entities as shown in Figure 3.1.

Perfect forward secrecy (PFS): A re-keying method that can guarantee confidentiality of a session, even after an existing encryption key is compromised.

Public seeds (salts): These seeds are used in the generation of pseudorandom numbers, such as domain parameters (see Diffie–Hellman later on in this chapter) that are public and may be validated.

Re-keying: Deriving new keys to replace the previous keys for the same cryptographic process.

Secret seeds: These seeds are used in the generation of pseudorandom numbers that should remain secret, such as secret keying materials.

3.1.2 Types of Cryptographic Algorithms

It is neither the intent nor within the ability of the authors of this book to provide detailed description of various cryptographic algorithms. The interested reader is referred to the large number of cryptography books and literature that can be found today. But for understanding the discussion in this book, it is useful to become familiar with the various types of cryptographic

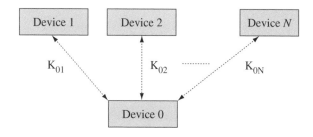

Figure 3.1 Usage of pairwise keys in dealing with multiple devices

algorithms, to understand how the key management mechanisms for each of these algorithms differ. For that purpose, we borrow the concept of classification of cryptographic algorithms provided by NIST documentation [KMGNIST1]:

Hash algorithms: Hash functions and their use for providing security services were described in earlier chapters. We feel that description is enough to understand the key management issues related to hash algorithms.

Symmetric key algorithms: Symmetric key algorithms use a single key to transform or encrypt data and require that the key remains a secret between the entities sharing the key. Symmetric algorithms are used for both data confidentiality (encryption), authentication and integrity services, and key establishment, mechanisms.

For encryption, American Encryption Standard (AES) and Triple Data Encryption Standard (3DES or TDES) are approved for Federal Government use. Typically symmetric keys are used as pairwise keys. In a system where each node or user has to engage in a secured communication with other nodes, the node must share a pairwise key with each of those nodes. A system compromising of N nodes therefore needs to create, distribute, and manage $N-1$ keys for each node and $N*(N-1)/2$ keys overall. In other words, an order of N^2 keys must be managed. Another issue that makes management of symmetric keys difficult is that these keys must remain secret (to anybody but the owning pair) during distribution and while in storage.

Asymmetric key algorithms: Asymmetric key algorithms are most commonly referred to as public key algorithms. We explained the concept of public key cryptography briefly in Chapter 2; we provide this paragraph to complete our discussion on cryptographic algorithms. Instead of having both of the two parties sharing the same key as they do for symmetric key algorithms, in public key cryptography the two parties each have two keys: a private key that only the owner knows about, and a public key that is passed along for use by the other party. Encryption is typically done with the receiver's public key, so that only the receiver can with her private key decrypt the message. Message signing is done with the sender's private key, so that any receiver can verify the authenticity of the signature, as long as they have access to the sender's public key. Public key cryptography solves the scalability problem associated with management of symmetric pairwise keys. Regardless of how many peers a given client is dealing with, the client is able to use the same key pair for dealing with all of them (2N keys are needed for N nodes). Public key cryptography, when combined with digital certificate technologies, will remove the heavy burden of shared key configuration and thus is closely aligned with the vision of zero-configured future network architectures.

3.1.3 Key Management Functions

A key management infrastructure provides the framework and services for generation, distribution, control, record keeping, and destruction of all cryptographic material, including symmetric keys, as well as public keys and public key certificates. Some of the major key management issues are as follows:

- *Selection of appropriate of cryptographic algorithms and key sizes*: This topic requires in-depth knowledge of cryptographic and crypto-analysis methods; the mathematically savvy readers with financial mean may possibly take off from work and spend the next couple of years learning

cryptography. The advice for the rest of us is to simply follow the word from NIST or other highly trusted sources on the cryptographic strength of the various algorithms and key sizes. Even then, the appropriate suite of algorithms and key sizes must be chosen after a careful study of the scenario at-hand and a trade-off between implementation complexity and security requirements.
- *Key management policy*: Policies for key management deal with defining key usage, key lifetimes, and key recovery procedures, and most of the time depend on the network management and security requirements for the scenario they are being deployed. For that reason, such policies are typically defined at the discretion of network designers or administrators. Because of the practical nature of this topic, we will be brief here. During our discussion of public key infrastructure and certificates in Chapter 9, we will provide more examples for such policy definitions.
- *Key establishment schemes*: Key establishment typically refers to generation and distribution of keys and is explained in more details in the following section. Since a variety of methods and protocols are being used or suggested for key establishment, the remainder of this chapter will be devoted to explaining key establishment issues.

3.1.4 Key Establishment Methods

As mentioned earlier, key establishment methods are used to set up keys between communicating parties. Methods of establishment of symmetric keys can be categorized as follows:

1. Key transport
2. Key agreement
3. Manual key establishment

In the following, we will go through the definition of each category briefly. Later, we describe some prominent examples of each category.

3.1.4.1 Key Transport

The term key transport is used to describe the act of distribution of keying material from one entity to another and typically refer to an electronic transfer using some sort of communications protocol. Keying material must be protected during transit. This means that the keying material must be encrypted for confidentiality and authenticated for integrity protection, while in transit. Integrity protection for key transport ensures that information is not tampered with during transit. Integrity protection in key transport becomes important when keying materials are being transported in the clear. Even though some key materials are not as sensitive as the keys and can be carried without encryption, protection against modifications is still required. Integrity protection can be provided by using HMAC or digital signatures. Both symmetric and public key cryptography methods can be used for encryption of keying material:

- Symmetric key encryption methods are sometimes referred to as key wrapping methods. These methods use a symmetric key called key encryption key that must be known by both parties.

- Public key encryption methods can also be used for protection of keys in transit. In such cases, the sender encrypts the keying material with the receiver's public key, while the receiver decrypts the received material with its own associated private key.
- Another method for providing confidentiality for keying material is to parse it into several components and handle each component individually and separately. Some components are transported to the other side in the clear, some in encrypted form, depending on their cryptographic value. The actual keys are then derived at each party by combining the transported components with other secret information that each party holds (typically a pre-shared secret). This way the more sensitive materials are never transported over any media, while the rest of the key material can be transported over less secure environments without strong encryption capabilities. An example of this method is provided in the Mobile IP–AAA signaling discussion in Chapter 8. The Mobile IP–AAA signaling mechanism allows a mobile node and network mobility agents to establish trust relationships and keys based on the fact that they both trust an AAA server. The AAA server can send the nonces to a mobile node over a clear channel, while the mobile node can, by using a key it shares with the AAA server, arrive at the session keys needed for interaction with mobility agents. This way the scalability problem of having to have the mobile node to establish keys with every single agent is also avoided. Another example of key management using an AAA server is the extensible authentication protocol (EAP) key management framework, explained in a later section of this chapter.

The last example provided deploys a widely popular design philosophy: use a third party trusted (AAA server) by both communications end points to perform key management. This is a typical case for key establishment by key transport. Unless the two parties have a trust relationship to protect the generated keys, no key transport can happen. This, in a way, indicates that key establishment by transport typically does involve a trusted third party. By contrast, the key agreement methods (described below) have mostly a peer-to-peer nature.

3.1.4.2 Key Agreement

As opposed to key transport, where the keys or key materials are generated at one end and transported to the other end/s through secure means, key agreement methods allow both sending and receiving parties to participate in the creation of shared keying material in a peer-to-peer manner. The most famous example of key agreement methods is the Diffie–Hellman (DH) mechanism, accomplished by public key techniques as explained later on in this chapter. Another example is the famous Internet Key Exchange (IKE), which is also based on DH concepts.

Regardless of the method used for key establishment, some form of key labeling is required to guide the involved parties on the usage of the established keys. Different applications may require different labels for the same key type. It is the responsibility of the implementer to select a suitable label for a key.

3.1.4.3 Manual Key Establishment

Even though the previous two key establishment methods use electronic communications methods for either transport of keys or agreement on the same, it is not always guaranteed that, at some point or another, a manual intervention of an administrator is not required. Here are some examples:

- When symmetric key wrapping methods are used, the two parties must have already established some sort of trust, since they both share the key wrapping key. The key wrapping key may well be a secret that is manually established and stored at both machines by the administrator.
- When parts of keying materials are transported so that they could be combined with other secrets to generate the session keys, those "other secrets" may be long-term secrets established by the administrators.
- When DH and IKE methods for key agreements are used, as we will see later on, some sort of authentication is required between the two parties to ensure that both parties are actually who they say they are, so the keys are established between legitimate peers. This authentication is often performed on the basis of knowledge of a pre-established key, called pre-shared secret that is sometimes established by administrators.

Only when public key methods are used for key transport, it is possible to avoid any manual key establishment as long as acquisition of public key pair and related certificates has not involved any manual key distribution (such as use of passwords to reach a server, as we will see later on). For these reasons, do not ignore the widespread practice of manual key distribution and include this method as one of the categories, even though this method is a highly hands-on one that cannot be covered in a book of this type!

The most prominent form of manual key distribution is configuration of passwords to be used for authentication signaling. As mentioned in Chapter 1, a user may present the password as is (such as in PAP), or use the password to generate a response to a challenge. The user may also use the password to generate further keying material or to fetch keying material as we said above.

Pre-shared keys are generally delivered or entered off-line, such as through phone, hard copy on a CD ROM, and so on. They are very administration-intensive, since the keys should be manually entered or delivered to each user/device. Generally, manual keys are also static due to the administrative burden associated with key updates. This entails higher likelihood of security breaches once the static keys are compromised.

The scalability problems typically lead to security breaches; for instance, it has been observed that some administrators use the same shared secret for all the NASs (see Chapter 1) dealing with one central server (such as a RADIUS server), to avoid the hassle of dealing with one key per NAS. Another prominent example of such security breach is the IEEE 802.11 wired equivalent privacy (WEP) security that uses the same secret for every host that interacts with the same access point. From a user standpoint, manual delivery methods do not scale well either, since a user or node dealing with many peers needs to manually establish one shared secret with each of those peers in order to protect itself/herself from spoofing.

Unfortunately, establishment of the pre-shared secret generally is considered outside scope for many security protocol designers, and hence often gets a simple hand-waving treatment. Nevertheless, advising on establishment of pre-shared secrets or working hard toward methods that avoid the need of pre-shared secrets all together (such as certificate-based methods) must be treated as an important issue in designing network security architecture. When left to the administrators or implementers, this issue may lead to either security breaches or non-interoperable end products, since network security is only as strong as its weakest link regardless of how sophisticated the rest of the security procedures are. For instance, if a network authentication is based on use of user certificates that are fetched by using passwords. When a password is compromised, the certificates do not bring any extra value.

3.2 Management of Symmetric Keys

In this section, we provide examples of some of the most prominent key management procedures being used today. It is of course not our intent to cover every imaginable method for symmetric keys. Other examples of key management procedures such as Mobile IP–AAA and public key infrastructure (PKI) certificates are provided in later chapters. It should also be mentioned that the tasks of key management and peer authentication are closely related and we now are witnessing a trend in combining the two. Since authentication is an expensive procedure, whenever possible, keys or keying materials for the following secure communications should be established in conjunction with the authentication process:

- When authentication of a peer is performed by a central server, it is common that authentication and key generation happens at the server simultaneously and then the keys are transferred from the server to the client (peer) along with the indication of the successful authentication.
- When key management happens through a peer-to-peer key agreement and independent of a main server, care must be taken so that neither peer establishes a trust relationship or keys with unknown or untrusted entities. For this reason well-designed key agreement methods also include an in-band mutual peer authentication.

3.2.1 EAP Key Management Methods

As mentioned earlier, combining authentication and key management procedure is an efficient exercise. A new trend is emerging in the design of wireless access technologies: access network authentication mechanisms are being used to create dynamic security associations between the wireless end client (in the hands of the user) and the network edge devices. In many ways, the IEEE 802.11 group and its business counter part, the Wi-Fi alliance [WIFIWEB], who deal with the design of wireless local area network (WLAN) technologies have been in the forefront of this movement. The main driver behind the movement may have been the massive critic that WEP, which was the first proposal for WLAN link security and authentication has been received for the weak protection that it provided. One of the main problems with WEP was the way it handled the key management for secure communications between the WLAN access point and the end clients. One of the main weaknesses of WEP is that it only supports the use of one single static pre-shared key between a WLAN access point and all the end clients that interact with that access point. Different clients authenticated with the access point all own a copy of that same encryption key and hence cannot privately communicate with the WLAN AP, while other authenticated clients are present. Furthermore, the WEP security mechanisms were easy to crack, which meant that as soon as an attacker discovered that universal key within the WLAN coverage area, the communications for all the clients within that area would be compromised. In order to solve these problems, the community has been looking into ways to be able to resolve the key management problem in a secure and scalable way. The result has been the design of new security methods that produce dynamic session keys in conjunction with the initial authentication procedure. Since EAP provides a generic authentication framework capable of providing native support for many authentication methods and their interactions with a backend authentication server, it has been used for key management procedures as well.

In Chapter 2, we presented an overview of the EAP as a generic framework for use of a variety authentication and access control algorithms. We will devote an entire chapter of this book (Chapter 10) to describing the details of many of these authentication mechanisms within EAP framework. Therefore, we will not go through the use of authentication methods within EAP framework here. However, the newer application of EAP as a key management framework is rapidly gaining popularity, and hence it is only responsible thing to do for the IETF EAP community to take the task of standardization of EAP as a key management framework. This work will hopefully not only supersede the popular IEEE 802.1x frameworks and clear the limitations and confusions experienced in the implementation of that protocol, but also provide a guideline for all the future instantiations of EAP-based authentication and key management protocols.

In this section, we provide an overview of the EAP key management framework [EAPKEYID]. However, it should be mentioned that as opposed to the EAP authentication framework, the use of EAP as a key management framework is still at a toddler stage of the standardization process and should be considered as such. In this treatment, we do our best to cover the important fundamentals that we deem rather stable and shy away from less stable details.

In Chapter 2, we described the way EAP is fitted for carrying the messaging for a generic authentication method in a three-party authentication model: EAP was originally designed to support network access and authentication mechanisms in environments where IP messaging was not available, such as over link layer access protocols. Hence, the EAP messages are meant to be carried over a link layer specific protocol between the end client and the edge device that acts as an authenticator. The EAP messaging between the authenticator and the backend authentication server, which is usually an authentication server, over a AAA protocol. We will describe encapsulation of EAP inside AAA protocols when we describe the two prominent AAA protocols, namely RADIUS and Diameter and will not go over any details here. Figure 3.2 shows this model, which can be used for providing a combination of access control and link security setup for an unknown end client (peer or user) perfectly. As the end entity (peer) requests for access and submits her/its credentials to an edge device (authenticator in the model), the authenticator

Figure 3.2 Elements of EAP key management for three-party authentication model

forwards the request to the backend server. The server authenticates the end peer and at the same time provides the necessary key material for the end peer and network edge device (authenticator) to be able to engage in a secure communication over the otherwise insecure link.

Key management according to the EAP framework comprises three distinct stages as described below: (shown in Fig 3.3)

- *Authenticator discovery phase (phase 0)*: This is the phase where the peer locates the authenticators and their capabilities, such as the type of access technology they provide (802.11, 3G, etc.), the bandwidth rates, the authentication algorithms, and cipher suites they can support. The key management framework requires that an EAP method that supports mutual authentication be chosen during this process. Note that the capabilities discovered during this phase are often based on the advertisements made by the authenticators. Thus, fully trusting such advertised capabilities, specifically those regarding authentication methods and cipher suites is not a recommended course of action by the EAP peer.
- *Authentication phase (phase 1a)*: Once the peer has discovered a network edge device (authenticator) and has decided to request access to the network represented by the authenticator, it starts an EAP conversation to perform a mutual authentication with the network. To keep our discussion simple, we only consider three-party authentication models, where the authenticator only acts as a pass-through for the authentication exchange with the AAA server, which also acts as an EAP server. The authenticator and EAP server encapsulate the EAP messaging inside AAA protocol messages. Note that the mutual authentication is between the peer and EAP server and happens concurrently and independently in the two directions. A normal authentication mechanism would stop here, but in an EAP key management framework, where authentication and key management is combined, during this phase authentication is accompanied by derivation of keying material between the peer and the EAP server. There are two types of keying materials that are produced during this phase. First, the keying materials that are used for protection of EAP conversation itself, such as secure confirmation of the capabilities discovered in the first phase and protection of the following exchanges within EAP conversation. The key material derived for this purpose is called transient EAP keys (TEK). As phrase "transient" suggests, the TEKs are used to create a secure channel that protects the EAP conversation between the peer and the EAP server. However, the term TEK itself is a sort of transient, since the IETF is still debating the terminology at this point! The second type of key material is the one that will be used to establish a secure channel between the peer and the authenticator and is called the AAA key. Note that, in a two-party model, the authenticator is also the EAP server and hence the same association protects the communications of the peer with both the EAP server and authenticator. However, in the three-party model, the secure channel between peer and EAP server does not provide a direct security relationship between the peer and authenticator (NAS).
- *Key transport (phase 1b)*: Since the key material was only generated between the peer and EAP server and no keys between the peer and authenticator exist, the link layer between the peer and the authenticator cannot be protected yet. Since the authenticator knows nothing about the peer, it does not have any information to build any sort of relationship with the peer. In order to help the authenticator with this process, the EAP/AAA server forwards part of the keying material generated (AAA key) in phase 1a to the authenticator

(NAS) through a secure channel it shares with the NAS. Using this keying material, the NAS can later trust the peer to start building a security relationship with that peer. We will provide more detail on that process later, but for now let us state the obvious: if the authenticator acts as the EAP server (two-party authentication model), no key transport is required.
- *Secure association (phase 2)*: This is the magic part. Based on the keying material (such as the AAA key) received from the EAP server in phase 1b, the authenticator (NAS) can now interact with the peer directly and without the involvement of the EAP server (Figure 3.3) establish security associations for secure communications with the peer without the involvement of the EAP server. Transient session keys (TSK) are the keys that are created during this stage and are used to protect data between the peer and the authenticator according to the cipher suite that is negotiated between the two. Security association protocols may be in place to refresh the generated keys before these keys expire. It should be mentioned that since the authenticator acts as a pass-through during the authentication (phase 1a), the peer and authenticator cannot trust each other during phase 1a. The only thing they have in common is the knowledge of the AAA key, which was derived by the peer and delivered to the authenticator from the EAP server. Hence, before going on and creating session keys, the peer and authenticator needs to establish trust by proving to each other that both possess the same AAA key.

A final but very important note is that the mutual authentication between the peer and the EAP server happens based on the *long-term credentials* for both parties. In other words, either they both share a pre-configured secret or they use their private keys in conjunction with their certificates. It should also be mentioned that in this book we try to follow the IETF terminology as closely as possible and therefore could not avoid the unfortunate misnomer that exists around the AAA key. Even though the EAP community refers to the key material created by the EAP server and used by the authenticator to establish a security association with the peer as the AAA key, the Mobile IP community, as we will see later on, calls the long-term secret that the mobile node shares with the AAA server as the AAA key.

Now that we described the three phases of the EAP key management process, we should mention the reason behind this phased design approach: the EAP key management process is being designed as a generic framework that needs to be

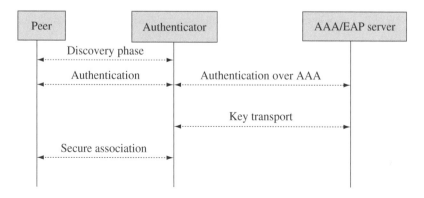

Figure 3.3 EAP key management process in a three-party model

- *Media independence*: The key management process includes a capability discovery process that is highly dependent on the specific link technology (PPP, 802.11, Ethernet) being used. Keeping the specifics of each link technology, such as discovery messaging (probe request/responses), information elements, host identifiers (such as MAC address) out of the EAP design, keeps the design of the EAP key management framework a manageable task and makes the specification a long-lasting one. Furthermore, secure association phase may involve negotiation of parameters and cipher suites are specific to a media and have no context for outsiders.
- *Cipher suite independence*: EAP key management framework involves two different family of cipher suites: (1) Cipher suites used by EAP authentication methods to protect the EAP conversation between the peer and EAP server; (2) Cipher suites negotiated by the secure association protocol between the peer and authenticator and used for protection of data between the peer and the authenticator. The cipher suites negotiation during secure association phase is performed in a manner that is out of band for EAP conversation and possibly without knowledge of the EAP server. The reason for this approach is to keep the code implemented at the EAP server simple. Otherwise, the EAP server code not only would have to include the implementation of every considered cipher suite but also would have to be upgraded every time a new cipher suite was introduced for media communications protection. Hence, cipher suite independence is another reason for having a phased EAP key management approach, in which the authentication and key materials generated as a result of this authentication, do not depend on the cipher suites implemented between the peer and the authenticator.
- *Handover and multi-port support*: With the increasing demand on fast handover, it is important for the peer to move quickly from one authenticator to the next and maintain the proper security standards despite its mobility. In order to fulfill these needs, the EAP server may, after performing a single authentication of the peer, transport the AAA keys to multiple authenticators to help proactive key management.

The strength of the EAP key management framework in deriving over-the-air communication channel keys, based on pre-existing trust with the AAA server, has been recognized by several standards communities. This strength combined with the flexibility of EAP as an authentication framework to provide support for a generic authentication mechanism has made EAP a powerful candidate for many emerging wireless technologies such as 802.11 security standards [IEEE80211i] and its robust secure network (RSN) [EDARB80211]. We will provide a much more detailed discussion on EAP authentication framework in Chapter 10.

3.2.2 Diffie–Hellman Key Agreement for Symmetric Key Generation

Diffie–Hellman (DH) is a public key-based method that allows two peers arrive at a shared secret after a series of exchanges and calculations, without ever transmitting the shared secret over the untrusted media. The DH agreement does however require each party to possess a public key pair and have knowledge about the public key of the other party. The idea of the DH agreement has been extended in many different ways, leading to many different variants, some of which are standardized by the American National Standards Institute (ANSI X9.42). IETF has also standardized the DH agreement in RFC 2631 [DH2631].

The basic idea of DH is based on the fact that each party combines its own private key with the public key of the other party to arrive at a shared secret (called ZZ in the RFC). That shared secret (ZZ) is later used to derive keying material for the security algorithms that protect the communications. The mathematics for calculating the shared secret ZZ is as follows:

- Each party holds a private–public key pair (x, y), i.e. y_a is the public key of party A and y_b is the public key of party B. The relationship between the private and public key is as follows

$$y_a = g \wedge x_a \bmod p$$

$$y_b = g \wedge x_b \bmod p$$

- *DH domain parameters*: g, p, and q are called DH domain parameters. They are either communicated out-of-band or negotiated and agreed upon during a signaling protocol that embeds DH. For instance, IKE defines six predefined groups of (p, g), with names group 1, group 2 till group 6 and sends only the number for the group, not the actual p and g values. The initiator proposes a group number. The responder agrees or counters the group number. The following DH exchange is then performed using the parameters out of the negotiated group number.
- *Shared secret (ZZ) generation*: RFC 2631 assumes that both the sender and the recipient of the messages uses a private key corresponding to the public key she has given to the other party, so that they both arrive at the same ZZ independently. To arrive at the shared secret, the sender A calculates y_a and sends it to party B.
- Party B after receiving the public key of A, y_a, performs the following calculation using its private key x_b to arrive at ZZ_b.

$$ZZ_b = y_a \wedge x_b \bmod p = (g \wedge x_a) \wedge x_b \bmod p = g \wedge (x_a \cdot x_b) \bmod p$$

- In the meantime, the recipient B also calculates y_b and sends it to A.
- Party A after receiving the public key of B performs the following calculation (using its own private key x_a) to arrive its own version of ZZ (ZZ_a)

$$ZZ_a = y_b \wedge x_a \bmod p = (g \wedge x_b) \wedge x_a \bmod p = g \wedge (x_a \cdot x_b) \bmod p$$

As we can see $ZZ_a = ZZ_b$, i.e. both parties arrive at the same shared secret.
- Next, the two parties convert the shared secret (ZZ) to other keying material. Using a hash algorithm such as SHA1 and zero-padding of the ZZ, an arbitrary amount of keying material (KM) with a variety of lengths (such as 1024 bits) can be created:

$$KM = H(ZZ \mid OtherInfo)$$

Examples of other information are: key information, algorithm for data encryption, type of content to be encrypted, information on party A, and so on.

For this discussion, we care about how DH is used to provide keys for encryption of traffic between two parties. Hence, we need to know how ZZ is used to calculate the key that is used to encrypt the actual session keys. You may say: "Ha?". Okay, let us repeat that again: the actual encryption key that is used to protect the data traffic is called content encryption key (CEK). In order to pass this key to the other side, this key is encrypted before transport using another key that is called key encryption key, KEK. The main thing to recognize is that KEK is special cases of the generic key material (KM) that was shown in the formula above.

The key size (n) for encryption of another key depends on the algorithm used for key encryption (encrypting the session key, CEK). For each algorithm, the KEK of size "n" is generated by using the n leftmost bits of the KM. Examples of common KEK sizes are 192 and 128 bits. Note however that the effectiveness of KEK depends on the original size of ZZ, since ZZ is the only source of entropy in KM generation.

It seems that the IETF specification of DH stops here. In other words, the specification only explains how the KEK that wraps the session key (CEK) is defined, without specifying a method for deriving CEK. So in a way the IETF specification [DH2631] defines a key agreement method for the KEK, and if we want to be pedantic we can say that this specification actually provides a key transport mechanism, since it defines a method on how to wrap the actual session key (CEK) for transport. The original DH method provided the researchers with the idea of arriving at the first shared secret (ZZ) above without having a pre-configured trust relationship. From then, it was up to the designer to use this shared secret directly as a session key or as a method to derive the session key or in the case of IETF specification, as we saw, even arrive at a KEK.

3.2.2.1 Problems with Diffie–Hellman

Although DH is a powerful key agreement concept, by itself it does not provide adequate security protection for a key exchange in a hostile environment. This is evident from the fact that many communities including the IETF have defined a variety of key exchange mechanisms such as IKE and transport layer security (TLS) that are developed on the basis of and around the concepts of the DH key agreement. In the following we describe the main problems with the raw DH approach in a way that motivates us in studying the more sophisticated methods such as IKE later on.

1. The two peers in the DH agreement do not authenticate themselves to one another. An attacker with a public–private key pair can engage in the DH agreement with other peers under a false identity, performing communications on behalf of a legitimate user. The attacker can receive authorization for service or join communications groups and so on instead of the legitimate entity. Using public key certificates that include the owner's identity as part of the exchange will eliminate this problem.
2. DH is vulnerable to IP spoofing attacks, where a peer can launch many DH agreements, each with a different and spoofed IP address towards another peer. This may bring the responder down through engaging the latter in the expensive computations involved in a DH exchange and that is why it is referred to as a flooding attack. This is not necessarily cured by authentication. Generally, the solution is to add cookies to verify that the peer actually has an interface identified by the presented IP address.
3. DH allows for man-in-the-middle (MITM) attacks, and this is due to the fact that public keys are not public enough. An attacker can insert itself in the middle and replace its own public keys with those of another peer during the exchange of public keys [IKEBORE].
4. DH does not prevent replay attacks. Another party can record the messages including a peer's response and replay them later on to arrive at a key with the second peer. This is

why it is important that one-time random numbers should be used in the process. As we will see later on, IKE uses nonces to provide anti-replay protection.
5. DH lacks the cryptographic method negotiation capabilities. If both peers do not suggest the same cryptographic parameters, the exchange fails. A remedy would be to have each party suggest the parameters, but that require support for negotiations in cases where they do not initially suggest the same parameters. Furthermore, transmission of all the parameters can be costly. If public key parameters had been published in well-known groups and the two peers followed a negotiation process, the problems mentioned above could be avoided.
6. There is no inherent reliability mechanism in a DH exchange. The exchange either needs to be overlaid on a reliable protocol or include inbuilt reliability mechanisms. The second version of IKE (IKEv2), currently being specified by IETF, provides some degree of reliability: the initiator is responsible for retransmissions until it receives a response. The responder retransmits the message only when it retransmission request.
7. Each DH negotiation leads to only one key, one time. Security requirements often dictate periodic refreshing of keys. A DH exchange does not provide facilities for refreshing the original keys. Key refreshes require new DH exchanges.

3.2.3 Internet Key Exchange for Symmetric Key Agreement

IKE [IKE2409] is developed to support the key management needs of IP security protocol, commonly known as IPsec. We will describe IPsec and provide more details on IKE in Chapter 4, since understanding those details require a better grasp of IPsec. However, to complete our key management discussion, we highlight some of the improvements that IKE has provided to the DH key agreement process.

From 10,000 feet above IKE can be seen as a refinement of the DH agreement, eliminating many of the shortcomings of the DH agreement:

1. IKE uses cookies to prevent flooding attacks; both initiator and responder create cookies in the beginning of the exchange and all the following messages include both cookies, to which the two parties associate the IP addresses of each other. A message including the wrong IP address–cookie association will be discarded. The cookie exchange can also be used to distinguish between multiple IKE exchanges between the two hosts (otherwise both IP addresses and UDP ports would be the same).
2. IKE prevents the man-in-the-middle problem of the DH exchange by requiring the participants to authenticate themselves to each other, using their IDs along with the proof of a secret. However, it should be noted that IKE performs the DH exchange after the cookie exchange, but before the authentication exchange, this way the identity of the two parties can be protected using the keys established by the DH exchange. IKE provides the opportunity for the peers to perform authentication in multiple ways. Both pre-shared secrets and public key certificates can be used for authentication purposes. We will discuss the details of each method in Chapter 4.
3. IKE defines various cryptographic groups, each with a set of well-known DH parameter group (domain parameters). IKE also prevents the problem of public key spoofing, by publishing well-known groups of DH parameters. This not only means the parameters themselves do not need to be transferred but also allows for negotiation of group

numbers; each peer suggests a number, until they both agree on the number. The IKE peers only need to pass the number of the group as part of the messaging.
4. Instead of using the exchanged key as shared keys, as the DH exchange does, IKE uses the key generated during the DH exchange to create a secure channel that can be used for creation of any number of IPsec keys. This way the encryption keys can be refreshed as often as required and perfect forward secrecy (secrecy in case some keys are compromised) is better guaranteed than with pure DH.

We will suffice with this short discussion on IKE for now and this will be detailed further in Chapter 4, where IPsec is being described.

3.2.4 Kerberos and Single Sign on

All the key management methods that we described so far have one thing in common and that is they mainly deal with authentication and access issues. Another common theme was that, regardless of whether a peer deals with an authentication server or with another peer, the keys were pairwise. In other words, the keys were established between only the two entities that were to establish a one-on-one trust relationship. However, in many real life applications, a user or more generally a client does not request access to the network just for access; often the client actually needs to receive services from a particular server. Mere authentication to an authentication server may not automatically authorize the user to receive those services. As mentioned in Chapter 1, in a majority of networks, the user is authorized for a service offered by a server based on authentication of the user to that server and checking the user's profile against some database including a list of privileged users. A mindset in tune with AAA philosophy would state that authorization for various services should be handled by an AAA server in conjunction with various service equipments as we explained in Chapter 1. However, in the pre-AAA era, each of these servers would individually require to authenticate the user. A great example was when all these servers were on a university campus but each would belong to a different department and was run under a different authority. Hence, the user needed to log in to each of these servers individually to receive their services. When the client needed to deal with multiple servers (as shown in Figure 3.4), the key management problem quickly became an unscalable one:

- First, the client needed to authenticate to each server individually. This meant the client had to have established authentication secrets with each server individually (S_{ASi}).
- Second, after the client authenticated to each server, it needed to establish separate session keys (S_{SSi}) with each server if privacy was required for communications with that server.

As shown in Figure 3.4, a user dealing with N servers had to have N authentication credentials (typically passwords) and establish N shared keys for encryption. This seemed rather tedious. Researchers at MIT came up with a clever idea to solve this problem and called the solution Kerberos. The name Kerberos, after the three-headed dog guarding the entrance to Hades in Greek mythology, provides a good analogy, since a single server watches and facilitates the interaction between a client and many other servers and this way handles the scalability problem associated with a client dealing with many servers at a time.

Key Management Methods 63

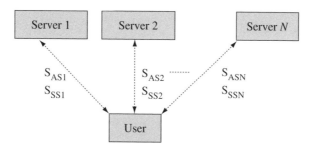

Figure 3.4 User dealing with multiple servers and using pairwise shared secrets for authentication and encryption

For establishing trust relationships between the clients and the servers, Kerberos employs a key distribution center (KDC) that acts as the third party involved in the interaction between the client and the servers. The KDC (acting as the three-headed dog) authenticates the client first and then issues a ticket granting ticket (TGT) to the client. For this reason, the KDC is sometimes also referred to as a ticket granting server (TGS). Anytime the client needs to access a server, it sends this TGT back to the KDC and requests a specific ticket to that specific server. The KDC issues the client a new ticket for authentication to that server. The KDC also generates a session key for the client and that server as part of this process. As we see, the Kerberos KDC builds the trust between the client and other servers based on the initial trust relationship it has established with the client during the initial client authentication. If a user is using a device that communicates to the Kerberos system, the user does not have to present her credentials every time it needs to deal with a new server. The user only presents her credentials once and after that the system and device that the user is using for access handle all other security procedures. This concept is now called single sign on (SSO) and is often marketed as an attractive network capability. Even though the concept of single sign on is nowadays broader than Kerberos it has its roots in Kerberos. The user notes that the KDC does not maintain any states regarding the sessions between the client and individual servers. The details of the procedure are described in the following.

In order to establish the initial trust relationship with the KDC, the client first needs to authenticate to the KDC. A Kerberos system may deploy a separate Kerberos authentication server (AS) to authenticate the user and provide the user with credentials for initial authentication to the KDC. It should be noted that the exchange below simply does not provide any identity verification by itself. The reader is referred to RFC 1510 [KERB_1510] for further details of client authentication in Kerberos systems. The authentication to the authentication server is performed using the following two messages:

1. *Kerberos authentication service request (KRB_AS_REQ)*: It is sent from the client to the authentication server to obtain credentials for the KDC and other servers that do not trust the mediation by the Kerberos system. This request is usually generated at the initiation of the user login process. The user password is typically used to derive a secret master key (K_{cp}) for encryption and decryption of the communications (the subscript "cp" stands for client password, to indicate that the key is generated from the client password). The client sends its own identity along with the identity of the server to which it wishes to authenticate along with any options that the client requires the ticket to have, such as possibility to renew or forward the ticket.

2. *Kerberos authentication service reply (KRB_AS_REP)*: It is sent from the authentication server to the client following successful processing of the KRB_AS_REQ. The authentication server looks the client name up and generates the KDC credentials for the client as follows:

   (a) S_c: A login session key between KDC and the client. The idea behind generation of this key is to discard the master key (K_{cp}) that was generated from user's password, as soon as a TGT is generated for the user. From that point on, the session key S_c will be used for encryption of future messages between the user and KDC within that login session. This eliminates the need of having the user type her password every time she is requesting a new ticket for a new server and allows the KDC to recover from a crash without maintaining any states.

   (b) *A TGT for the user*: The TGT includes information about the user identity, the login session key issued to the user (S_c), ticket expiration time, and the current time encrypted with the user's login session key (S_c). Note that the information in TGT is encrypted with a key that only KDC holds (K_{KDC}), so that nobody (not even the client that will own the ticket) can modify the TGT contents.

$$TGT = K_{KDC} \{user\_ID, expiration\_time, S_c[time]\}$$

3. The AS encrypts the KDC credentials using the secret key, K_{cp} (derived from client password), and sends the encrypted credentials to the client:

$$K_{cp}[S_c, TGT]$$

4. The user decrypts the session key S_c and then stores the TGT for future use with variety of servers.

5. When the client wants to access the services of a server, the client sends a ticket request to the KDC to receive a ticket for the new server. As we mentioned earlier, the KDC is called a ticket granting server for this reason and that is also why this request message is typically called Kerberos TGS request (KRB_TGS_REQ), which includes the client's own identity, the server's identity, the TGT and the current time encrypted with its login session key (S_c). The latter is used for two purposes: first, since the information is encrypted with the user's session key S_c, the user proves its possession of the key shared with the KDC. Second, inclusion of the current time provides a measure of anti-replay protection by adding the non-repeatable element (time) to the message. However, this requires that the user and the KDC have somewhat synchronized clocks, possibly through using protocols such as network time protocol [NTP1305]. In general, usage of time critical messages may also require adequate provisioning of quality of service in routing of such messages. For those reasons, it is recommended that the Kerberos system should tolerate a lack of synchronization up to a few minutes.

$$User\_ID, server\_ID, TGT, S_c[time]$$

6. When the KDC receives the ticket request from the client, it decrypts the TGT (using its own K_{KDC}). After checking the validity of the TGT and its expiration time, the KDC recovers the client login session key (S_c) from the TGT. The KDC then generates a specific session key between the client and that server (K_{cs}) and includes the key inside a ticket for the user to use for authentication with that server. Beside the session key, the ticket also includes the client name, the server name (server_ID), the session key

for that server (K_{cs}), and ticket expiration time. Here is another fundamental assumption in Kerberos (the third head of the dog if you will): *The KDC shares a secret (K_{Ks}) with the server to which the client is trying to authenticate.* Using this secret (K_{Ks}), the KDC encrypts the ticket that has just generated for the client and sends the ticket to the client.

$$\text{Ticket} = K_{Ks}[\text{ID, server\_ID}, K_{cs}, \text{expiration time}]$$

This makes sure that neither the client nor any other middleman can alter or view any of the information inside the ticket. Only the server, for which the ticket is generated, can decrypt the information provided by the KDC. The KDC also sends the session key (K_{cs}) to be shared with that server to the client. This key along with server identity and the ticket is encrypted with the login session key (S_c) to the client inside a Kerberos TGS reply message (KRB_TGS_REP):

$$S_c[\text{server\_ID}, K_{cs}, \text{ticket}]$$

The trust relationship as a result of using the Kerberos system is shown in Figure 3.5.

7. Later on, when the client intends to authenticate to the server, the client presents the ticket that it just received from the KDC to the server. The user also encrypts the current time with the session key it shares with the server, K_{cs}, as a measure to provide proof of key possession and anti-replay protection and includes that in a Kerberos authentication request (KRB_AP_REQ).
8. Kerberos provides mutual authentication by having the server authenticate to the user as well. Once the server receives the ticket from the user and decrypts it with the key it shares with the KDC (K_{Ks}), the server retrieves the key with the user (K_{cs}) from the ticket and decrypts the time value submitted by the user. The server then increments the time value and re-encrypt it with K_{cs} and sends it to the user to authenticate itself to the user (by showing that it possesses the correct key: K_{cs}).

3.2.4.1 Kerberos Issues

As seen above, Kerberos provides both confidentiality and anti-replay protection for the authentication procedures by providing encrypted authentication. Kerberos also provides

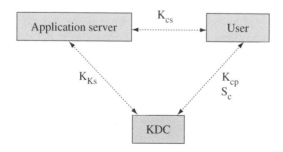

Figure 3.5 The three-way trust relationship in Kerberos-based authentication

good scalability properties when it comes to dealing with multiple servers. However, there are still several issues with usage of Kerberos:

- Kerberos is a client–server protocol that does not make a clear distinction between device and user, since the Kerberos was originally developed to support users that came inside a trusted facility and use fixed and trusted terminals.
- Since KDC holds all the session keys for all the clients and servers, it becomes a single point of failure. Having an alive and always available KDC is crucial to the operation of Kerberos. Although backup KDC servers can be deployed, this is still an inherent weakness of Kerberos. In comparison, when deploying certificate-based authentication, the certificate authorities (CAs) do not have to be available in every protocol exchange.
- Again, since the KDC holds all the keys, it is a very attractive attack target. The KDC must be secured properly, otherwise a compromised KDC leads to a compromised system.
- The initial authentication with the KDC or the authentication server generally requires a pre-shared secret. Many systems use a password for this purpose. A way to get around the shared secret problem is to have the client use certificates for authentication to the KDC. A recent technology, currently being developed in IETF, provides a way to get around the shared secret problem by having the client using smart cards including public key certificates for the initial authentication. The protocol, called PKINIT, short for "public key cryptography for initial authentication in kerberos" is still not standardized (20th revision as of this writing) [PKINIT] but is also used in Windows 2000. PKINIT allows the client to provide a signature along with its public key certificate for the purpose of authentication to the KDC, while leaving the rest of Kerberos messaging in tact. The downside is that the KDC must trust a certificate authority for verification of the certificate.
- As mentioned earlier, Kerberos emphasis on use of time values for authentication requires implementation of accurate clocks at clients and servers.

3.2.5 Kerberized Internet Negotiation of Keys (KINK)

Deploying DH in IKE means that IKE requires several expensive public key operations to establish the keys between the two parties. That makes IKE computationally expensive and slow. Also, when pre-shared keys must be configured at the two peers to perform the authentication required by IKE, scalability issues are also added to the performance problems. Another issue is that since IKE is a peer-to-peer protocol, it does not provide facilities for central management of keys. Every time a secure channel is needed with a new or existing peer, the same set of IKE messages and processes must be repeated, and furthermore the authentication within the IKE requires another pre-shared secret or set of certificates to exist for this IKE exchange. When n clients deal with m servers, the number of keys required are in order of $O(n \times m)$.

As mentioned earlier, a Kerberos KDC can be used to establish trust between a user and many unknown servers, as long as both user and the servers trust the KDC. The initial authentication to the KDC can be performed using either symmetric keys or public keys. Even though the pre-shared keys may still be required for authentication with the KDC, this would be only with the KDC and not every other server. A working group in IETF, calling itself Kerberized internet negotiation of keys (KINK), is chartered to create a protocol to facilitate a centralized key management based on Kerberos architecture [KERB1510] as an alternative to IKE. This way, not only a central and consistent security policy and key management procedure can be employed, but also the difficulty in managing pre-shared keys is avoided and computational burden of using public key cryptography in IKE is reduced. The protocol will also provide security protection for a key exchange message themselves.

The aim of the working group is to produce an easily managed, but cryptographically sound protocol that does not require public key operations. The working group will not require changes to either IPsec or Kerberos specifications. We will not provide any further details on this protocol, since it is still a work in progress, but like to warn the reader about its potential viability in the industry when it does become a standard.

3.3 Management of Public Keys and PKIs

As seen numerous times before, many encryption mechanisms and even some key exchange mechanisms, such as IKE rely on pre-shared symmetric secrets. Usage of symmetric keys for trust establishment introduces serious scalability problems, since it requires that each node pair in the system share a unique secret. A network consisting of n nodes, may have to potentially deal with $O(n^2)$ keys. In a large network, such a large number of keys can pose a serious key management problem. On the other hand, using public–private key pairs solves many of the scalability problems related to manually distributed pairwise secrets: each node often needs to only hold one key pair, leading to n key pair for a network of n nodes. However, using public keys has its own set of issues:

- When Alice sends a message to Bob, Alice uses Bob's public key to encrypt the message, so that only Bob can, with his private key, decrypt the message and get hold of its content. However, this assumes that Alice knows Bob's public key beforehand. Either, Bob must have sent his public key to Alice, or, Alice, must have downloaded it from some place. At any rate, Alice must have a way of being able to trust the fact that the public key she has received does actually belong to Bob.
- When we say that Alice sends a message encrypted with Bob's public key, one important fact that is usually forgotten is that this by itself does not protect the integrity of the message. Anybody could have sent an encrypted message on behalf of Alice, since Bob's public key is accessible to anybody who has access to the public key database. This means that Alice needs to also add her signature to the message, signing some part of the message with her own private key. Bob can verify Alice's signature using Alice's public key, trusting that nobody else can forge Alice's signature, since Alice's private key is a well-kept secret. However, again, Bob needs to either acquire Alice's public key out-of-band or from Alice directly.

We see that when using public key methods two main problems need to be solved:

1. The public key of each party must be made available to the other parties. People need to know Bob's public key to be able to send him encrypted messages or verify his signatures. Either Bob needs to send his public key to people in advance or there needs to be a trusted authority where people can go and download Bob's public key.

2. There needs to be a way to tie the public key to the identity of its owner in a trusted fashion. When people get Bob's public key from some source, they need to make sure that it actually belongs to Bob. Even when they receive the key from Bob himself, they need to make sure that it is actually Bob who sent it.

Both problems can be solved with introducing certificates for any entity that presents its public key to other entities in the network. A public key certificate includes the owner's identity and public key, so that the recipient of the certificate can determine who the public key belongs to. In order to assert the authenticity of the certificate, it must be signed by an authority that both the presenter and recipient of the certificate trust. This entity is called certificate authority. The CA is actually the entity that issues the certificate for all the clients within the network, regardless of where the private–public key pair for that entity is generated. To ensure both identity and public key in the certificate belong to the same client, the CA authenticates the client thoroughly (often physical presentation of identity rather than electronic one).

Unfortunately, managing certificates is not a simple task. Beside the authentication, many other issues, such as safeguarding the private keys, managing certificate lifetimes and verifying their validity, revoking certificates if needed, are involved. For this purpose, use of certificates for authentication and security purposes typically require an entire infrastructure, typically referred to as PKI. Since we are devoting an entire chapter (Chapter 9) to PKIs, we will not continue with management issues for public keys and end this chapter at this point.

3.4 Further Resources

Bruce Schneier has written a comprehensive book, that includes a wealth of information on various cryptographic concepts, methods, and their implementations [SCHNEIER].

Going through the mathematics in Schneier's book can however be tedious at times. Charlie Kaufman *et al.* have written a book on network security [KAUFMAN] that explains the mechanics of many security algorithms such as Message Digest (MD) and Data Encryption Standards (DES) in an algorithmic way. This book also explains Kerberos in a comprehensive manner.

More information about DH standards can be found in the document entitled "PKCS #3 Diffie–Hellman Key Agreement Standard" at RSA Security® web page: http://www.rsasecurity.com/rsalabs/pkcs/.

Edney and Arbaugh have written a book on 802.11 security [EDARB80211] explaining key management framework within 802.11 networks which can be used as an excellent material on real-life applications of EAP key management framework.

A good source of information on the progress of 802.11 WLAN (known as Wi-Fi in the business world), the reader can check the weekly news brief, called "Blue-Print Wi-Fi" at http://www.arcchart.com/. Unfortunately, this publication is not free. However, information on Wi-Fi Alliance can be found for free at Wi-Fi Alliance website [WIFIWEB].

For more information on EAP framework, IKE, and PKI, the reader is referred to Chapters 4, 9, and 10 of this book.

For more information on KINK and its applications in the packet cable industry, the reader is referred to [PACKCABLE].

3.5 References

1. [KMGNIST1], "Key Management Guidelines, Part 1: General Guidance", second draft, NIST, June 2002.
2. [EAPKEYID], Aboba, B., *et al*. "Extensible Authentication Protocol (EAP) Key Management Framework", IETF work in progress, Internet draft, draft-ietf-eap-keying-03.txt, July 2004.
3. [IEEE80211i], Institute of Electrical and Electronics Engineers, "Supplement to Standard for Telecommunications and Information Exchange Between Systems – LAN/MAN Specific Requirements – Part 11: Wireless LAN Medium Access Control (MAC) and Physical Layer (PHY) Specifications: Specification for Enhanced Security", IEEE 802.11i, July 2004.
4. [EDARB80211], Edney, J. and Arbaugh, W., *"Real 802.11 Security, Wi-Fi Protected Access and 802.11i"*, Addison-Wesley, March 2004.
5. [KERB1510], Kohl, J. and Neuman, C., "The Kerberos Network Authentication Service (V5)", IETF, RFC 1510, September 1993.
6. [DH2631], Rescorla, E., "Diffie–Hellman Key Agreement Method", IETF, RFC 2631, June 1999.
7. [IKE2409], Harkins, D. and Carrel, D., "The Internet Key Exchange (IKE)", IETF, RFC 2409, November 1998.
8. [IKEBORE], Borella, M., "Methods and Protocols for Secure Key Negotiation Using IKE", *IEE Network Magazine*, July/August 2000.
9. [PKINIT], Tung, B. *et al.*, "Public Key Cryptography for Initial Authentication in Kerberos", IETF work in progress, draft-ietf-cat-kerberos-pk-init-20.txt, July 2004.
10. [PACKCABLE], Mihai, B., "Packet Cable Security Architecture", Helsinki University of Technology, November 2000, http://members.fortunecity.com/burli2/packetcable0.html.
11. [SCHNEIER], Schneier, B., *"Applied Cryptography: Protocols, Algorithms and Source Code in C"*, 2nd Ed., John Wiley & Sons, 1996.
12. [KAUFMAN], Kaufman, C., Perlman, R. and Speciner, M., *"Network Security: Private Communication in a Public World"*, 2nd Ed., Prentice Hall PTR, 2002.
13. [WIFIWEB], Wi-Fi Alliance website, http://www.wi-fi.org/OpenSection/index.asp.
14. [NTP1305], Mills, D., Network Time Protocol (Version 3), IETF, RFC 1305, March 1992.

4

Internet Security and Key Exchange Basics

4.1 Introduction: Issues with Link Layer-Only Security

In previous chapters, we discussed some of the problems associated with security provisioning for links providing access to networks. Often, the link security mechanisms are developed specifically for the physical media on that link. We also provided some examples on how protocols such as EAP can be used to provide authentication and secure channels between a client and the network over a generic link technology. Still, methods such as EAP must be fitted tightly to the link technology. A bigger problem with such link security mechanisms is that they only provide security mechanisms for the link between the client and a device at the edge of the network (network point of presence) and do not extend beyond that single hop. The following example intends to demonstrate the problem that arises when link layer security mechanisms are deployed for generic communications scenarios.

Consider the case, when the two communicating hosts A and B are four hops away, i.e. separated by three routers. When only link layer security mechanisms are available, security protection can be provided over a single link (between two neighboring routers) at a time. In our example, this would mean that each of the hops 1, 2, 3, and 4 in Figure 4.1 must be secured individually based on the trust relationships between the routers on each end of that hop. When a packet traverses from host A to host B, each receiving router on the path must first decrypt the packet from the previous hop, and re-encrypt the packet before forwarding it over the next hop toward the following router. This decryption/encryption of each packet would not only significantly increase the processing burden on each of the intermediary routers, but also leaves the host A and host B at the mercy of those routers: they simply have no choice but to trust all of these routers with their traffic. For that to be acceptable, the two hosts must in advance know how their traffic is being routed through the Internet and trust all the security relationships across the path. We can see how this approach can quickly bring the Internet speed and scalability down!

All the impracticalities described in the previous example point to the fact that the two hosts are better off with establishing an end-to-end secure channel between themselves in

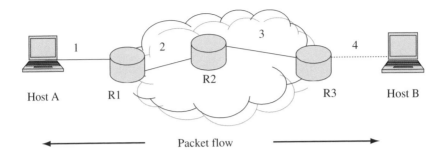

Figure 4.1 Securing communications between two entities that are multiple hops away

a manner that is transparent to the intermediate networks. That way the intermediary routers do not have to carry the burden of security processing and can focus their processing power on routing and forwarding. The two hosts can be sure that no intermediaries had access to the content of their packets.

Another case, where link security provisioning becomes impractical, is a popular scenario that is shown in Figure 4.2. An enterprise wishes to connect two parts of its network at two remote locations (site 1 and site 2) in a way that the nodes and users residing at each location can communicate with each other securely and transparently, as if they belonged to a single network. The enterprise leases a high bandwidth link from a commercial provider to connect the two sites, and is assured about the link performance by the provider, but cannot trust the provider with the content that is being transferred between the two sites. Furthermore, the enterprise may wish to handle both sites as if they were part of one site by for instance using addresses from the same IP address pool or applying the same network policies at both sites.

One prominent method for solving the problem described above is to deploy a so-called virtual private network (VPN). The VPN establishes a secure channel between the two network parts (hence the term "private") and routes the traffic in a way that the users at each end feel they are part of the same network (hence the term "virtual"). Each of the two network portions is considered as a part of the secure and trusted domain. Specific routers that also act as security gateways and are often called VPN gateways are deployed at the edge of each site and define the boundaries of each site. All traffic going in or out of each site must go through these gateways and most often in encrypted format, while the traffic within the boundaries does not have to be secured. When host A needs to communicate with host B, the traffic is encrypted by security gateway 1 (SG1) before going over the untrusted link to security gateway 2 (SG2), which in turn decrypts the packets and forward them over the trusted topology in site 2.

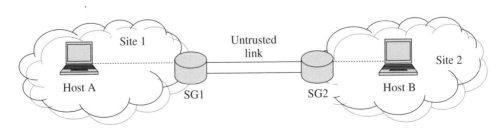

Figure 4.2 Use of secure channels to connect networks at remote location

Deploying security between the security gateways at IP layer allows the application traffic travel between any host A and host B without any regard to how the traffic is being routed.

As we see, it is often practical to provide key exchange and security at a layer higher than link layer. There are several choices on which layer to apply the security protection to the ongoing traffic. Looking at the TCP/IP stack, applying security at any of the network layer (IP layer), transport layer and application layer comes to mind. In this chapter, we will discuss the most prominent security protocols at network and transport layer, namely, internet protocol security (IPsec) and transport layer security (TLS). Out of those two, TLS provides its own key exchange mechanism, while IPsec relies on internet key exchange (IKE) for key exchange. For that reason, we also devote a section to IKE as well. The reader may even experience an imbalance on the amount of text dedicated to IKE compared to that for IPsec. The reason for that conscious effort is that we believe there is a large amount of literature (including books) devoted to IPsec, while there are very few comprehensive sources on IKE. Also, we do not provide any text on application layer security for several reasons. First, application layer security can be directly applied by using algorithms that are described in previous chapters and other cryptographic text. Second, providing security at application layer requires the application designer to possess the know-how on cryptographic procedures. Using IP and transport layer security mechanisms allows the application designers to focus on the application functionality and simply use standardized lower layer mechanisms to provision security.

4.2 Internet Protocol Security

As mentioned before, when two entities or networks attached to two different parts of a private network or public Internet, the exchanged packets may have to traverse over multiple hops. In such cases, it is useful to provide a security mechanism that is negotiated and established in an end-to-end or gateway-to-gateway manner at the IP layer or above without any regard to the communications path.

IPsec protocol is an IP layer security mechanism that can be used to protect the entire path between two entities or only the untrusted part of the path. IPsec allows the two parties establish a secure channel capable of providing data integrity, data confidentiality, anti-replay protection, and a number of other security services. A variety of network scenarios for which IPsec can be used are specified in the IETF security architecture document [SECARC2401]. Together with its key exchange mechanism (IKE), IPsec allows the two entities to negotiate and select the required protection mechanism, such as "authentication only" or "authentication and encryption", select proper cryptographic transform to use for the chosen protection, and exchange the keys required for those transforms.

IPsec uses two protocols to provide security protection for IP traffic: authentication header (AH), specified by [AH2402], and encapsulating security payload (ESP), specified by [ESP2406]. In the following sections, we provide more details on AH and ESP. However, we like to point out that AH and ESP *per se* do not provide any security protection. AH and ESP must be used along with cryptographic transforms to provide the security protection they are intended for. In fact, as we will see shortly, they are designed to be algorithm-independent to allow more choice in the selection of cryptographic processes of various strengths and functions.

4.2.1 Authentication Header

AH protocol provides data integrity and replay protection for the whole IP datagram (Table 4.1) and is an effective measure against IP-spoofing and session-hijacking attacks. The AH protocol can use a hash function, such as MD5, that takes the IP packet and an authentication key as input and computes an authenticator field called the integrity checksum value (ICV). The sender includes the ICV as part of the AH header. The receiver, after receiving the packet, re-computes the ICV on the received packet and compares with the value included in the header. This way the receiver can detect any modifications performed by a possible middleman. It should be noted that the authenticator field must not be calculated over IP header fields that may be modified by the intermediary routers. For instance, in order to prevent infinite loops in the packet forwarding process, the header of each IP packet includes a so-called time to live (TTL) field that determines the maximum number of hops a packet can go through, before it reaches its destination. As each router forwards the packet to the next hop, the router reduces the value of TTL by one. Hence, as the TTL field within the IP header keeps changing, the AH ICV calculation should not include the TTL field value. The AH protocol specifies a set of IP header fields and flags, similar to TTL, that should not be used to compute the ICV.

As it may be obvious, IPsec AH protocol does not provide any confidentiality, since no encryption algorithm is deployed. For those reasons, when data confidentiality protection is required, use of AH alone is not sufficient.

4.2.2 Encapsulating Security Payload

IPsec AH protocol, despite being a powerful mechanism for data integrity protection, does not offer encryption services. This means when encryption is required in a communication that is deploying AH, another mechanism must be deployed alongside the AH protocol to provide the encryption service. IPsec ESP protocol provides both encryption and message authentication services. This may be one of the reasons behind the fact that the security implementers are now shying away from AH. If you have a multipurpose protocol, such as ESP, that accomplishes both encryption and authentication, why use a single purpose

Table 4.1 IP header followed by the AH header

Sub-field	Description
IP header	IP heder prior to IPsec AH header
Next header	Identifies the type of the next payload after the authentication header
Payload length	Showing the length of the entire message
Security parameter index	Showing an arbitary 32-bit value that, in combination with the destination IP address and security protocol (AH), uniquely idenfifies the SA for this datagram
Sequence number	It is an increasing value and is mandatory and is always present even if the receiver does not elect to enable the anti-reply serice for a specific SA
Authenticator	A variable length field containing an ICV computed over the ESP packet minus the authentication data

Diagonally shaded fields are authenticated.

protocol, such as AH that necessitates the use of yet another protocol for encryption and incur the "double" processing costs? In the same manner as the AH protocol, ESP protocol adds an ESP header after the IP header and provides authentication protection for the information following the ESP header (Table 4.2).

Now that we established a firm argument for use of ESP, we need to be a bit more honest with the reader and state that the claim by ESP on providing authentication is not as strong as that by AH. The ESP authentication is not quite as comprehensive as AH. As shown in Table 4.2, unlike the AH protocol, the ESP protocol does not provide protection for the IP header preceding the ESP header. Still, even though an IP header unprotected against tampering can be an argument against ESP versus AH, the additional cost of AH seems to not justify deployment of both AH and ESP.

As in AH, ESP uses secure hash algorithms to compute the authenticator (ICV) that provides the data integrity protection service. For data encryption, ESP uses symmetric key algorithms. When the encryption algorithm requires initialization data, these data will be pre-pended in an unencrypted form to the payload data that is encrypted. A variety of hash and encryption algorithms can be used with ESP. However, there is a minimum set of mandatory-to-implement algorithms that all standard-compliant ESP implementations must support. HMAC-MD5, HMAC-SHA1, and DES are few examples of such algorithms.

4.2.3 IPsec Modes

In the introduction at the beginning of this chapter, we mentioned two cases for which implementation of security at network layer would make more sense, namely an end-to-end communication between two nodes at distant parts of the network and a VPN connection between two remote sites. These two scenarios represent the need for two different methods in using IPsec protocols or as the IPsec community calls them: two IPsec modes, namely tunnel and transport modes. In the following, we describe each mode briefly.

Table 4.2 An IP header followed by ESP Header

Sub-field	Description
IP header	IP heder prior to ESP header
Security parameter index	A random value identifying the security association for this packet
Sequence number	It is an increasing value and is mandatory and is always present even if the receiver does not elect to enable the anti-replay service for a specific SA
Payload data	A variable length field containing data described by the next header field
Padding (0–255 bytes)	Padding for encryption
Pad length	Padding indicates the number of pad bytes immediately preceding it
Next header	Identifies the type of data contained in the payload data field
Authenticator	A variable length field containing an ICV computed over the ESP packet minus the authentication data

Diagonal shaded fields are authenticated, while Trellis shaded fields are authenticated and encrypted.

| IP header | IPsec header (AH, ESP, or both) | TCP/UDP header | Payload |

Figure 4.3 IPsec in transport mode

4.2.3.1 Transport Mode

In transport mode, the IPsec protocols provide protection primarily for data at the transport layer and higher. As shown in Figure 4.3, this would mean that the transport layer header (such as TCP header) immediately follows the ESP header.

Security architecture specification [SECARC2401] allows transport mode to be used by end hosts. Security gateways can use this mode only when they act as an end node for the communication, not when they act as an intermediary. This means, transport mode is generally used for securing the communications between two parties in an end-to-end manner. By only protecting the transport layer information, the IP headers are sent in the clear and the IP infrastructure can perform its normal routing and forwarding duties based on the information in the IP headers.

4.2.3.2 Tunnel Mode

When tunnel mode is used with an IPsec protocol, the entire original IP packet is covered by the IPsec protection. As the name suggests, the initial IP packet is encapsulated inside another IP packet that includes an IPsec header (such as ESP or AH header) (shown in Figure 4.4).

Looking at this configuration, a legitimate question would be: if the original IP header is hidden by encryption, how is the routing performed? This question can best be answered by considering the VPN scenario shown in Figure 4.2. The two hosts in each of the remote locations are not aware of the security gateways at the boundary of each domain. The packet from host A simply takes the normal route till SG1. As long as the security gateway SG1 intercepts the packet going from host A and encapsulates the packet inside another IP packet destined to SG2, there is no need for any of routers between SG1 and SG2 to know the contents of the original IP header. SG2 simply removes the outer header and IPsec header and decrypts the contents, which also include the inner IP header. Using the inner IP header, SG2 is now able to forward the packet towards host B. The routing is accomplished and the contents of the original IP packet are protected from outsiders while the packet is traveling from SG1 to SG2. The contents of the packets do not need to be encrypted while traveling inside the enterprise network.

| IP header (outer header) | IPsec header | IP header (inner header) | Transport header | Payload |

Figure 4.4 IPsec in Tunnel mode

Another use case for tunnel mode is when the user and the enterprise are concerned about revealing the IP address of their hosts. By having the IP address information, an attacker may not only be able to find out the user's location but also lay down a map of enterprise routing topology. By encapsulating the initial IP header inside an outer IP header which only bears the IP addresses for SG1 and SG2, no information about internal addressing topology is revealed to the outside world.

Security architecture specification mandates the use of tunnel mode when a security gateway is the end point for the IPsec tunnel (not the end point for the communications).

4.2.4 Security Associations and Policies

The term "security association" (SA) has been used loosely in many different contexts and has caused confusion at times. Up to this point, we have been trying to avoid the use of the term "security association" and used the term "trust relationship" instead. The reason for exercising such caution is that the IPsec SAs are very strictly defined. An IPsec SA includes a well-specified set of information and is looked up and referred to in a specific way. Furthermore, the term "IPsec SA establishment" has a well-specified meaning:

> When it is said that a security association is established between two nodes, it means that the two nodes have agreed to use IPsec as a means for securing their communications, agreed on what kind of protection to use, what protocol to use (AH and/or ESP), what cryptographic transform to use, have established keys to perform the transforms, and agreed on the life time of these keys.

As one can guess, the IPsec SA is a data structure that includes all the aforementioned information as well as some practical details, such as what IPsec mode to use, sequence numbers, and so on.

One important fact is that if the two ends decide to use both AH and ESP protocols or the scenario dictates security protections both in tunnel and transport mode, such as a case when two hosts decide to use IPsec encryption regardless of whether a VPN tunnel exists or not, separate SAs must be established for each protocol and each mode. Another fact about IPsec SAs is that they are unidirectional. In other words, upon establishing a secure connection, the two nodes establish one SA in each direction. The receiver side of the SA assigns a security parameter index (SPI), which is a 4-octet long bit-string, to each SA and sends it to the sender side. The SPI has only local significance to the destination, even though the sender receives the SPI from the receiver. The sender only passes the SPI along with the IPsec protected packet (see AH and ESP packet formats in Tables 4.1 and 4.2) to aid the receiver with the processing of the packet. We will discuss the use of SA and SPI in IPsec processing of traffic in a later section. For now, we will suffice by saying that each SA is uniquely identified using a triplet of SPI, destination IP address, and a protocol identifier (whether it is ESP or AH).

In a node that is using IPsec for secure communications with other entities, security policy is an important element that determines how the communications with those entities are handled. The security policy dictates what sort of security service is implemented for each packet as it is being received or transmitted. Three polices are applied to both inbound and outbound traffic: discard, apply IPsec (protect the packet), and bypass IPsec (send without IPsec protection).

Since IPsec SAs are directional and security can be applied in an asymmetric manner in the two directions, the security policies are defined differently for inbound and outbound traffic.

Before going through the discussion on IPsec processing, we need to describe the main databases used by IPsec.

4.2.5 IPsec Databases

Two main databases are defined by the security architecture specification [SECARC2401] provided by the IETF IPsec working group [IPSECWEB]:

- Security association database (SADB): IPsec SAs are maintained at this database. Each SA has an entry in this database. As mentioned earlier, the IP destination address of the outer IP header, the SPI, and the IPsec protocol identifier (ESP or AH) make up the triplet that is used to look up the SA within the SADB.
- Security policy database (SPD): When processing any inbound or outbound traffic, the receiving or sending entity needs to know which security policy applies to the traffic. SPD is where all the security policies, to be applied to data traffic, are stored as entries. The communicating entity consults the SPD to fetch the policy regarding the traffic. When the security policy is to apply IPsec, the specific SAs, that show the details of the security protection to be applied, must be located. Since the SAs are directional, the SPD needs to support separate entries for inbound and outbound traffic. Also, as mentioned earlier, a communication link may be protected by more than one SA, and this means, a security policy entry may return one or multiple SAs. For those reasons, SPD contains an ordered list of policies. The lookup in the SPD is performed by using selectors. The selector is formed on the basis of IP source address, IP destination address, name (for user or system), protocol (can be used when not encrypted), and transport ports.

Note that the SPD does not include specific information on how the security is implemented. That information is included in the SAs that are stored in SADB. When there are SAs in place, the SPD needs to return an SPI for SA to allow the processing entity to perform yet another lookup at the SADB as described in the following.

4.2.6 IPsec Processing

IPsec distinguishes between the ways in which inbound and outbound processing of packets are handled.

4.2.6.1 Outbound Processing

On processing packets that are going out (outbound), the transport layer at the sender sends the packet down to the IP layer, which in turn consults the SPD to determine what security policy applies to the packet. As mentioned earlier, in order to perform a lookup in the SPD, the processor needs to build a selector from the information included in the transport and IP layer headers. Based on the selector, the policy is retrieved from the SPD. The IP layer will drop the packet or add the IP header to the transport packet, if the policy dictates a drop or a bypass without security protection.

If on the other hand, the security policy that is fetched dictates an IPsec processing, a pointer (SPI) to the SA should be returned through the SPD lookup. If no SAs are returned, it may mean that no SAs are established and an IKE may have to be invoked to create the necessary associations. We will explain IKE process in detail later on. When the policy dictates IPsec processing, no packets can be transmitted until the SAs are established. After the SAs are established, it processes the packet by adding the appropriate AH and ESP headers, as was shown earlier.

4.2.6.2 Inbound Processing

On the receipt of an incoming IP packet (inbound traffic), the IP layer processing first needs to compare the provided security protection for the packet against those required by the security policies.

If the packet contains an IPsec header, first the IPsec processing is performed. The receiver extracts the SPI from the included IPsec header and simply builds the triplet of SPI, IP destination address, and protocol to identify the IPsec SA in the SADB. The processor then processes the payload according to the protocol (AH and/or ESP) that is specified. Once the processing is complete, the selector for lookup in SPD is built and the SPD is consulted to fetch the policy that should have been applied to the packet. The policy would also indicate whether the SA that was used was the correct one.

If on the other hand, the packet does not contain any IPsec headers, the fields for building the selector for SPD lookup has to be extracted from the packet IP and transport headers. Once the selector is built and the security policy is fetched, the receiver will know whether it needs to drop the packet (discard policy) or pass the packet to the transport layer (bypass policy). Note that if the policy was to apply IPsec, the packet is still dropped since no IPsec protection was added to the packet.

4.3 Internet Key Exchange for IPsec

During the discussion of key management mechanisms in Chapter 3, we briefly mentioned the IKE. IKE is a method that is based on the Diffie–Hellman key agreement concepts, even though it is the result of a great many improvements to the Diffie–Hellman agreement as a protocol. Like the Diffie–Hellman agreement, IKE is a peer-to-peer key agreement mechanism; no central servers are directly involved in the IKE conversation. However, compared to the Diffie–Hellman agreement, IKE has a less generic deployment base, since its main purpose is to facilitate the use of IPsec. With high likelihood, any IPsec implementation bundle includes IKE as the key exchange mechanism to establish SAs. Another difference between DH and IKE is that, as we will see later on, the purpose of the DH exchange within IKE is to establish shared secrets, that will be used to merely protect the IPsec SA negotiation, following the DH exchange. Those shared secrets are not used for encryption of the actual session data. The actual session data are protected by the IPsec SAs that are established during the IKE conversation.

4.3.1 IKE Specifications

IKE is one of those protocols that, despite their large deployment base, are very rarely treated in books and literature. This may be due to the complexity of its documentation;

just understanding the relation between IKE documentations will take a while, let alone understanding the details of each. We will explain that relation to the extent that helps the reader with referencing those documentations, instead of sending the reader's head into a spin.

The IETF specification is a mixed pot of multiple documents:

- Secure key exchange mechanism for Internet (SKEME) defines a method to establish an authenticated key exchange and together with Oakley (below) provide the theoretical background for IKE. SKEME was first published in the form of a research paper for an IEEE symposium [SKEME].
- Oakley is a newer mechanism than SKEME and provides an entire key management framework. The framework defines a family of the actual key exchange techniques and their modes. Oakley also specifies the cryptographic strength in various DH groups that can be negotiated between the IKE peers. Oakley is described in RFC 2412 [Oakley2412].
- Internet security association and key management protocol (ISAKMP), on the other hand, provides a generic framework with protocol details, packet formats, and payload types for different methods of authentication and key exchange and hence is the basis of IKE as a conversational protocol. ISAKMP also defines the two phases that have been later on known as IKE phases. During phase 1, the two ISAKMP peers authenticate each other and establish a secure channel that is called ISAKMP security association (ISAKMP SA). This SA is now widely known as IKE security association (IKE SA). ISAKMP phase 2 uses the secure channel in phase 1 to negotiate SAs for the upcoming IPsec services. We will provide much more details on ISAKMP later on.
- Domain of Interpretation (DOI) for ISAKMP is another specification that bundles security protocols and cryptographic transforms into groups that are used for a variety of situations. Protocols and transforms are given specific identifiers and group-belongings. This way, the protocols and transforms within a DOI can have a common understanding of the semantics of payload contents, attributes, and their identifiers. DOIs also set security policies and may define additional DOI-specific key exchanges and notification messages. DOI specification is provided in [DOI2407].
- And, there is the IKE itself that is specified in a separate RFC [IKE2409] that describes the IKE protocol using part of Oakley and part of SKEME in conjunction with ISAKMP payloads and procedures.
- Finally, a newly formed working group in the IETF [PKI4IPSECWEB] is working on defining the use of public key certificates for the IKE authentication. This specification has just started at the time of this writing and no standards are available yet. We will provide more details on this specification in Chapter 9, when we describe PKIs.

The relationship between Oakley and ISAKMP is rather straightforward: Oakley defines the cryptographic details such as key exchange modes, while ISAKMP defines the protocols details, such as the two phases and their signaling details.

The reader might feel better if she or he knows that the IETF implicitly admits the complex relationship between the documents to the point that the IPsec community has provided a document roadmap just for IPsec [DOC2411].

4.3.2 IKE Conversations

Even though complete understanding of IKE conversation requires understanding of ISAKMP, we stop mentally torturing of the readers interested only in IKE highlights and provide an overview of the IKE conversation. Those interested in the details can read the following sections on the ISAKMP and the IKE authentication process.

IKE is a peer-to-peer protocol, but it distinguishes between the party that starts the conversation (initiator) and the party that responds to the initiator (responder). It may or may not be intuitive to the reader that even though IKE serves the purpose of key management for IPsec (a layer-3 protocol), IKE conversations run above transport layer: IKE runs over UDP port 500.

IKE conversation has a two-phase nature, imitating the two phases of ISAKMP: in the first phase of IKE, the two parties establish a secure communications channel that is also called the IKE SA. During its second phase, IKE uses the IKE SA to securely negotiate the IPsec SAs, including the keys and transforms. It should be noted that unlike IPsec SAs, the IKE SA (the ISAKMP SA) is a bidirectional SA that is identified by the initiator cookie (CKY_I), followed by the responder's cookie (CKY_R). As mentioned earlier, IKE is a peer-to-peer protocol, which does not segregate between the initiator and the responder except the fact that the party that starts the conversation is the initiator and hence that party's cookie is the initiator cookie. For that reason, it is important that the cookies do not swap places during the signaling, so that each party can easily identify the IKE SA. In the following, we will provide a brief overview of the two IKE phases.

4.3.2.1 IKE Phase 1

The purpose for IKE phase 1 is establishment of an ISAKMP security association (IKE SA). As shown in Figure 4.5, the basic mode of IKE phase 1 (called main mode) involves a total of six messages between the two peers, requiring three exchanges (three round trips).

The first round involves exchange of initiator and responder cookies, CKY_I and CKY_R, respectively, and SA transform proposals, such as Diffie–Hellman groups and so on. Note that peer identifiers are not exchanged at this point to protect their privacy. In the picture, the cookies are called C_i and C_r for simplicity. The details of cookie generation are implementation-dependent, but the cookie must be unique to the party generating the cookie (possibly based on some local secrets). In other words, it must not be possible for anyone other than the issuing peer to generate the cookie.

The second round accomplishes the Diffie–Hellman exchange itself. However, the DH exchange within IKE uses the cookies exchanged previously (to prevent an attacker launching many DH negotiations with the responder) and nonces, N_i and N_r, to prevent replay attacks. X and Y in the figure are the DH half keys that were explained in Chapter 3. Once the two peers performed the DH exchange, they now share a secret that can be used to encrypt the following traffic.

During the third exchange of IKE phase 1, the two peers engage in a protected mutual authentication, including exchange of both their identities and their authentication credentials. This way, the identity of the peers are protected by the encryption provided for this exchange. Several authentication mechanisms are suggested for IKE as described later. For that reason, we keep the generic notation "Auth" within the figure. When the third exchange of IKE phase 1 is complete, the two peers share an IKE SA that can be used to protect the conversations during phase 2.

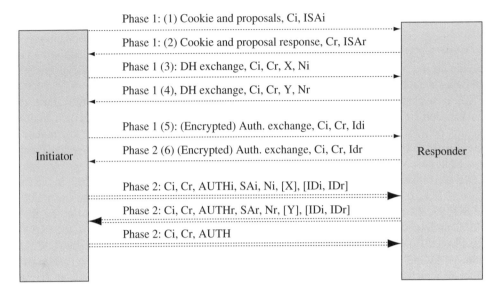

Figure 4.5 Basic IKE conversation consisting of two phases with total of nine messages. A main mode phase 1 and a quick mode phase 2 exchange are shown

4.3.2.2 IKE Phase 2

During phase 2 conversations, the two peers negotiate the details of IPsec SAs to be used to protect the data traffic. The basic mode of operation in this phase is called quick mode. (No, there is no slow mode!) Except the message headers (explained later on), all payload carrying the exchanged information are protected by the IKE SA negotiated during the phase 1. The IPsec SA details that are negotiated in this phase include keys, transforms (algorithms), and IPsec protocols (ESP and AH) for protection of the session. Note that IPsec SAs are unidirectional, and therefore a pair of IPsec SAs must be established for each IPsec transform during this phase.

Once the IPsec SAs negotiation is complete, the original IKE SA can terminate after the phase 2 exchanges, while the IPsec SAs remain in existence as long as their lifetime allows. However, the original IKE SA typically is kept alive for future refreshment of IPsec SAs.

It is implicitly assumed that the identities passed during phase 2 are the IP addresses of the peers. Phase 2 exchanges must always follow a phase 1 performed earlier and cannot be initiated prior to completion of phase 1. However, once the non-ISAKMP SA (e.g. IPsec SA) are established as a result of the first phase 2 exchange, these exchanges can be repeated at later times to refresh the keys for those SAs. This is done by producing new nonces (N_i and N_r) and running the exchange again. This is one of the major advantages of IKE over plain Diffie–Hellman exchange that does not provide key refreshes. Periodic re-keying is a desired feature for protection of long-lived sessions.

4.3.2.3 Round Trip Optimizations

As we can see, the number of round trips required by the two IKE phases is rather large, making the key exchange process very lengthy and CPU consuming. For those reasons, people have been looking for ways to reduce the delay associated with these round trips.

The IKE specification [IKE2409] provides a faster alternative for phase 1, called aggressive mode, which requires only three messages (1.5 round trips) as opposed to the three round trips required by the main mode. As shown in Figure 4.6, aggressive mode combines the cookie and proposal exchange with the Diffie–Hellman exchange. Also as one can see in the figure, the peer identities are carried in the clear, which means in the aggressive mode is less secure when it comes to protecting the privacy of the clients.

IKE specification provides two modes for phase 2 as well. Besides the quick mode, there is another mode called new group mode, which can be used to change the cryptographic group for future negotiations without establishing any new SAs on its own.

4.3.3 ISAKMP: The Backstage Protocol for IKE

To go further into the details of the IKE, one needs to understand the details of ISAKMP better. ISAKMP is described in a fairly thorough manner in RFC 2408 [ISAKMP2408]. First, we can see the resemblance between the ISAKMP and IKE modes. The main mode in IKE phase 1 is an instantiation of the ISAKMP identity protect exchange, which consists of a policy exchange, followed by a Diffie–Hellman exchange and an authentication exchange. The IKE aggressive mode is also an instantiation of the ISAKMP aggressive exchange.

Understanding details of IKE messaging and especially the newer profiling work that is being done to fine-tune the certificates for IKE (as explained in Chapter 9) requires better understanding of ISAKMP message and payload formats.

4.3.3.1 ISAKMP Message Format

Each ISAKMP message has a header with a fixed format, followed by a number of ISAKMP payloads (Figure 4.7).

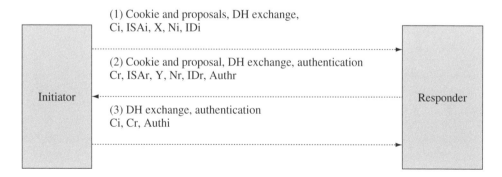

Figure 4.6 IKE phase 1 conversations according to aggressive mode

Figure 4.7 ISAKMP message format

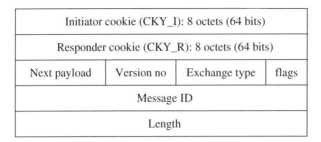

Figure 4.8 ISAKMP header format

The ISAKMP header, which is referred to as HDR in IKE documentation and in pictures in the rest of this chapter, is shown in Figure 4.8. The ISAKMP header is sent unencrypted even after ISAKMP SA (IKE SA) is established. As mentioned earlier, since ISAKMP SA is identified by the initiator and responder cookies, these cookies are included in the header for ISAKMP exchanges.

- The "next payload" field in the ISAKMP header indicates the type of the first payload included in the ISAKMP message. Table 4.3 shows some examples of the payloads that will be discussed in the following text. These payload types are also used for nesting of multiple payloads in a single ISAKMP message. The "next payload" field identifies the type of the following payload in the ISAKMP message and uses the types shown in Table 4.3.
- The field "exchange type" in the ISAKMP header indicates the type of exchange in progress. These exchange types translate into the IKE modes, such as main and aggressive modes (as explained earlier). Several ISAKMP exchanges types are defined such as, "base", "authentication only", "aggressive", and so on.

Each ISAKMP payload (all following the ISAKMP header in the message) begins with a generic payload header. The payload header has a simple format as shown in Figure 4.9.

Table 4.3 Examples of ISAKMP payload types

Next payload type	Value
None	0
Security association (SA)	1
Proposal (P)	2
Transform (T)	3
Key exchange (KE)	4
Identification (ID)	5
Certification (CERT)	6
Certificate request (CERTREQ)	7
Hash (HASH)	8
Signature (SIG)	9
Nonce (NONCE)	10

Figure 4.9 ISAKMP generic payload header

Now that we explained the building blocks of the ISAKMP messages, let us go through the semantics of a few important payload types that are important for understanding the details of IKE messaging and authentication.

- *SA payload*: This payload is used for negotiation of security attributes and to indicate the DOI that helps the two parties interpret the context of the other ISAKMP payloads. For instance, the DOI defines how the cryptographic algorithms and certificate authorities (CAs) should be named in a consistent manner or how a proposal for protection of traffic under a given situation is formatted.
- *Proposal payload*: This payload is used by each party to propose a security protocol for protection of the data. The protocols are identified by protocol identifiers and may include IPsec ESP, IPsec AH, and TLS depending on the type of protection needed for the data communications (authentication, encryption, or both). In order to facilitate the negotiation, each party may include multiple proposal payloads along with each exchange message. The proposal payload includes a proposal number that is used in a clever manner: when a party needs to use a bundle (multiple) of security protocols to protect the data, for instance both ESP and AH, then that party uses the same proposal number within the proposal payloads for each of those protocols. In other words, using the same proposal number in multiple payloads has a logical-AND meaning. The message to the other peer is "I need to use ESP AND AH to protect the data". On the other hand, using different proposal numbers for each proposal has a logical-OR function and means "I need to use either ESP or TLS". The proposal payload also includes a "number of transform" field and a "sending entity SPI", whose size and type depends on the protocol that is being offered.
- *Transform payload*: Each proposal payload is a proposal to use a specific security protocol such as IPsec ESP. As we know, each security protocol can use a number of transforms (security algorithms). The transform payload is used within the security association negotiation exchange to propose a transform such as DES or SHA1. The transform payload includes a "transform number" field. The responder chooses a transform within those proposed in conjunction with each proposal number to indicate its choice of security algorithm.
- *Key exchange (KE) payload*: As mentioned in Chapter 3, to avoid overuse of the limited bandwidth resources, IKE facilitates the negotiation of Diffie–Hellman (or Oakley) parameters between the two parties, without the need for transfer of the complete parameter data. IKE accomplishes this by using the ISAKMP KE payload defined in [ISAKMP2408]. This way only the group numbers, rather than the parameters that define the group, need to be sent. It should be noted that the actual group parameters are however defined by IKE specification [IKE2409] rather than by ISAKMP specification.
- *Identification payload*: This payload is used to provide the identity of the communicating peer. We will refer this payload again when describing the use of certificates for IKE in Chapter 9.

4.3.3.2 ISAKMP Payloads in IKE Conversations

As mentioned earlier, SA negotiation takes place in the first phase of IKE and is started with SA offers by the initiators. Among the payloads that are shown in Table 4.3, the security association, proposal, and transform payloads are very important, since these are the payloads that are used to build ISAKMP messages used for negotiation and establishment of IPsec SAs.

Each SA establishment message includes a single SA offer with one or more proposals, each of which may include one or more transforms. For instance, one peer may propose the use of AH and ESP as protocols to accomplish both authentication and encryption by using two proposal payloads. A user that needs to use MD5 for authentication with AH and 3DES for encryption with ESP and sends a transform payload indicating MD5 along with a proposal including AH as protocol and a transform payload indicating 3DES along with a proposal including ESP as protocol. All this together forms the IKE SA proposal that is sent by each party during the first exchange of IKE (shown as ISAi and ISAr in Figures 4.5 and 4.6). The offer is included in "transform" payload/s encapsulated in "proposal" payload encapsulated in SA payloads.

4.3.4 The Gory Details of IKE

The IKE specification, specifically [IKE2409] tends to be very brief on the lingual side. Following the details and mathematics of the key management process requires going through the ISAKMP and other specifications. Another important and at the same time difficult topic is the IKE authentication process. The specification provides several authentication alternatives, but it is short on the explanations or providing pros and cons. To make matters worse, it seems that there may be a few typing errors in the specifications and none of the material offered on the Internet seem to have discovered or admitted this. There are very few articles explaining such details, with [IKEBORELLA] being a great exception. Our main intention in this section is to make a humble attempt at shedding more light on the IKE key management and authentication methods. The mathematical descriptions are mostly borrowed from [IKE2409].

4.3.4.1 Derivation of ISAKMP Short-Term Keys

As mentioned earlier, the Diffie–Hellman exchange in IKE phase 1 creates a shared secret between the two peers. We explained the Diffie–Hellman process in Chapter 3, but to be consistent with the IKE terminology we replace some of the notations in Chapter 3 with those used in [IKE2409]:

- $g^{\wedge}xi$ and $g^{\wedge}xr$ are the public keys for the initiator and responder, sometimes referred to as DH half keys for initiator and responder. It should be noted that the "modulo p" operation is omitted from the notation, since it is understood that a DH (g, p) group indicates a modulo p operation.
- The shared secret that is the result of the Diffie–Hellman exchange is denoted as $g^{\wedge}xy$, which for simplicity we sometimes call DH key and denote DHK.

The DH shared secret (DHK) created during the second exchange of IKE phase 1 is not used directly as the key that protects the second-phase negotiation. Instead, IKE uses the

DHK in deriving a master key that is called SKEYID, which in turn is used to derive further keys for protection of second phase of IKE conversations:

- SKEYID_d is the keying material used to derive keys for non-ISAKMP SAs (such as IPsec SAs) from the SKEYID during main or aggressive mode as follows

$$SKEYID\_d = prf(SKEYID, (g^\wedge xy \mid CKY\_I \mid CKY\_R \mid 0))$$

where prf stands for pseudorandom function. Typically, a keyed hash function prf(key, information) is used.

- SKEYID_a is the keying material used by the ISAKMP to authenticate the messages during the IKE phase 2 conversation and is derived from SKEYID as follows:

$$SKEYID\_a = prf(SKEYID, SKEYID\_d \mid g^\wedge xy \mid CKY\_I \mid CKY\_R \mid 1)$$

- SKEYID_e is the keying material used by the ISAKMP to protect the confidentiality (encryption) of the messages during the IKE phase 2 conversations and is derived from SKEYID as follows

$$SKEYID\_e = prf(SKEYID, SKEYID\_a \mid g^\wedge xy \mid CKY\_I \mid CKY\_R \mid 2)$$

Values 0, 1, and 2 are used as single octets.

Again, we like to emphasize that SKEYID_a and SKEYID_e are used to create an ISAKMP SA for authentication and encryption of IKE phase 2 conversations, i.e. the conversations that carry the negotiation of IPsec SA and keys. These are different from keys used by IPsec protocols later on. The IPsec keys are derived from the SKEYID_d as follows:

$$protocol\_key = prf(SKEYID\_d, protocol \mid SPI \mid Ni \mid Nr)$$

where protocol and SPI are provided by the ISAKMP proposal payload described earlier.

At this point, an observant reader might ask: How the SKEYID, itself, is derived? We simply said that SKEYID is derived from DHK. What we did not mention is that derivation of SKEYID requires more than DHK and the reason (aside from all the mathematics) may be obvious:

> We did mention in Chapter 3 that Diffie–Hellman exchange must be accompanied with authentication, otherwise the two peers cannot know the identity of the other party and consequently do not know whom they are engaging in a key exchange with.

For the same reason, the mere establishment of a DHK is not enough for an ISAKMP SA that in itself will be the source of trust for all the IPsec SAs that are established later on, since any node could have spoofed the legitimate party's identification in the exchanges prior to DHK and thereby during SKEYID derivation. Hence, the calculation of SKEYID needs to draw from the authentication credentials that are used for the authentication exchange following the DHK. Such authentication credentials are typically long-term credentials that are either statically configured in the form of pre-shared secrets or public-private key pairs. On the other hand, the ISAKMP SA keys (SKEYID, SKEYID_a, SKEYID_e, and SKEYID_d) derived in the IKE process are called short-term keys, since they are derived only for the IKE conversations.

For all this added security, there is a catch: since IKE provides multiple authentication alternatives (as we will see later), derivation of SKEYID is different for each authentication method. We will show how SKEYID is derived when we describe each of the authentication methods in the following section.

4.3.4.2 IKE Authentication Alternatives

We mentioned earlier that IKE phase 1 conversation includes a mutual authentication process. This authentication is required to ensure each peer that the other party is legitimate before the two peers proceed with the generation of IKE and IPsec key materials. We also mentioned that to perform this authentication a set of long-term cryptographic secrets are required. These secrets are typically established out-of-band either as pre-shared keys or public key-based credentials. Depending on what types of secrets are used for authentication, a different IKE authentication mechanism is used. In the following section, we briefly explain each of the authentication alternatives.

Authentication with a Pre-shared Key

For simplicity, we call this key a password for the intuition that the term password provides. According to this method, the peer proves its identity based on knowledge of a shared key. For instance, by calculating an SKEYID based on the shared key and then using this SKEYID, the peer signs some information known by the other party. For pre-shared key authentica SKEYID is calculated as below:

$$SKEYID = prf(preshared\_key, (Ni \mid Nr))$$

The signature provided by initiator is called HASH_I

$$HASH\_I = prf(SKEYID, (g^\wedge xi \mid g^\wedge xr \mid CKY\_I \mid CKY\_R \mid SAi \mid IDii))$$

where IDii stands for the identity of ISAKMP initiator.

The signature provided by the responder is called HASH_R and calculated as follows

$$HASH\_R = prf(SKEYID, (g^\wedge xr \mid g^\wedge xi \mid CKY\_R \mid CKY\_I \mid SAi \mid IDir))$$

where IDir stands for the identity of ISAKMP responder.

As we saw, the initiator and the responder each calculate the SKEYID based on the shared secret and by sending their respective HASH value to the other party during the authentication exchange (messages 5 and 6 of main mode) they provide adequate proof for authentication. The SKEYID_e that is derived from SKEYID (see previous section) is used to protect this IKE exchange.

A note is now in order: an observant reader might have already noticed, both HASH_I and HASH_R use the SAi in their calculation. We believe that this was an unfortunate typing error in specification (HASH_R should be calculated using the SA of the responder, SAr), but IETF decided not to correct the error to preserve RFC-compliant implementations.

IKE authentication using pre-shared keys has two problems: scalability and lookup key issues. The scalability and administration problems stems from the fact that a node that engages in IKE exchanges with N parties requires $N-1$ passwords.

The second problem arises from the fact that the password is used to calculate the SKEYID and the SKEYID_e, used for encryption of authentication exchange of phase 1.

This means the responder needs to look up the initiator's password in its database to be able to calculate the SKEYID_e and decrypt the contents of the messages in the authentication exchange. Since the identity of the initiator (IDii) is within the encrypted part of the message, IDii cannot be used as a lookup key. This means using pre-shared keys and IKE main mode, the IP address must be used for password lookup. This defeats the purpose of hiding the identity in the main mode the first place. Using aggressive mode, that does not hide user identity, resolves this problem.

Using IP address as identity creates a problem for nodes with dynamically assigned IP addresses (DHCP) or roaming nodes using Mobile IP as we will see later on. We do not recommend using pre-shared secrets for IKE authentication.

Authentication with Digital Signatures

In this method, for the purpose of authentication during the IKE phase 1 (messages 5 and 6 of main mode), the contents of these messages are hashed with a keyed hash function and signed with the sender's private key. The hash functions HASH_I and HASH_R are calculated on the basis of the SKEYID in a manner similar to that explained for IKE with pre-shared keys. The other details are as follows:

The SKEYID is calculated using the DHK from previous IKE exchange:

$$SKEYID = prf((Ni \mid Nr), g^{xy})$$

We see that in this method, calculation of SKEYID does not involve any long-term secrets. However, each peer signs its own HASH with its own private key, where the private key acts as long-term credential in this case:

$$SIG\_I = hash(Priv\_I, HASH\_I)$$

$$SIG\_R = hash(Priv\_R, HASH\_R)$$

where the hash function itself can be negotiated in first IKE exchange. Each of the initiator and responder then attaches its signature (SIG_I and SIG_R, respectively) to its authentication message exchange (messages 5 and 6 in main mode). The main advantage of using signatures is that the peers cannot deny participation in an IKE exchange.

The issue is that the receiver must have knowledge of the public key of the sender to be able to verify the signature and thereby the authentication. Certificates can be used to attach a public key to an ID and verify the authenticity of the attachment. However, sending certificates is only optional and has its own set of problems with IKE as we will see in Chapter 9.

Authentication with Public Key Encryption

Two public key encryption methods are suggested for IKE authentication: a basic and a simpler revised method. For this authentication method, the SKEYID is calculated as follows

$$SKEYID = prf[hash(Ni \mid Nr), (CKY\_I \mid CKY\_R)]$$

The use of public key encryption for authentication is quite different from the other methods. In fact, all the three exchanges within the IKE phase 1 are different. The two parties no longer send their nonces in the clear. The initiator encrypts its own nonce (Ni) and ID (IDii) with the responder's public key, PubKr, and adds these encrypted fields into its DH exchange message (message 3) as shown in Figure 4.10. In the figure, this encryption is shown as <Ni>PubKr.

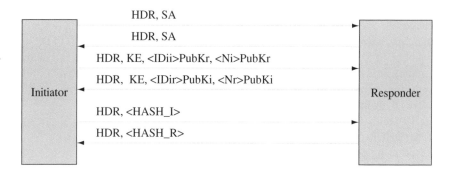

Figure 4.10 IKE phase 1 exchange when public key encryption is used for authentication

The responder decrypts the data with its own private key (PrvKr) and in turn adds an encrypted version of its own ID (IDir) and nonce (Nr) using the initiator's public key (PubKi) and adds the result to its DH exchange message (message 4).

The HASH_I and HASH_R are still calculated as before, however, here calculation of these hashes requires the knowledge of both identities and nonces from both sides. Encrypting these values with the recipient's public key ensures that only the recipient can decrypt the values with its private key and this implicitly ensures authentication, when each party receives the HASH value from the other side.

Again the issue is that each party must have access to the other party's public key and must be able to look for it without receiving that party's identity before. This in turn means that the initiator must be loaded with public key (or certificates) of the responders with whom it is planning to engage in IKE exchange and know which one of its public keys the responder has used to encrypt the message.

The main disadvantage is that this authentication method requires a total of four public key calculations, which are very intensive. Therefore, a revised simpler method was added to IKE to reduce the numbers of calculations to two. Here, only the two nonces are encrypted with public key encryption (as explained above), while the IDs are only encrypted using a key derived from the nonce and the symmetric encryption algorithm negotiated in the first message pair of phase 1. This way, only two public key encryptions are performed.

Another disadvantage of using encryption instead of digital signatures is that each side can potentially completely reconstruct its own side of exchange and the other party can completely deny that the conversation ever took place.

4.3.4.3 IKE Deployment Issues

IKE suffers from a variety of problems:

- IKE specification is very complex, consisting of an entangled web of IETF and IEEE specifications, combined in a complicated manner: ISAKMP is a generic negotiation protocol that was not specifically designed for IKE. The entire SKEME is not implemented in IKE and not all of cryptographic groups in Oakley are implemented in IKE. All this makes IKE a difficult specification for many implementers to understand, leading to potential interoperability problems between implementations. For those reasons, care was taken to specify the new version of (IKEv2) in one single document [IKEV2DR]. But that did not help the document from being a scary one. Once IKEv2 draft is standardized, it will obsolete [DOI2407], [ISAKMP2408], and [IKE2409].

- IKE exchange is lengthy and computation-intensive, especially since the expensive DH operation is completed before mutual authentication, significant computing power can be wasted if IKE is started with unauthorized users. This by itself is a potential security threat.
- Usage of IKE does not guarantee elimination of the problem of having to deal with pre-shared secrets. The fact that IKE is a peer-to-peer protocol, that inherently does not use the services of a centralized key database, does not make the process easier. Usage of certificates in IKE is neither straightforward nor standardized. Profiling PKI and transport of certificates by IKE messages are still being debated in a newly formed IETF group called PKI4IPsec [PKI4IPSECWEB]. IKEv2 is planning to provide support for EAP-based authentication methods to allow more choices for IKE authentication process [IKEV2DR].
- IKE cannot handle peer IP address changes due to mobility or multi-homing. IKE negotiations in both phases typically use the IP address as identifier but both IKE and IPsec SAs are based on static IP addresses. The problem is not being addressed for IKEv1 as a protocol. The Mobile IP community is looking at architectural ways to provide interoperability between Mobile IP and IPsec-based devices such as VPN gateways. On the other hand, a newly formed group called Mobike, short for "IKEv2 for mobility and multi-homing" is going straight for adding support for mobility and multi-homing for IKEv2 [MOBIKEWEB].

As we see, the IETF is working on multiple avenues to evolve IKE specifications, both in the form of augmenting IKEv1 with certificate support and in the form of developing new IKE versions. Unfortunately, IKEv1 and IKEv2 are not designed to be compatible, which means a transition will be abrupt and painful. This might make people less receptive toward IKEv2, unless Mobike provides a compelling solution for the hard-to-solve problem of providing IPsec for mobile users and multi-homed gateways.

4.4 Transport Layer Security

Implementing security mechanisms at network layer for end-to-end communications can run into practical problems, since the network layer needs to take care of the details of packet routing between the two communicating end parties. For IP networking, this means the routers and other network agents such as mobility agents need to access the information inside the IP header to perform routing functions. As we saw earlier, when IPsec is used to protect the packets at IP layer, the information inside higher layer headers (such as TCP headers) is hidden from outside world. Interim network management entities or middle boxes, such as network address translators (NAT) or quality of service (Qos) policing entities that need to look at transport layer ports to perform address translations or traffic shaping will not be able to do their jobs. Significant amount of design and standardization effort has been devoted to solving interoperability issues between IPsec and a variety of middle boxes and network management entities.

Another serious problem for network layer security protocols such as IPsec and IKE is that they typically establish trust relationships (SAs) using IP addresses as identifiers. When a node

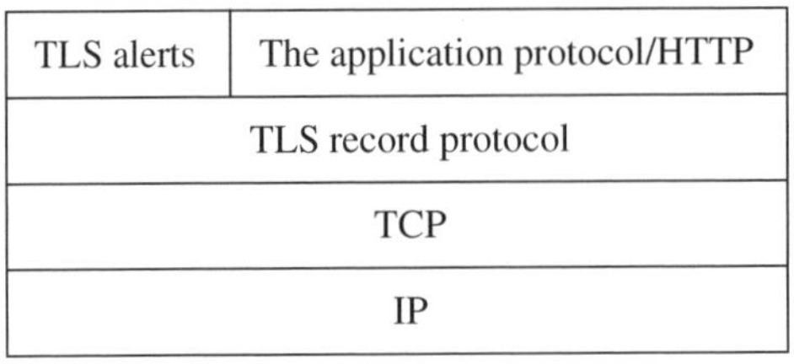

TLS alerts	The application protocol/HTTP
TLS record protocol	
TCP	
IP	

Figure 4.11 TLS Protocol stack

changes its IP address due to using Mobile IP or other methods, the original SAs are no longer valid.

Finally, IPsec and its key management procedure (IKE) were not designed to incorporate use of certificates as a means of dynamic key management procedures and hence IPsec implementers may suffer from key management scalability issues.

For the reasons mentioned above, implementing security at transport layer has been seen as an attractive solution. TLS is a protocol that has been standardized a few years ago [TLS2246]. Despite its rather recent standardization, TLS has its roots in secure socket layer (SSL) used for a long time for e-commerce.

SSL was designed by Netscape to support Internet shopping. Prior to SSL, the user needed to register with a website and then receive a password through the mail later on. Using SSL, the website could prove its legitimacy by sending out a certificate issued by a CA to the user. If the user decided that it could trust the certificate, she could type her credit card number on the web interface and complete her shopping.

Even though TLS gets its fundamentals from SSL, at this point SSL and TLS do not directly interoperate. As shown in Figure 4.11, TLS is designed to run on top of a reliable transport such as TCP and therefore is specially suited for application protocols that run over TCP. Examples are HTTP, TELNET, FTP, and SMTP (simple mail transfer protocol).

TLS is an asymmetric client–server protocol, which means in general the server authenticates itself to the client using a server certificate, while client is not always required to present a certificate. This was of course done to promote the spirit of impulse buying. Otherwise, by the time the user realized she had to register with a CA to have her own digital ID and certificate she may have either changed her mind or simply did not know how to proceed. Despite all our sarcasm, mutual authentication based on certificates for both client and server is supported in TLS, when so is required. This will eliminate the need for shared key configurations.

Besides providing security at transport layer and avoiding the problems associated with IP layer security, TLS has another attractive feature: TLS handles its own key exchange negotiations in protected manner and without requiring another key exchange protocol (as IPsec does with IKE).

TLS handles the trust relationship establishment and security provisioning based on its two layers:

- The TLS handshake protocol uses public key cryptography for authentication, cipher suite negotiation, and key exchange to establish a secure channel.
- TLS record protocol carries the actual session data in a secure fashion, using the symmetric keys created during the handshake stage. The data are encrypted as well integrity protected

(using hash-based digests) and compressed. The secure channel is based on use of symmetric keys as using public keys for session data would be too expensive.

Note that as we mentioned, TLS also provides compression. However, as explained later, this feature is not used in all security mechanisms that are derived from TLS, such as EAP-TLS.

Another note is that TLS handshake uses the record protocol for communications, which in turn uses the keys established during the handshake. To solve the chicken-and-egg problem, handshake data goes over an unencrypted channel.

4.4.1 TLS Handshake for Key Exchange

The handshake presented here includes some optional elements that are not always used, but the order in which the messages are exchanged is important. Also for final integrity checking, both client and server need to maintain a copy of all the messages that are exchanged during the handshake. The handshake is performed as shown in Figure 4.12 and described as follows:

- The client, who is typically the initiator of the session, sends a *client hello* message to the server. This message includes an ordered list (based on preference) of cipher suites the client can support. In TLS context, cipher suite is the certificate types and encryption and integrity checking methods the client can understand. The client also includes a nonce to provide liveliness and anti-replay protection.

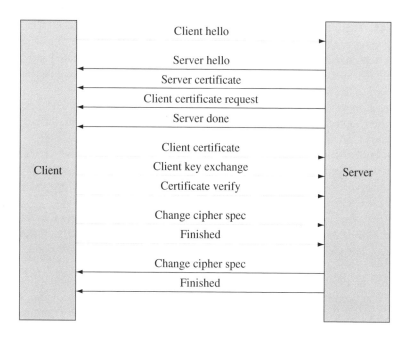

Figure 4.12 TLS message exchange for a full handshake

- After checking the cipher suites supported by the client, the server in turn sends a *server hello* message and possibly counters with its own cipher suites. The message also includes a nonce from the server and a session ID assigned by the server for the session to be established. By now the client and the server have agreed on the cipher suites to use. The session ID can be used by the client at any later time when resuming a TLS session with that server, if desired.
- To accommodate authentication to the client, the server sends its certificate in a *server certificate* message. Often, to save signaling overhead, this message is piggybacked with the server hello. The name and public key of the server are included in a certificate, which is signed by a CA. The CA signature proves that the public key and name actually belong to the server.
- The client uses the public keys included in the server certificate to verify the server certificate and authenticity. The client extracts server's public key for encryption of future messages to server as well.
- If the server requires client authentication, it asks for client certificate, using a client certificate request, otherwise the server sends a *server done*, which can also be piggybacked over the server hello and server certificate messages.
- The client then generates a random number called *pre-master secret* and encrypts the number with the public key of the server (obtained from the certificate) and sends it inside a *client key exchange* message to the server. This number is called pre-master key, since it is later used to create the master key between client and server and therefore is hidden from public. Only the server is able to decrypt the pre-master key using its own private key. Note that without client authentication, nothing prevents a rouge client to create the same pre-master secret.
- If required by the server, the client sends its own signing certificate using the *client certificate* message. To prove to the server that it is owner of the certificate, the client hashes all the messages exchanged up to this point (except the client hello) and signs the hash with its private key. The client sends the signature in *certificate verify* message. The client also sends a *change cipher spec* message to the server to indicate that the following messages will be protected using the cipher suites and keys that were just negotiated and established. The client also sends a *finished* message to confirm the use of the new master (calculated as described in the following).
- The server decrypts the pre-master secret with its own private key. Also if client authentication was required and certificate verify message was provided by the client, the server verifies the authentication provided by the client.
- Now that both client and server have the pre-master key, they both can create the master key by hashing the client nonce, the server nonce, and the pre-master secret. Now both the client and server share this master key. At this point, the handshake process is complete and keys are exchanged.
- The server sends a *change cipher spec* message to indicate to the client that the following messages will be protected using the cipher suites and keys that were just negotiated and established. The server also sends a *finished* message to the client at this point.

As we see, the handshake protocol not only provides a means for negotiation of security mechanisms and exchange of cryptographic parameters, but also provides for mutual authentication and establishment of a secure channel for the following communications. In a way, the handshake protocol within TLS provides for the record layer what IKE does for IPsec without requiring a separate protocol.

The master key generated during TLS handshake protocol is used for generating future symmetric keys to provide encryption and message authentication services for data traffic exchanged through the TLS record protocol.

4.4.2 TLS Record Protocol

The TLS record protocol provides confidentiality and message integrity using shared keys established by the handshake protocol. TLS defines record protocol to transfer application and TLS signaling information, such as alerts and so on.

The TLS record protocol is a layered protocol, which means that at each layer, messages may include fields for length, description, and content. The record protocol takes messages to be transmitted, fragments the data into manageable blocks, optionally compresses the data, applies a message authentication code (MAC), encrypts, and transmits the result. Received data at the receiver are decrypted, verified, decompressed, and reassembled, then delivered to higher-level clients. Figure 4.13 shows the steps in the record protocol.

4.4.2.1 TLS Alert Protocol

One of the content types supported by the TLS record layer is the alert type. Alert messages are used to send TLS-related alerts to peers. Like other messages, alert messages are encrypted and compressed. In the following, we describe two examples of the fatal messages, which would result in immediate termination of the connection.

- *The closure alert*: The client and the server notify each other that connection is ending by sending the closure alert message. Either party may initiate the exchange of closing messages.
- *The error alert*: Any time an error is detected during a TLS handshake, the detecting party sends a message to the other party. Anytime a party sends or receives a fatal alert message, it immediately closes the connection and deletes the related parameters such as session ID, keys, etc. Examples of error alerts are: certificate revoked, certificate expired, decryption failed, and so on.

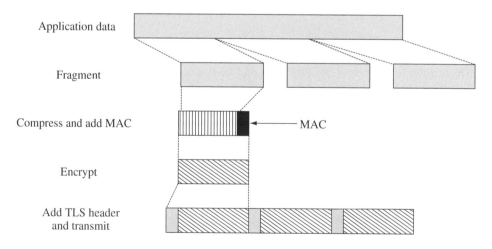

Figure 4.13 Processing within TLS record protocol

4.4.3 Issues with TLS

One of the issues with TLS handshake and record layers is that they must be carried over a reliable transport protocol such as TCP. Hence, TLS suffers from the overhead associated with TCP compared to UDP. First, a TCP connection is set up; second, a TLS connection is set up over the TCP connection, then the client and the server start communication using TLS over TCP.

Another issue is that there are three main security-related operations that introduce overhead to TLS: MAC computation, compression, and encryption.

Another issue is that TLS handshake does not guarantee that the client and server choose the strongest cipher suite available. The system is only as strong as the weakest key exchange and authentication algorithm supported, and only trustworthy cryptographic functions should be used. Implementers and users must be careful when deciding which certificates and CAs are acceptable; a dishonest CA can do tremendous damage.

4.4.4 Wireless Transport Layer Security

In case of mobile devices, an application layer protocol called wireless application protocol (WAP) was developed by an international organization called WAP forum. WAP defined a set of protocols in transport, security, transaction, session, and application layers for mobile services. A mobile user would be able to use the WAP services all around the world regardless of the local mobile network technology.

To protect WAP traffic, people suggested the use of TLS. However, when both bandwidth and processing power (battery consumption) are issues for the client, as it is the case for small wireless devices, TLS can have problems. The large number of exchanged messages and certificates along with the intensive public key operations can be too expensive. For those reasons, a trimmed down version of TLS called wireless transport layer security (WTLS) was developed for this purpose. The WTLS protocol was suggested as the security layer underneath WAP. In the same manner as TLS, WTLS is a shim security layer between the application layer and transport layer. A number of modifications and changes were made to the TLS to help WTLS adjust to the nature of wireless networks.

- The protocol should be able to cope with long round trip time. Also the protocol needed to be able to deal with low bandwidth associated with the wireless links at the time.
- The handset sets requirements for the algorithms because of the limited processing power and memory. Also restrictions on exporting and using cryptography were considered.

The goal of WTLS was to provide privacy, data integrity, and authentication for WAP applications that were being suggested at the time, such as online banking and e-commerce through the wireless phone. The WTLS did not get too much traction since the suggested cipher suites did not seem to cut the security muster [WAPWTLSM].

4.5 Further Resources

IP security document roadmap [SECARC2401] is a great source of information for understanding the details of how an architecture based on IPsec should be designed. This

RFC is being modified at the time of this writing [SECARCDR] and will be soon be replaced by another RFC.

For more information on IPsec, IKE standards, and progress on IKEv2, the reader is referred to the IETF IPsec working group official website [IPSECWEB].

Readers interested in open source implementations of IPsec may refer to the FreeS/WAN project web site [FREESWAN].

The IETF web page for TLS working group can be found at [TLSWEB], while information on open source implementations of SSL can be found at [OPENSSL].

4.6 References

1. [SECARC2401], Kent, S. and Atkinson, R., "Security Architecture for the Internet Protocol", IETF, RFC 2401, November 1998.
2. [AH2402], Kent, S. and Atkinson, R., "IP Authentication Header", RFC 2402, November 1998.
3. [ESP2406], Kent, S. and Atkinson, R., "IP Encapsulating Security, Payload (ESP)", IETF, RFC 2406, November 1998.
4. [Oakley2412], Orman, H., "The Oakley Key Determination Protocol", IETF, RFC 2412, November 1998.
5. [SKEME], Krawczyk, H., "SKEME: A Versatile Secure Key Exchange Mechanism for Internet", from IEEE Proceedings of the 1996 Symposium on Network and Distributed Systems Security.
6. [ISAKMP2408], Maughan, D. *et al.*, "Internet Security Association and Key Management Protocol (ISAKMP)", IETF, RFC 2408, November 1998.
7. [DOI2407], Piper, D., "The Internet IP Security Domain of Interpretation for ISAKMP", RFC 2407, November 1998.
8. [DOC2411], Thayer, R. *et al.*, "IP Security Document Roadmap", IETF, RFC 2411, November 1998.
9. [IKE2409], Harkins, D. and Carrel, D., "The Internet Key Exchange (IKE)", IETF, RFC 2409, November 1998.
10. [IKEBORELLA], Borella, M., "Methods and Protocols for Secure Key Negotiation Using IKE", *IEE Network Magazine*, July/August 2000.
11. [IKEV2DR], Kaufman, C., "Internet Key Exchange (IKEv2) Protocol", IETF work in progress, draft-ietf-ipsec-ikev2–17.txt, September 2004.
12. [IPSECWEB], IETF IPsec working group website, http://ietf.org/html.charters/ipsec-charter.html.
13. [SECARCDR], Kent, S. and Seo, K., "Security Architecture for the Internet Protocol", IETF work in progress, draft-ietf-ipsec-rfc2401bis-04.txt, October 2004.
14. [PKI4IPSECWEB], IETF website for "Profiling Use of PKI in IPsec", working group, http://ietf.org/html.charters/pki4ipse-charter.html.
15. [MOBIKEWEB], IETF website for "IKEv2 for Mobility and Multi-homing", working group, http://ietf.org/html.charters/mobike-charter.html.
16. [FREESWAN], Linux FreeS/WAN project website, http://www.freeswan.org/.
17. [TLS2246], Dierks, T. and Allen, C., "The TLS Protocol", RFC 2246, January 1999.
18. [TLSWEB], IETF website for Transport Layer Security working group, http://ietf.org/html.charters/tls-charter.html.
19. [WAPWTLSM], Saarinen, M.J., "Attacks Against the WAP WTLS Protocol", University of Jyvaskyla, September 1999.
20. [OPENSSL], Open SSL project web site, http://www.openssl.org/.

5

Introduction to Internet Mobility Protocols

5.1 Mobile IP

Mobile IP is widely proposed to be the protocol of choice to provide routing services to nodes that need to change their points of attachment to the network. The concept of Mobile IP is a lot like mail forwarding. While Joe Smith lives in Chicago and has a permanent address of "1000 Industrial Drive, Chicago", when he moves to live with his mother in Seattle at "10 Easy Street, Seattle" for two months, he simply notifies his local post office in Chicago about his temporary address in Seattle as a "care of" address. None of the people who need to send mail to Joe Smith during those two months need to know his temporary address. They simply send his mail to his permanent address. The local post office intercepts the mail and, based on the stored temporary address for Joe Smith, forwards his mail to his mother's house in Seattle. In the same manner, a host that is otherwise assigned an IP address by its local network may require further IP addresses and receive forwarding services while away from home:

- *Home address*: The home address (HoA) is a "permanent" IP address that is assigned to a mobile node (MN) by its home network, which is fixed and known to associated with the mobile node. The home address is attached to the home network, i.e. the network prefix of the HoA matches that of the home subnet. The mobile node should use this address and network prefix only when residing at its home subnet. If the mobile node moves to another subnet and still insists on using its HoA, the routers on the path of the mobile node's traffic would have to create special entries in their routing table for the mobile node. If the routers in the network had to do this for every mobile node that roams into their subnet, this would essentially make the routing in the Internet fall apart, since this not only makes the routing both non-scalable and slow (due to potentially large routing tables), but also makes the job

of routing protocols through which the routers update each other about the happenings of the network much more complicated.
- *Care of address*: Instead, when the mobile node moves into a new network subnet, the routers in the new subnet do not serve the MN through its HoA. Thus, the mobile node needs to acquire a new temporary IP address that is topologically correct within those networks' subnets. This temporary address is called the care of address (CoA).

Regardless of the mobility pattern of the mobile node and its current point of attachment to the network, its HoA can remain unchanged. Any correspondent node (CN), i.e. a node that is engaged in communications with the MN and needs to send packets to the MN, can be unaware of the MN's CoA and simply continues to send the packets to the MN's HoA. The standard IP routing mechanisms will deliver data destined to a mobile node's HoA to the mobile node's home network. However, just as in the mail-forwarding example, a local authority must exist in the home network to intercept the packets and forward them to the MN's current location, i.e. CoA. In Mobile IP architecture, this local authority is the home agent (HA). However, when the MN moves into a foreign network and acquires a CoA, the MN first needs to register its new CoA with the HA through a Mobile IP registration process. The HA then creates a temporary *mobility binding* between the mobile's HoA and the CoA. When the correspondent node sends a packet to the MN's HoA, the packet arrives at Mobile IP HA, which intercepts the packet and forwards it to the CoA based on the mobility binding for the MN.

An observant reader will think back to the mail-forwarding example: what happens on the way between the Chicago post office and the Seattle post office? The Chicago post office will put a stamp on Joe's mail with his new temporary CoA in Seattle as the new receiver address. If the post office wanted to be a bit neater, it could put each mail envelope inside a new envelope with the Chicago post office as sender address and the temporary CoA for Joe Smith in Seattle as the receiver address. This is closer to what Mobile IP does. IP routing achieves the effect of double enveloping by the process of **IP in IP encapsulation** or tunneling. The HA intercepts the packet destined for the MN's HoA and encapsulates the original IP packet, originally destined for the MN's HoA, inside a new IP packet destined for the MN's CoA as shown in Table 5.1. The original header is called the inner header, while the header added by the HA is called the outer header. The outer header includes the new source and destination address required for the routing of the traffic to the MN's current location.

Since the HA is the starting end of the tunnel, like every other tunnel, we expect that there is another entity at the end of the tunnel to perform the de-tunneling (de-capsulation), i.e. to remove the outer header and forward the inner packet in its original form. This entity is typically another Mobile IP agent, called the foreign agent (FA), since it is at a subnet that is foreign to the MN. We will talk about cases where the FA is not deployed shortly, but for now let us assume an FA exists. In such cases, the CoA belongs to the FA and hence the FA is the entity that forwards the original packet to the MN. Figure 5.1 describes the operations on the packet in the forward direction, i.e. from a correspondent node to a mobile node, as the packet moves through the mobility agents.

Table 5.1 Forwarding of packets destined for the MN's temporary address by the HA

HA IP Address	MN's Care of Address	Correspondent Node IP Address	MN's Home Address	Payload

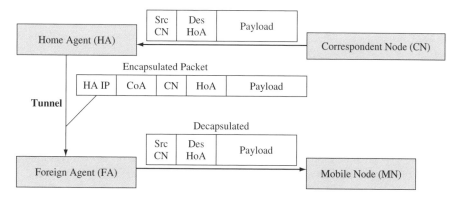

Figure 5.1 Mobile IPv4 processing in the forward direction

Considering how the packet was generated from the CN and arrived at the MN, we will see the value that Mobile IP agents provide: hiding the mobility of the mobile node from the correspondent node and the applications that it and the MN are running.

When Mobile IP was first being designed, security and firewall concerns were not as pronounced as today. Routing was the main driver for Mobile IP design. Hence, Mobile IP specifications simply stated that the MN would use its CoA when sending packets to the CN and would not involve any mobility agents in the reverse direction. This is shown in Figure 5.2. However, as we will see later, the presence of firewall and security mechanisms in the foreign network may hinder this process. For these reasons, Mobile IP reverse traffic routing is not performed as shown in Figure 5.2.

Now that we have completed the description of Mobile IP forwarding in both directions, we might ask why not simply forward the tunneled packets directly to the MN itself as opposed to an FA? The local post office in Seattle does not have to know that Joe has moved into the neighborhood, it simply sees his mother's address and takes Joe's mail to that house. The answer lies in the problems that IPv4 has with its limited address space. Giving each mobile node a new CoA during its moves seems to be a wasteful approach to using IP addresses. Thus, many IPv4 networks implementing Mobile IP protocol deploy specific mobility agents, called foreign agents (FA) to deal with visiting mobile nodes to allow the network to save on its pool of IPv4 addresses. Depending on

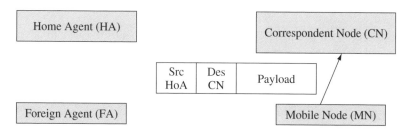

Figure 5.2 Mobile IPv4 messaging in the reverse direction

whether there is an FA at the foreign (visiting) network, there are two different types of CoA configurations:

- *Foreign agent care of address (CoA)*: When a foreign agent is present at the visited network, the FA can provide the MN with a care of address that is essentially the same as the IP address for one of the FA's own interfaces instead of giving the MN a "personal" address. In this way, the foreign agent care of address forces many mobile nodes to share the same care of address without placing unnecessary demands on the already limited IPv4 address space. As we will see later, the MN must register its new CoA with the HA, which means the MN must be made aware of this address when it moves in. The FA announces this care of address through a so-called Agent Advertisement message. In this mode, when the HA tunnels forward packets to the MN's CoA, the FA is the endpoint of the tunnel. Upon receiving tunneled data, the FA de-capsulates and delivers data to the mobile node using local-link routing mechanisms.
- *Co-located care of address (CcoA)*: When foreign agents are not present or cannot be deployed, the mobile node needs to acquire its own "personal" temporary CoA through other means, such as through DHCP. Alternatively, the mobile node may own a long-term address for its use only while visiting some foreign network. Since this CoA is associated with one MN's local interfaces, it is referred to as a co-located CoA (compare to the previous case, where the CoA is the address of one FA's interfaces). When using a co-located care of address, the mobile node serves as the endpoint of the tunnel and itself performs de-capsulation of the data tunneled to it. The advantage of using a CcoA is that it allows a mobile node to move to networks that have not yet deployed a foreign agent.

Regardless of what sort of CoA is used, it should be noted that the CoA is the endpoint for Mobile IP tunnel that started from the HA. In the collocated care of address case, the MN itself is the end of the tunnel. On the other hand, in the case of an FA-based CoA, the FA handles the de-tunneling and routing of the packet to the MN.

This and many other differences between IPv4 and IPv6 have led to separate design specifications for Mobile IPv4 and Mobile IPv6. In this book we will focus mostly on Mobile IPv4 specifications due to the far more wide-spread use of IPv4 and would refer the reader to the references at the end of chapter for more information on Mobile IPv6.

5.1.1 Mobile IP Functional Overview

The following steps provide a rough outline of the operation of Mobile IP protocol:

1. Foreign agents and home agents send Agent Advertisement messages to advertise their presence and ability to support various Mobile IP functionalities.
2. A mobile node, receiving these Agent Advertisements, determines whether it is on its home network or a foreign network or has moved from one foreign network to another. The mechanism for this detection is part of what is called agent discovery and is discussed in detail later on.
3. When the mobile node finds it is located on its home network, there is no need to use mobility services.

4. If the MN detects that it has moved to a foreign network, it needs to acquire a care of address in order to use Mobile IP services. The CoA may either be a foreign agent care of address, or a co-located care of address assigned by some external assignment mechanism such as DHCP.
5. By exchanging a registration request and a registration reply message, the mobile node registers its new care of address with its home agent.
6. Once the registration process is complete, data sent to the mobile node's home address are intercepted by its home agent, which tunnels the data to the mobile node's care of address. As mentioned earlier, these data are received at the tunnel endpoint (either at a foreign agent or at the mobile node itself), and finally delivered to the mobile node. In the reverse direction, the mobile node sends data directly to the correspondent node, i.e. data sent by the mobile node are delivered to their destination using standard IP routing mechanisms. Due to security concerns, this may, however, not work in many networks and reverse tunneling must be employed. We will discuss reverse tunneling later.
7. If an MN that was residing on a foreign network detects that it has returned to its home network after being registered elsewhere, the mobile node de-registers with its home agent, through the exchange of a registration request and registration reply messages.

For more information on reverse tunneling the reader is referred to the IETF standard specification [REV3024].

5.1.1.1 Mobile IP Registration

As mentioned earlier, Mobile IP registration is the procedure by which the mobile nodes inform their home agent of their current care of address and request forwarding services while visiting a foreign network. Mobile IP registration is also used when the mobile node needs to inform the home agent that the mobile node has returned home.

The concept of Mobile IP registration is very simple: there are only two control messages: registration request and registration reply, the main purpose of which is to allow the MN to communicate its latest IP address and possibly point of attachment to the HA. We will explain how these messages are used in a variety of circumstances shortly. But first we will show the exact format of these two messages and how they can be extended.

As can be seen, both of these messages can be carried over UDP and they both carry lifetime fields in order to limit the extent of the period of time in which the registration (the binding created at the HA) is valid (see Table 5.2 and Table 5.3). This is sometimes called a registration lifetime. Mobile IP allows certain extensions to be added to these messages and as we will see shortly, these extensions serve important purposes, such as providing security and interaction between the Mobile IP and AAA signaling.

Depending on the topology of the foreign network, FAs may or may not be deployed, which means the registration process may not involve an FA. For that reason the Mobile IP defines two different registration procedures, one via a foreign agent that relays the registration to the mobile node's home agent, and one directly to the mobile node's home agent. Both registration procedures involve the exchange of registration request and registration reply messages. Also, when the mobile node returns home, it sends another registration request to the HA to indicate that it no longer requires forwarding services from the HA. This process is called de-registration.

In the following we describe each of these procedures briefly. For each case, we show the exact content of the source and destination addresses in the IP header of the registration request. This

Table 5.2 Mobile IP registration request highlights, including some examples of registration extensions

Field name	Sub-field	Description
IP header		
UDP header		
Mobile IP registration fixed payload	Type	= 1 for registration request
	Flags	Showing simultaneous bindings, presence or absence of FA (use of collocated CoA), reverse tunneling, and so on
	Lifetime	Life-time in seconds before the registration expires
	Home address	Home address of the mobile if configured, if NAI is used, cleared to zero
	Home agent	IP address of home agent
	CoA for the mobile	CoA the mobile acquires either directly (CcoA) or from the foreign agent
	Identification	To match registration requests and response as well as to provide anti-replay protection for messaging to HA
Registration extensions		

Table 5.3 Mobile IP registration reply and some related extensions

Field name	Sub-field	Description
IP header		
UDP header		
Mobile IP registration fixed payload	Type	For registration reply = 3
	Code	The result of registration request, e.g. = 0 registration accepted, = 67 mobile node failed authentication
	Lifetime	If successful registration, seconds before the registration expires
	Home address	Home address of the mobile
	Home agent	IP address of home agent
	Identification	To match registration requests and response as well as to provide anti-replay protection for messaging to MN
Registration extensions		

will help the reader understand how the mechanics of Mobile IP routing works. In each case the MN uses the IP address of its own physical interface as the source address and the IP address of the entity, that it is sending its registration request to, as the destination address. The specifics that the HA needs to create the mobility binding are included in the registration request fields.

When registering a foreign agent-based CoA, the MN sends its registration request via a foreign agent, which means the MN sends the packet to FA first (see Figure 5.3). The registration procedure requires the following four steps:

1. The mobile node sends a registration request to the prospective foreign agent to begin the registration process. The MN uses its home address as the source address, since the CoA

belongs to the FA. The MN uses the IP address of the FA as the destination address to send the registration request through the FA.
2. The foreign agent processes the registration request and then relays it to the home agent.
3. The home agent sends a registration reply to the foreign agent to grant or deny the registration request.
4. The foreign agent processes the registration reply and then relays it to the mobile node to inform it of the disposition of its request.

When a mobile node is using a co-located care of address (obtained from a DHCP server), the mobile node registers directly with its home agent as described in the following steps (see Figure 5.4).

1. The mobile node sends a registration request to the home agent by using the IP address of the HA as the destination address in the IP header.
2. The home agent sends a registration reply to the mobile node, granting or denying the request.

If a mobile node has returned to its home network and is (de-)registering with its home agent (Figure 5.5), the mobile node MUST register directly with its home agent.

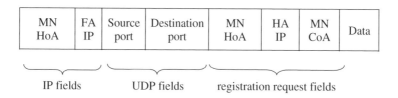

Figure 5.3 Mobile IPv4 registration packet, when an FA is in place

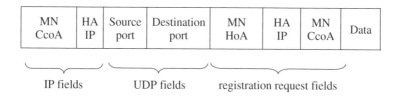

Figure 5.4 Mobile IPv4 registration packet, when the MN is using a CcoA

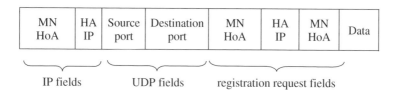

Figure 5.5 Mobile IPv4 de-registration

5.1.1.2 Mobile IP Reverse Tunneling

As mentioned earlier, Mobile IP allows the correspondent node (CN) to communicate its traffic to the mobile node without knowing the instantaneous whereabouts of the mobile node. When Mobile IP is implemented, the CN simply sends the packet towards the MN's HoA. When the MN is away from home, these packets are intercepted by the home agent, which in turn tunnels the packet towards the MN's CoA.

In the reverse direction, Mobile IP is not needed for routing purposes, since IP routing mechanisms do not use the source addresses in the IP packet. Theoretically, the mobile node could simply use its own home IP address as the IP source address in packets it transmits towards the CN. This means, in the reverse direction, no tunneling needs to be implemented and the triangular routing through the HA that exists in the forward direction can be avoided. This was how it was done in the original Mobile IP design.

With the increasing security threats to the enterprise networks, the security designers began to realize that conventional firewalls that filtered only inbound traffic would not protect the network from internal threats. Ingress filtering is now an important security component that prevents attacks from malicious nodes that physically reside inside the network boundaries. When ingress filtering is implemented on an edge router interface, the router will not forward IP packets received on that interface unless the packet source address matches the interface network prefix.

The implication of implementing ingress filtering is that, when the MN resides on the foreign subnet and transmits packets using its own HoA as a source address towards the CN, the packets will be dropped, since the HoA is not topologically correct within the foreign subnet. To avoid this problem, the MN must use its topologically correct CoA as the source address for any packets it is sending from its current location. However, the packets going to the CN must include the MN's HoA, in order to hide the routing and mobility complexity from the application running at the CN. In order to comply with these two conflicting requirements, the MN will tunnel those packets in the reverse direction, using its CoA as the source address in the outer packet to lead the packet through the ingress filtering, while using its HoA as the source address of the inner packet directed to the CN.

So far, we have not discussed where the outer packet is destined. To keep the forward and reverse tunnels symmetric, the MN may opt to send the packet to the HA, which decapsulates the outer packet and sends it to the CN. When a foreign agent is present, the MN may use the services of the FA for reverse tunneling. When a mobile node arrives at a foreign network, it listens for agent advertisements and selects a foreign agent that supports reverse tunnels. It requests this service when it registers through the selected foreign agent. At this time, and depending on how the mobile node wishes to deliver packets to the foreign agent, it also requests either the Direct or the Encapsulating Delivery Style.

In the Direct Delivery Style, the mobile node designates the foreign agent as its default router and proceeds to send packets directly to the foreign agent, that is, without encapsulation. The foreign agent intercepts them, and tunnels them to the home agent.

In the Encapsulating Delivery Style, the mobile node encapsulates all its outgoing packets to the foreign agent. The foreign agent decapsulates and re-tunnels them to the home agent, using the foreign agent's care of address as the entry-point of this new tunnel.

5.1.2 Mobile IP Messaging Security

Experience tells us it is better not to trust strangers completely, regardless of how nice they may seem. Mobile IP is a protocol that was originally designed to deal with mobile nodes connecting through networks other than their home network. However, trust was something that was not discussed when designing network routing protocols a few decades ago. However, with the increased need for mobility and connectivity through unknown networks, especially wireless ones, trust establishment has become a major component of network signaling protocol design. The signaling required for management protocols must be protected from active attackers. In the case of Mobile IP, a rogue mobile device can use its own or a fake IP address as CoA and create false registration requests on behalf of a legitimate Mobile IP client and cause the traffic for that client to be diverted to different location. Another example is that a rogue mobile device can intercept a legitimate registration request from another mobile device and replace the CoA in that packet with its own IP address. This would cause future traffic to that MN to divert to the rogue MN. A third example is a rogue user that simply creates a large number of registration requests towards a foreign network, causing a denial of service attack.

For those reasons and many others, integrity protection for Mobile IP registration requests is a required security measure:

- The MN must authenticate its registration requests to the HA and the HA must authenticate its registration reply to the MN.
- The foreign network and foreign agents must be protected from false registration requests or replays of previously emitted registration requests.
- The foreign agent and the home agent should be able to trust each other and authenticate control messages to each other, when needed.

The basic Mobile IP standards [MIP3344] do recognize the need for message integrity protection for registration requests and specify methods by which the MN and HA can authenticate registration requests and replies. The MN performs this authentication by calculating a signature called an authenticator and including the authenticators within specific authentication extensions to the registration request. RFC 3344, however, only mandates authentication of registration requests to the HA, while leaving an authentication to the FA optional. However, the HA must also use an authentication extension to authenticate its reply to the MN.

The authenticator value is calculated using a previously agreed upon hash algorithm over pre-specified message contents as follows:

$$\text{authenticator} = \text{HMAC\_MD5}(X\_Y\_key, UDP\_payload, prior\_extension, type, length, SPI)$$

For instance, the mobile may add a mobile-home authentication extension to authenticate its message to the HA. In the same way the mobile may optionally add a mobile-foreign authentication extension to authenticate its registration request to the FA. Note that the UDP payload above includes the registration request or reply message (please see registration signaling format shown earlier in Figures 5.3 and 5.4).

In order to make the processing of authenticator values at the sender and the receiver side more straightforward, the exact format of the authentication extension carrying these authenticators is defined. The mobile-home authentication extension is defined in RFC 3344 for authentication of the MN and its messages to HA, while mobile-foreign authentication extension is defined for authentication of the MN and its messages to the FA. For authentication of the FA and its messages to HA, a foreign-home authentication extension is used. Note that the extension can be used for authentication in the reverse direction as well, e.g. the mobile-home authentication extension can be used for HA to MN authentication as well, although the exact value of some fields (such as SPI) will be different, as explained later.

The exact format of these extensions is shown in Table 5.4. However, since the format for all these extensions is very similar, we use the notation of X-Y authentication extension. As we can see, the extension not only includes the calculated authenticator values, but also includes information that helps the receiver to verify the authenticator value, such as the length of the authenticator, and the SPI (security parameter index).

The SPI helps the receiver locate the security context, i.e. the context that is required to verify the authenticator value and thereby authenticate the sent message or the sender. The context can include algorithms and the shared key between the two entities (X and Y), an agreed hash algorithm, and so on. RFC 3344 does not provide further guidance on this SPI and the security context it represents. We will discuss these SPIs in more detail in Chapter 8.

As mentioned, the base Mobile IP specifications (RFC 2002, 3344) only mandate the use of mobile-home authentication extension, while they consider the other authentications to be only optional. However, an efficient network security design should require all the authentications.

Another security problem is that, although the registration request includes an identification field that provides anti-replay protection for the HA, the authentication extensions as designed do not provide any protection for the foreign network against replay attacks. Random one-time values, such as nonces or challenge/response mechanisms with non-repeatable challenges, should be included in the messaging to ensure the messages are not replayed at a later time.

IETF RFC 3012 [MIPCHAL3102] and its draft revision [3102bis] specified a method to provide protection for the foreign network against replay attacks: the FA issues a temporary challenge towards the mobile node and expects the mobile to append that challenge to its registration

Table 5.4 Authentication extensions for Mobile IPv4

X–Y Authentication extension	
Field name	Description
Type	32 for mobile–home, 33 for mobile–foreign, 34 for foreign–home
Length	4 bytes (for SPI) plus the number of bytes in the authenticator
SPI	Security parameter index (4 bytes), which is used by the receiver to select the algorithm, mode and secret that was used for computing the authenticator. For instance: When MN is sending a message to HA, this field includes SPI allocated by the HA to locate MN-HA MSA When HA is sending a message to MN, this field includes SPI allocated by the MN to allocate HA-MN_MSA
Authenticator	Calculated on (UDP payload, all extensions prior to the current Auth. Extension, type, length, and SPI)

Table 5.5 Mobile IP Agent Advertisement Challenge extension

Field name	Description
Type	24
Length	Length of challenge
Challenge	The challenge value issued by the FA

Table 5.6 Mobile–foreign challenge extension

Field name	Description
Type	132
Length	Length of challenge
Challenge	The challenge from latest registration reply or from Mobile IP agent advertisement challenge extension of the latest agent advertisement by the FA

request. Since the challenge has a finite lifetime and can only be used once, the FA can be sure that the registration request is not replayed by illegitimate nodes later on. The details are as follows.

According to RFC 3012, the FA includes the challenge within its periodic or solicited Mobile IP agent advertisement messages. The challenge is added inside a "Mobile IP Agent Advertisement Challenge Extension" as shown in Table 5.5.

When the MN is ready to send a registration request towards the HA through the FA, the MN includes the latest challenge value inside a so-called mobile-foreign challenge extension and adds this extension to its registration request message. The format for the mobile-foreign challenge extension is shown in Table 5.6.

5.1.2.1 Caveat: Key Establishment

Almost all the security protection mechanisms mentioned above require the establishment of trust relationships and secret keys between the mobile node and mobility agents. We devote an entire chapter (Chapter 8) to describing the use of AAA infrastructure for this trust establishment process for Mobile IP and hence will not go into the details here.

5.2 Shortcomings of Mobile IP Base Specification

Mobile IP specification has developed over many years of evolution. The IETF RFCs have been revised several times to cover issues that have been pointed out for Mobile IP operation. Furthermore, the specification for IPv4 is different from the specification for IPv6. We do not cover Mobile IPv6 specification in this text and refer the reader to the end of this chapter for more references on Mobile IPv6. Also, as mentioned above, the security procedures for Mobile IP may require interaction with AAA servers. The support for this interaction is being standardized outside the scope of Mobile IP base specification as will be described in Chapter 8.

Issues related to full mobility support are many and complicated and cannot be covered in a small text such as this. To name a few things that we do not cover in this chapter, we would mention interaction of mobile and VPN gateways, traversal of Mobile IP traffic through network address translation boxes and network mobility and refer the user to the end of the chapter and the related IETF work in the area.

In the remainder of this section we will discuss many Mobile IP issues such as address, HA and key bootstrapping. We also briefly discuss the performance issues for Mobile IP, when handover latency can be an issue and cover the fast handover methods that have been suggested to alleviate these problems. In the rest of this chapter we describe some protocols that are complementary to Mobile IP in providing seamless mobility experience for moving users.

5.2.1 Mobile IP Bootstrapping Issues

The original design of Mobile IP assumed that, each mobile node would be assigned a fixed home address (HoA) tied to a pre-specified home agent. Since then experience has shown that these assumptions place several restrictions on the deployment of Mobile IP and for a variety of reasons that will be explained shortly, it may be more practical to perform the HoA and HA configuration dynamically.

According to the ongoing standardization work in IETF [MIPBOOT], *bootstrapping* is defined as the process whereby the mobile node obtains enough information so that it can successfully register with a Mobile IP home agent. This information generally includes the mobile node's home IP address, the home agent address and the security credentials required to protect Mobile IP registration signaling. Depending on the amount of information the mobile node has available, the bootstrapping process may include any or all of the following:

- dynamic home agent assignment
- dynamic home address assignment
- dynamic key establishment.

As mentioned above, the deployment scenario dictates what information is needed and hence influences the bootstrapping details. A variety of bootstrapping solutions are being proposed in IETF or used in the industry. Furthermore, the IETF Mobile IPv4 and Mobile IPv6 communities seem to be going their separate ways at the moment. However, one common theme is that bootstrapping is done in conjunction with Mobile IP registration process, even though it may add additional round trips between the mobile node and the network or between Mobile IP agents and other entities within the network (such as AAA servers).

In the following we provide a brief overview of the dynamic home agent and the home address assignment for Mobile IPv4, but leave the dynamic key establishment to Chapter 8.

Before going into details, we would mention that as with any added flexibility, the addition of mechanisms such as dynamic assignment of home agent and home address will bring additional complexity to the deployment. For example, not only must the mobile node be made aware of its new home agent and home address, but it also must be authorized to use that IP address. Furthermore, the new addresses must be entered in the proper authorization tables and security policy databases, wherever the network traffic is being secured or firewalled.

5.2.1.1 Dynamic Home Address Assignment

Instead of configuring both the mobile node and its home agent with a fixed home address, the mobile node's home address can be obtained dynamically from the DHCP servers to permit a more flexible use of the allowed pool of addresses. Dynamic assignment of the home address also opens the possibility of involving the AAA server in the process of address assignment and thereby directly authorizing the mobile node to use that address. Without going into the details of dynamic address assignment, we want to point out that when the MN does not have any IP address at the time of booting, it needs to have some sort of identifier, especially when it needs to authenticate itself to the AAA server first. The industry is moving towards an identifier called the network access identifier (NAI) in such cases. When presenting the NAI as an identifier, the mobile node needs to use a so-called NAI extension along with its Mobile IP signaling. We will not go into the use of NAI for MNs and other entities within Mobile IP here and refer the reader to [MIP3344] and [MIPNAI3846].

5.2.1.2 Dynamic Home Agent Assignment

As mentioned earlier, the original Mobile IP design assumed that each mobile node would be initially assigned a home agent and would be configured with the IP address of that home agent. However, a variety of deployment conditions make a static assignment of home agent less desirable:

- In larger networks, the pre-configuration of mobile nodes and home agents is not scalable and hence makes the administrative cost increase. Dynamic assignment of HAs during the registration process eliminates the need for such configurations.
- When multiple home agents are available, it may be desirable to perform load balancing between these HAs, so that arriving MNs are assigned to HAs with a smaller load.
- In geographically large networks, when the MN is roaming far from its home network (for instance, in a different country), having the traffic for the MN go through a HA that is located in a home network far from the current location of the MN creates necessary large latency and network traffic. To eliminate the large latency due to the traffic going through a distant HA, it is desirable to assign an HA locally in the network that the MN is currently visiting. This is called the local home agent assignment.

Mobile IPv4 group [MIP4WWW] is standardizing a procedure for dynamic assignment of Mobile IPv4 home agents [MIP4DYNHA]. This procedure is not yet finalized and may or may not become the de facto standard due to the intellectual property rights associated with it. However, it provides a good demonstration of how a home agent can dynamically be assigned to a mobile node as part of the registration process. It should be noted that the specification provides a process for assigning an HA to the MN, but does not describe how this HA is selected. The HA selection process can be done either by a designated HA or a AAA server or another designated mobility management entity.

The process for dynamic home agent assignment according to [MIPDYNHA] is as follows: the MN sends a registration request to a mobility agent and indicates that it needs a HA assignment. The network rejects the original registration, but in the process assigns an HA to the MN and sends the address for the HA to the MN inside the registration reply. The MN attempts the registration request again, but this time directed towards the assigned HA. The details for the process can, however, be tricky:

- Where does the mobile node send its registration request to?
- How does the mobile node build the registration request without having the HA and how does the MN indicate that it does not have an HA and needs one?
- How is the HA information conveyed to the MN?

When the MN sends the registration request for the first time, it does not know the address of its HA, so it needs to send the registration either to an FA (if deployed) or to a symbolic HA. Now a good question is what is the symbolic HA and how does the MN find out its address? We use the term "symbolic HA", since this is simply an HA, which receives the MN's original registration request and assists the MN with HA assignment, but does not act as a Mobile IP HA for traffic forwarding. Since the MN sends its first registration request towards this HA hoping that this HA will agree to act as a Mobile IP HA for the MN, this HA is referred to as the "requested HA". When an FA is deployed, the MN simply needs to send the registration request to the FA without needing to know any HA. It is then the job of the FA to find out which HA to send this registration request to. The FA may be configured with the address of some designated HA or may have to contact the AAA server to get this sort of information. However, when the MN is not registering through the FA, it needs to find the address of the "requested HA" on its own. Different methods have been suggested for this, such as extensions within DHCP signaling, requests made by the MN to DNS servers, or even the static configuration of the MN with the address of some designated HAs. In either case, the MN needs to use the IP address of the requested HA as the destination IP address for the registration request.

As described in Section 5.1.1.1, when the MN is sending a Mobile IP registration request, it needs to includes both the MN's home address (HoA) and its home agent address (the HA field). The dynamic HA assignment process in [MIPDYNHA] mandates that the MN uses its NAI and requests a home address as well. This makes sense, since if the MN does not have an HA and needs to be assigned an HA, the home address of the MN should also be assigned by that HA as well.

To indicate that the MN is requesting the dynamic assignment of an HA, the MN includes an ALL_ZERO_ONE_ADDR (either 255.255.255.255 or 0.0.0.0 depending on the location of the HA) inside the HA field of the registration request. Note that the MN does not include the address of the requested HA in the HA field, even when the MN knows this address. This is to ensure that the network is made aware that it needs to perform an HA assignment, even though in some cases the requested HA may actually agree to act as the HA for the MN and sends a registration reply indicating the success of the registration process. Instead the MN includes the address of the requested HA inside a "requested HA extension" (introduced by [MIPDYNHA]). The important fields within the first registration request sent to an FA including the request for HA assignment are shown in Figure 5.6.

Source IP = MN HoA	Destination IP = FA	MN HoA	HA address = ALL-ZERO-ONE-ADDR	CoA = FA CoA	Requested HA extension

Figure 5.6 First Registration Request during the dynamic HA assignment

| Source IP = requested HA | Destination IP = source IP of RRQ | MN HoA | HA address = requested HA IP | CoA = FA CoA | Redirected HA extension (assigned HA IP) |

Figure 5.7 First registration reply indicating a failure and the address of the redirected HA

If the MN is using a co-located care of address, then the source and destination IP addresses for the registration request would be the MN CcoA and the IP address of the requested HA, respectively.

If the requested HA agrees to become the HA for the MN, it sends a registration reply indicating successful registration and includes its IP address inside the HA field of the registration reply. In the majority of cases, the requested HA will not agree to become the HA for the MN, but the process in [MIPDYNHA] is based on the idea that the requested HA knows which HA the MN needs to be redirected to (regardless of which entity performs this selection). IETF has decided to call this HA the re-directed HA. As mentioned earlier, upon receiving the registration request, the requested HA rejects the request by creating a registration reply that includes an error code "REDIRECT-HA-REQ" and includes the address of the assigned HA inside the "redirected HA extension". An example of this message for scenarios where the FA is present is shown in Figure 5.7. The FA forwards the message to the MN, which then creates a second registration request, this time towards the redirected HA.

5.2.1.3 Dynamic Key Establishment

As mentioned, the bootstrapping solution depends on the deployment scenario. For instance, when the mobile node is roaming outside its home domain, it needs to authenticate itself to a service provider network prior to receiving Mobile IP services. At any rate, it is safe to assume that the mobile node shares a trust relationship (shared secrets or certificates) with the AAA server serving the domain to which Mobile IP home agent belongs. This relationship, along with the relationship between the visited domain and the mobile node's home domain, can be exploited to provide the security framework that is required for Mobile IP signaling. We will describe this process in a later chapter.

5.2.2 Mobile IP Handovers and Their Shortcomings

So far we have talked mostly about the use of Mobile IP for routing traffic for roaming mobile nodes. Sometimes roaming scenarios are when a mobile node boots up in a foreign network or initiates a session within the foreign network. By going through Mobile IP registration process during the boot-up or prior to the start of the session, the mobile node ensures proper routing procedures are set up for its traffic. However, true mobility support requires support for the handover of active sessions as well. In other words, a mobile node, involved in a communication session with a correspondent node, should be able to move from one network to another and still maintain that session. Although not emphasized so far in this section, it may be obvious to the reader that Mobile IP is also designed to support layer-3 handovers as well: a mobile node that moves to a new network acquires a new CoA and needs to register its new CoA with the HA. If the mobile node was already at another foreign network and had another CoA there, this registration means that the old CoA is no longer valid. Since the mapping between the home address and a CoA at home agent is called a binding, sending a registration for a new CoA is referred to as a binding update.

Despite its simple concept, Mobile IP as a signaling protocol for handover support has been criticized for its poor performance. The signaling has to go all the way to the mobile node's home agent that is typically in the mobile node's home network, even though the mobile may be moving between two foreign networks that are relatively close. Another reason for this poor performance is that completion of a Mobile IP handover relies on a number of other things that need to happen first. In the following subsections we describe each of these factors. But before we go on, we need to describe a few key terms related to mobility architectures:

- *Access point*: Many wireless access operators deploy so-called wireless access points (AP) as the front end or the point of presence (POP) for their networks. From a logical standpoint the AP is a layer-2 device that provides layer-2 functionalities such as media access control, layer-2 resource management, layer-2 framing, layer-2 authentication to the users and at same time deals with layer-3 functions. In practice, the AP can vary from cheap and dumb devices (what the industry has started calling light-weight access points) to rather complicated devices capable of performing many more advanced functions, such as sophisticated interactions with AAA servers or providing the system administrator with many added tools such as simple network management protocol (SNMP) facilities.
- *Access router*: Due to the scarcity of IPv4 addresses, Mobile IPv4 deployed the specialized FAs to offer a way of providing IP connectivity to mobile nodes and saving IP addresses at the same time. IPv6 does not face such issues and thereby FAs did not have to be introduced in to Mobile IPv6 procedures. Simply a router at the edge of the access network would be able to provide routing functionality for mobile nodes joining the network. The IETF community needed a name for this edge router that provided a set of functions within access networks. Hence the term access router (AR) was introduced.
- *Mobile-controlled handovers* are types of handovers where the mobile makes the ultimate decision on whether to go ahead with the handover or not. Note that such handovers could be network-assisted, meaning that the mobile decides to hand over and performs the handover possibly based on the information it had received from the network elements. This handover model is the preferred one when each of the neighboring networks is owned by a different operator or deploys a different technology.
- *Network-controlled handovers* are types of handovers where the network makes the decision on handover and instructs the mobile node to perform the handover. From the load balancing and network management point of view, this is the preferred model, especially if the handover is within a homogeneous network owned by the same operator.

Now that we have described the fundamental terminology for mobility, we can run through the factors that impact on Mobile IP handover performance.

5.2.2.1 Layer-2 Triggers and Fast Handovers

Most mobile nodes use wireless connections to ensure easy mobility. Wireless links typically use specific link layer (layer 2) technologies and access points, APs. This means handovers almost always involve a layer-2 handover first. In other words, each handover is the result of the mobile node's change of point of attachment from one AP to another. Depending on whether a change of APs leads to a change of subnets, the layer-2 handover may also lead to a layer-3 handover that requires Mobile IP signaling. However, since within each node the information flows up the stack, Mobile IP processing software will only find out about the need for handover, if it receives an indication that a layer-2 handover is performed or is about to be performed. Hence the performance of Mobile IP handovers (latency required for completion of handover) depends on the timing of the information received about layer-2 handovers. This information is usually referred to as layer-2 trigger, since it is usually information about some sort of event that has happened at layer 2, such as the disconnection of the mobile node from its old AP, or the reception of an acceptable signal quality from a new AP by the MN.

If layer-2 triggers are presented to the network layer entities in a timely manner, these entities may be able to start Mobile IP processing early on without having to wait for the layer-2 handovers to be completed. For instance, if the mobile node could start acquiring a new CoA for its new point of attachment as soon as it has received an indication that the link with its old AP is going down, the mobile node could do some of its layer-2 and layer-3 handover processing in parallel and hence save some time. These ideas provided the motivation behind the handover optimization processes that were called low-latency handovers (for Mobile IPv4 [LOWMIP4DR]) and fast Mobile IP (for Mobile IPv6 [FMIP6DR]).

5.2.2.2 Candidate Router Discovery Issues

Mobile IP registration signaling provides a mechanism for the mobile node to inform its mobility agents about its new subnet and IP address, so that the traffic for the MN can be properly routed. However, Mobile IP does not provide any guidance on how the MN determines which new subnet it needs to go to.

With the emergence of many contender access technologies, the MN may be presented with a situation where the MN is in a coverage area of several access network operators (multi-domain scenario) or multiple points of attachment from the same operator (single domain scenario). The MN must choose from a pool of networks, each represented by an edge device such as an access router. The problem of making a choice between all those so-called candidate access routers (CAR) to select a target AR (TAR) is referred to as access router discovery and selection and may be affected by the services and pricing that each of these networks offer or by the loads that each of these networks experience. When the MN makes a decision on the TAR and then realizes that the TAR is different from its current AR, if the MN supports Mobile IP, the MN initiates Mobile IP signaling to attach to the TAR.

Another related problem is that for economic and performance reasons, many operators, especially wireless ones, deploy a number of low-cost APs connected to much fewer layer-3 access routers. In such cases, knowing the address of an AP does not reveal the address of AR automatically. This has an effect on wireless network handovers since as the user moves across the APs, it needs to find out whether it has also moved between access routers. For this the MN must perform a so-called reverse address translation, meaning that a MN from the layer-2 address of the AP needs to find out the IP address of the AR, to which the AP is connected. If the MN finds out that it has also changed subnets, it again needs to perform a Mobile IP handover. The delay involved in address translation and access router discovery affects the overall delay involved in a layer-3 handover.

IETF Seamless Mobility group [SEAMOBY] has designed a protocol called Candidate access router discovery (CARD) protocol that attempts to solve both of the reverse address translation and router discovery problems for a single domain scenario. We will come back to this protocol in the next section.

5.2.2.3 Delay and Disruption Tolerance by Applications

Low bandwidth in many access networks generally imposes a limit on the amount of signaling that can be done over the air during a handover. Large amounts of signaling over a low-bandwidth link translate to a long delay associated with completion of handover signaling. The basic Mobile IP handover solutions introduce a disruption in traffic flow, since layer-3 handover signaling is in the timing critical path and must be completed before MN can access service at the new location.

In order to avoid such disruptions, several measures have been proposed. However, the effectiveness of these measures typically depends on the type of traffic as well. For instance when dealing with non-real time data applications, such as file downloads, a temporary disruption in the flow of the traffic during handovers is not crucial, as long as no packets are lost. To avoid packet loss, the data can be buffered at some point along the path prior to the handover. After the handover, when the new path is established, the buffered traffic is delivered to the MN in a way that the entire file can be reassembled again. For real time voice conversation, however, it is crucial that the packet loss does not lead to loss of voice over a time period that is discernible for average human ears. This time interval has been stated to be around 150 milliseconds. However, many wireless access technologies now use voice coders that may only tolerate loss of voice packets over much smaller periods. Such coders are not able to regain their state if packets over a larger period of time are lost. Regardless of the number tolerable by a voice coding technology, one thing is obvious: once the handover signaling delay is larger than the period tolerable by the voice application, no amount of buffering will help a disruption. Then session will simply fail (dropped call).

Another measure taken to reduce traffic disruption is to use fast handover methods that employ tunnels. One such method described in [LOWMIP4DR] is to deploy a so-called bi-directional edge tunnel (BET) between the mobile node's two points of attachments prior and after handover. In Mobile IPv4 this could be a temporary tunnel between the mobile node's old FA and its new FA, where all the mobile node's traffic arriving at the old FA can be forwarded to the MN through the tunnel from the old FA to the new FA until the handover signaling is complete and the routing path to the MN through the new FA has been established. Thus, the only disruption occurs before the establishment of the tunnel and after its break-down.

5.2.2.4 Establishment of Network Services

Establishment of new routing paths to the mobile node, as performed by Mobile IP, is not the only service the mobile node requires from the network during a handover, especially when the mobile node is engaged in traffic sessions. Applications running at MN may also require continuous support for various features such as data encryption, QoS treatment, authentication services, etc. after (and maybe during) handovers. Typically, each of these network features is managed by specific processes that maintain states regarding the mobile node and its traffic. Re-establishment of these features is a necessary process for continued service after handover so that the service at the new point of attachment is almost the same as it was prior to handover. This is, however, time-consuming and can add to latency caused by handover signaling and to the latency perceived by user. The problem is that some of these processes, such as authentication and establishment of security associations, require further interaction with the user following the handover. This not only adds to the latency of the handover, it makes the handover visible to the user. Hence, a method that reduces the delay associated with feature re-establishment and makes the handover signaling transparent to the end user, is important for providing a seamless mobility experience for the user. Context transfer is the method that has been proposed for this purpose. We will go through context transfer details in the next section.

5.3 Seamless Mobility Procedures

Many people from different communities and backgrounds came to the IETF mobility community, all with the aim of finding a protocol that solves their mobility issues for a variety of applications. One of the useful things that IETF did was to define a set of terminology for the field of mobility [MOBTERM3753], to make sure everyone could express their requirements in a homogeneous manner. The list of handover types defined in that document is almost endless. However, we borrow a few definitions from there almost verbatim. These definitions help the reader understand the meaning of a seamless handover, since this term, due to countless number of misuses, has a very convoluted meaning:

- *Fast handover*: A handover that aims primarily to minimize handover latency, with no explicit interest in packet loss.
- *Smooth handover*: A handover that aims primarily to minimize packet loss, with no explicit concern for additional delays in packet forwarding.
- *Seamless handover*: A handover in which there is no change in service capability, security, or quality. In practice, some degradation in service is to be expected. The practical definition of a seamless handover is that other protocols, applications, or end users do not detect any change in service capability, security or quality, which would have a bearing on their (normal) operation. As a consequence, a handover that is considered seamless for a less demanding application might not be seamless for a more demanding application.

We have already dedicated a lengthy section to describing the shortcomings of Mobile IP when it comes to providing the seamless handovers. As we mentioned, the IETF Mobile IP community tried to enhance the performance of Mobile IP type handovers by designing a variety of optimizations. Some of these optimizations were by modifying Mobile IP architecture to reduce the time involved with the registrations by doing more of the registration work at the local level. Examples of these efforts are the regional registration procedures for Mobile IPv4 [REGREQDR] and the hierarchical mobility management for Mobile IPv6 [HMIP6DR]. Other optimizations were made by improving the signaling by doing layer-2 and layer-3 handover signaling as simultaneously as possible. Examples of the latter group were fast and low-latency procedures that were briefly described in the previous sections.

However, to achieve the seamless experience during handovers, some measures can to be taken that are outside the scope of what Mobile IP can do. Some of these measures are beyond Mobile IP remit simply because of political reasons, such as limiting the busy charter of the IETF Mobile IP working group. The fact that Mobile IP working group has now been split into three IETF working groups attests to that. Some other measures could simply be harder to justify as inclusions in Mobile IP specification. Hence, one group (among others) was formed in IETF and called itself (chosen from among many less pleasant terms) the Seamless Mobility (abbreviated to seamoby) working group [SEAMOBY] that dedicated itself to defining procedures that are beyond Mobile IP but are important in providing the seamless mobility experience. The group ended up working on solutions for candidate access router discovery and context transfer problems among other things. In this section we will cover these two topics in more detail.

Before going on to discuss the engineering details of those solutions, we would like to make a point. When more than one access technology and operator are available, each of these operators may have different pricing, authorization, authentication and security schemes. Furthermore, they may require the user to use different types of devices and even identify herself differently (phone number, user-name, member ID, and so on). The user would typically also need to use a different set of credentials for authentication for each provider, since they typically do not and cannot share these credentials. Unless there are roaming and pricing agreements, seamless mobility is not possible, even if we optimize the mobility and handover protocols to death. Without roaming agreements that guarantee network security (trust relationships) and protection for revenue loss (accounting relationships), the visited networks will not even route the MN requests for any kind of access or service.

5.3.1 Candidate Access Router Discovery

We have described the issues associated with handover and candidate access router discovery earlier and promised to provide more details on the IETF CARD protocol, designed for this purpose. Before starting the protocol description, we need to explain the concepts of network discovery and selection:

- Network discovery is the process by which the mobile node (MN) discovers a number of candidate networks and their presence.
- Network selection is the process of making a choice between the available networks.

Network selection is done as a result of executing a selection algorithm that seeks to match the requirements of the MN with the services and capabilities offered by each CAR. The selection process and its algorithm can be complicated. Furthermore, making a selection is a control issue as well. Some network architectures only allow network entities to make this selection (although the choice may not be in the best interest of the end-user, but anything to keep a customer from going to a competitor operator!). Other architectures seek to give control to the MN and its user.

CARD protocol is designed more from the point of view that MN could perform the selection function, since it provides facilities for the MN to query each network about the capability of that network. However, since the selection process is not standardized by CARD, the MN could potentially forward the result of all the queries to another entity to perform the selection algorithm. When queried, each candidate network must convey its capabilities to the MN prior to the selection process. Although the first capability that comes to mind is the availability of link bandwidth and network resources to handle the MN's traffic, more advanced capabilities may be considered in the future. An example would be whether or not the network is able to offer encryption for the MN's traffic. The network may also advertise its policies and restrictions on what it expects from the MN. For instance, one example of those policies may be the choice of the minimum-security policy on authentication algorithms.

As previously mentioned, the candidate access router discovery protocol (CARD) [CARDDRF] was developed to deal with two problems:

1. Reverse address translation.
2. CAR capability discovery.

In cases where the network keeps information about the capability of its neighbors, the CARD protocol also provides facilities for each AR to keep a CAR table, including the capability of each of those ARs. The CARD specification is now finalized and at the time of writing is awaiting the RFC publication process. However, the specification is intended as an experimental specification rather than a standard one. The goal of the group was to provide a generic protocol that can be applied to a variety of media and architectures. Hence, as mentioned earlier, neither the actual capabilities nor the selection algorithm are specified. The protocol simply defines the signaling needed for query and delivery of the information requested by the selecting entity to run the selection algorithm that derives the so-called target AR.

In the following we provide only the highlights of the process in an example scenario, where an MN is moving across coverage areas of multiple WLAN 802.11 APs:

- The MN periodically scans signal quality on beacons possibly received from surrounding APs. As it moves out of the coverage area of the current AP (we will call this old AP the OAP), it starts sensing stronger beacons from the surrounders APs. The MN may detect one or more APs that can be "heard" better than the OAP. These APs are now candidate APs (CAP) for handover. Let us assume for our example that the MN has detected two CAPs: CAP1 and CAP2. By detection, we mean that the MN has at least measured signal strength and a layer-2 identifier (ID) for each of these CAPs.
- Now that the MN has the L2 ID of each of these CAPs, the MN needs to determine the IP address for the access router (AR) that each of these CAPs is attached to. The CARD protocol provides a CARD request/reply mechanism through which the MN can receive the IP address for each of the CAPs for which it provided the L2 identifier. If the MN decides to detach itself from the OAP and attach itself to any of these CAPs (a layer-2 handover), it needs to connect to a new AR, other than its old AR (OAR), and the MN then needs to perform a layer-3 handover as well. This why a layer-2 to layer-3 identifier translation is important. For simplicity, we assume that each of these APs is served by a different AR, i.e. CAR1 and CAR2 to see the full extent of the CARD protocol.
- When confronted with one or more CARs that are different from the AR that was serving the OAP, the MN needs to make an informed decision as to which CAR to choose for handover. As mentioned earlier, such a decision is made based on the capability of each of these CARs. This means the MN needs first to discover the capability of each CAR. The CARD protocol provides an enquiry mechanism to support this capability discovery. The MN, through the CARD request and the CARD reply message exchange with each of the CARs, can capture the capabilities of each CAR. Alternatively, as shown in Figure 5.8,

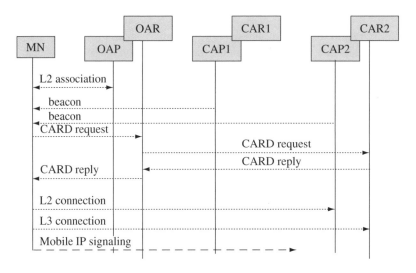

Figure 5.8 Determination of target AR for handover with the help of the CARD protocol. The MN interacts with its current AR (old AR) for CARD inquiries

CARD protocol also allows the MN to perform the capability discovery and the L2 translation through interaction with its current AR. In that case, the current AR in turn contacts the candidate ARs on behalf of the MN. The current AR may have mechanisms to acquire the capability information for its neighbors in a manner that is not on the time-critical path for the MN handover signaling. The CARD protocol suggests a mechanism for this as well.
- When the capabilities of all the CARs are collected, a selection algorithm is deployed to determine the best candidate, which in CARD terminology, is called the target access router (TAR).
- When the MN finds out that the TAR is different from its current AR, it needs to perform Mobile IP signaling with the TAR to complete the layer-3 handover.

The CARD protocol provides several other neat features as well. For instance, the MN can ask its current AR (we called this Old AR) to screen some of the candidate ARs for the MN to see whether their capabilities match the MN's minimum requirements. The CARD protocol calls this process capability pre-filtering and provides several options that can be carried out by the CARD request and reply messages to perform pre-filtering. For instance, an MN may have only an IEEE 802.11b interface card available and hence signals to the current AR that it can only connect to CARs that offer 802.11b services. Another example is that the MN can only connect to CARs that can satisfy a specific QoS level. The MN can signal its requirements and preferences by adding the so-called requirement and preference sub-options to the CARD request that it sends to its current AR. The reader is referred to the IETF specifications for more details [CARDDRF].

5.3.2 Context Transfer

As mentioned earlier, to minimize the period of disruption that the user experiences during a handover, some optimizations, such as fast and low-latency handovers, were on Mobile IP procedures. However, as mentioned earlier, these methods simply attempted to re-establish the layer-3 forwarding paths quicker. They did nothing for the re-establishment of network services, such as security or QoS treatment at the new point of attachment. When the MN is using a network feature such as QoS treatment for its packets, the network maintains some state relating to these features. These states are abstracted into what is called the MN's context. When the MN moves to a new network, the new network must recreate the states for a variety of features (context) to be able to provide the service to the MN after the handover and in a consistent manner. This is where context transfer comes in. Even though designing an efficient and secure context transfer process is a tricky job, the basic motivation for context transfer is rather simple: When the MN moves to a new network, if the mobile's context is transferred to the new network, the features can easily be installed in the network and if done in a time-efficient manner, the user's session can continue seamlessly following or even during the handover. Hence the context transfer is defined as follows:

> Context transfer is a process by which the old point of attachment transmits the feature states related to the MN to the new point of attachment, so that the MN and the new point of attachment do not have to engage in additional signaling (following the completion of handover signaling) to re-establish these features.

It should be noted that, initially, there was controversy over which entity would be responsible for the context transfer, since it is not always the case that the point of attachment (which may even be a dumb device) is the entity that holds the states. There were debates on whether the context transfer should be between more centralized management entities at each of the networks or between layer-2 APs or layer 3-ARs. Finally, it was decided that the context transfer would be performed between layer-3 entities at the edge of the network (ARs). This would, however, require that the state would have to be present at the AR prior to the context transfer.

There have been several proposals for context transfer, some of which (such as TEXT [TEXTDR]) have tried to accomplish simultaneous context transfer and traffic forwarding even during the handover. Finally, the IETF seamless mobility working group [SEAMOBY] came up with an experimental specification of a context transfer protocol (CTP) [CTP3374] that is a waiting RFC status at the time of writing.

As mentioned earlier, there could be many different network services (features in context transfer terminology) that have to be performed for the applications that the mobile user is running. Each of these features is defined by a specific protocol. For instance, if header compression is deployed on the packet headers, the robust header compression (Rohc) protocol may be used for this purpose. The context that would allow header compression operation to run seamlessly on a new processing entity within the new network would also have to be defined by the specifics of Rohc. In a similar manner, a QoS service that is deployed through a QoS-specific protocol needs to define its own feature-specific context. For handover, this would mean the context for each of these features must be transferred to the new AR and from there be forwarded to the entity that is responsible for processing that feature. Since each of these protocols is standardized by other standard authorities such as other groups in IETF, it is up to those standard authorities to define the exact context for the feature that their protocol provides. For instance, the working group responsible for the QoS protocol would have to standardize exactly which states could be transferred to establish QoS provisioning services at the new AR.

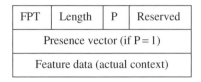

Figure 5.9 Format for context data blocks (each row represents 4 octets)

Context transfer designers did not have the authority to define the context for any specific feature. So they defined a standardized format with which the designers of each feature protocol need to comply in order to use CTP for their protocol. This format is called a context data block (CTB) and is formatted as shown in Figure 5.9.

The feature profile type (FPT) is a number assigned by IANA (Internet assigned number authority) to each feature protocol that needs to have its context data transferred through context transfer protocol. The Presence Vector provides the freedom of sending specific parts of context data if required.

By using CTB and FPT, CTP provides facilities for multiplexing context from a variety of different features, so that as many features as required can restart after the handover as quickly as possible. When sending a CTP message (request or response) in which context is relevant for multiple features, the multiplexing is simply done by stacking the context data block for each feature one after another. For requests, the CTBs would only include the first 4 octets, while for response messages, that carry the actual context data, the full CTB including feature data is included.

5.3.2.1 Design Considerations

As one can imagine, defining the data blocks and passing them back and forth is not rocket science and has been achieved by endless application and transport protocols. The main point with context transfer is how it is tied to the rest of the handover procedures so that context transfer by itself does not create additional latency, state mismatch, or security compromises. Hence timing of various actions and the entities that control them and the underlying trust relationships are very important:

- *Controlling entity*: Depending on the choice made for architectural or business reasons, either the network or the MN can control the handover. This has always been a source of contention in standard bodies. We will not go through the pros or cons here, since that discussion is usually never-ending. The choice of entity initiating the context transfer process may also depend on the affiliation of the designer with the network manufacturer or the MN manufacturer. Hence, we will not issue a verdict on our favorite alternative either. We simply mention that either the network or the MN can initiate the context transfer procedure and warn the designers that. their decision will definitely have timing and security implications, either way. In that discussion, the network is represented by either the previous AR (pAR) or the new AR (nAR) in the handover process.
- *Choice of triggers*: The timing of the context transfer is very important. Context transfer should start as soon as possible so that when Mobile IP signaling is complete and the new paths for both layer-2 and layer-3 forwarding have been established, as little time as

possible is wasted on re-establishing the service features. Starting too late in conjunction with possible retransmission would defeat the purpose of context transfer. The end user might as well establish the context directly with the new point of attachments. Choices of triggers are typically the same as those for fast handovers. Typically the best triggers are provided by layer-2 protocols and examples are those that provide information on the quality of the signal at the link level, such as those indicating when the link to the old AP is going down or is already down, or those indicating when a new link to a new AP is visible, or when a new link is established. Another trigger issue is security. To avoid denial of service attacks, the trigger must come from a trusted source, i.e. either from an internal driver (inside the entity receiving the trigger) or from a source that can prove its identity.

- *Feature type*: Various features and their protocols have different degrees of sensitivity to timing. Some features, such as those providing security, may need to refresh their state only once during short sessions (such as re-keying procedures), while others, such as header compression or accounting protocol, change their state based on every traffic packet that has been forwarded. This means various protocols have different life-times associated with the state that they want the CTP to transfer to. For instance, an old AR cannot forward an accounting state (such as the number of packets forwarded on behalf the MN) to the new AR and then keep sending more data packets to the MN until the connection goes down. This means that the timing of the transfer of context is also dependent on the feature type. At least, this means context transfer cannot be done too early either.
- *Reliability*: As with any protocol, whose purpose is transport of data blocks, reliability is important for CTP. Nobody can use the faulty state to re-establish a feature state machine. However, reliability has always been a great source of contention for design of CTP. One reason is that reliability often means retransmissions, and retransmissions take time and time is a very scarce commodity that nobody has during handovers. Retransmissions may not even be allowed or simply limited to once or twice. When no re-transmission is allowed, it is important for the receiver of the context to be able to know whether it can trust the reliability of the actual transfer or not.
- *Security*: Apologizing to unlikely portion of our reader with background in dentistry, now we are coming to what considered as the root canal operation in the eyes of network mobility designers. Nobody likes to deal with security, especially when things need to be done quickly, but down the road problems appear if we don't consider security issues. First, both the MN and the old AR must trust the new AR before they can send potentially sensitive material (such as keying material) to the new AR. This is partly accomplished in CTP by an authorization token. Second, even when the new AR can be trusted with the context, a sensitive context must be protected during transit from the old AR to the new AR. This would imply that security associations (such as IPSec SAs) must exist between the new AR and the old AR. However, establishment of these SAs can take a long time. For the secure context transfer to happen in a time-efficient manner, it is essential that the two ARs will have established these SAs (for instance, through IKE) prior to the handover. This will rule out CTP for time-sensitive applications between networks that do not trust each other. The final security implication is that, as mentioned above, the trigger source or the entity that requests a context transfer must be able to be trusted.

The CTP solution [CTPDR] provides several measures for dealing with these issues. Requiring the CTP messaging between the routers to be performed over SCTP (Stream Control Transport protocol) was a bold move. SCTP provides reliability and flexibility in the timing constraints of CTP. One can argue that SCTP is a bit extreme for a lightweight application layer protocol that must be nimble and quick. Some of these decisions can be attributed to the current hype in IETF to take extreme measures for congestion control. The authors argue that CTP and its small blocks can hardly be found guilty of causing congestion in the Internet, especially given that CTP is run at the edge of the network. The flip side of argument is that one can only go so far when designing home-brewed reliability at the application layer. We leave the readers to give their own verdict over their beverage of choice in their favorite gathering place. SCTP would be too heavy to use for messaging between the MN and the routers. Hence, ICMP was suggested to carry the messaging that involves MNs.

The TEXT procedure [TEXTDR] although it never made it as a standard track document, had proposed a method to combine traffic forwarding with context transfer and handover signaling to eliminate the impact of context transfer and handover latency on traffic flow. The reader is referred to the draft as well as [RELCT] for thorough discussions and solutions on trigger, timing, and reliability issues.

5.3.2.2 Messaging Overview

As mentioned earlier, one of the important design considerations is the trigger and the entity that receives it. This could be the entity that initiates the context transfer process, even though it may not be the entity that authorizes it from both the control and security standpoint. Depending on who initiates the trigger and who authorizes a context transfer, different scenarios were envisioned. The messaging for each scenario is slightly different.

Scenario 1: Proactive Context Transfer
The messaging for this scenario is shown in Figure 5.10. We like to call this scenario the proactive one, since the pAR is the entity that sends the context to the new AR (nAR) without the new AR asking for it. In this scenario, the previous AR (pAR) initiates context transfer, based on either an internal link layer trigger, or a request from the MN.

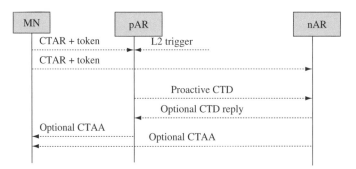

Figure 5.10 Proactive context transfer

- The MN makes the request by sending a Context Transfer Activate Request (CTAR) to the pAR. The CTAR includes information on the IP address of the nAR (so that the pAR knows where to send the context), the IP address that the MN is using at the pAR (to identify the MN to the ARs), and optionally a list of contexts to be transferred. MN also includes a so-called authorization token. The token is calculated based on a hash (using HMAC_SHA1) of MN's IP address at pAR, the context data block and some other information.
- Upon receiving the CTAR, the pAR sends a context transfer data message, including all the context blocks (described earlier) to the nAR.

For added reliability, the MN also sends the CTAR message to the nAR, in order to make nAR aware that a context transfer is happening. This is so that if the proactive process fails, the nAR can request context data according to the process in the second scenario. Again with the same aim, any of the entities that has received a CTAR may send a context transfer authorization acknowledgement (CTAA) to the MN to confirm the context transfer and may even include a list of FPTs for context blocks that have been successfully transferred, if this information is available.

Scenario 2: Reactive Context Transfer
We call this scenario the reactive context transfer, since in this scenario the nAR is the entity that requests the pAR to send the context. The nAR knows when to do this based on either an internal trigger or a CTAR message from the MN. Once the nAR receives either of these indications, the nAR sends a context transfer request (CT-req.) to the pAR, which in turn responds with a context transfer data (CTD) message including the context data.

5.4 Further Resources

For more information on Mobile IPv4 specifications, the reader is referred to the specifications that were completed by old Mobile IP working group, whose website is: http://ietf.org/html.charters/OLD/mobileip-charter.html.

Specitications are being completed by the IETF Mobility for IPv4 working group, found on the group's website http://ietf.org/html.charters/mip4-charter.html. IETF specifications on the interaction of Mobile IPv4 and VPNs can also be found on this page.

There is also an excellent book written on Mobile IP by one of the founders of Mobile IP, i.e. Jim Salomon. This book, called *Mobile IP: The Internet Unplugged*, provides a great overview for people who do not have an IP networking background.

For more information on Mobile IPv6 specification, the reader can look at the "Mobility for IPv6" working group website http://ietf.org/html.charters/mip6-charter.html and the "Mobile IPv6 signaling and handoff optimization" working group (mipshop) website at http://ietf.org/html.charters/mipshop-charter.html. More information on the interaction of Mobile IPv6 and IPsec architecture can also be found on the Mobile IPv6 working group website.

For more information on network mobility protocol for IPv6 networks, the reader is referred to the IETF "network mobility" (nemo) working group website at http://ietf.org/html.charters/nemo-charter.html.

5.5 References

1. [MIP3344], C. Perkins, "IP Mobility Support for Ipv4", RFC 3344, August 2002.
2. [REV3024], G. Montenegro, "Reverse Tunneling for Mobile IP", RFC 3024, January 2001.
3. [MIPCHAL3012], C. Perkins, and P. Calhoun, "Mobile IPv4 Challenge/Response Extensions", RFC 3012, November 2000.
4. [3102bis], C. Perkins et al. "Mobile IPv4 Challenge/Response Extensions (revised)", draft-ietf-mipv4-rfc3012bis-02.txt, June 2004.
5. [CARDDRF], M. Liebsch et al., "Candidate Access Router Discovery", IETF Internet draft, draft-ietf-seamoby-card-protocol-07.txt, IETF work in progress, June 2004.
6. [SEAMOBY], Context Transfer, Handoff Candidate Discovery, and Dormant Mode Host Alerting (seamoby), concluded IETF working group, http://ietf.org/html.charters/OLD/seamoby-charter.html.
7. [MIPBOOT], A. Patel, "Problem Statement for Bootstrapping Mobile IPv6", IETF work in progress, draft-ietf-mip6-bootstrap-ps-00.txt, IETF, July 2004.
8. [MIP4WWW], IETF Mobile IPv4 working group website, http://ietf.org/html.charters/mip4-charter.html
9. [MIPDYNHA], M. Kulkarni et al., "Mobile IPv4 Dynamic Home Agent Assignment", IETF work in progress, Internet draft, draft-ietf-mip4-dynamic-assignment-02.txt, June 2004.
10. [LOWMIP4DR], K. El Malki, "Low Latency Handoffs in Mobile IPv4", IETF work in progress, Internet draft, draft-ietf-mobileip-lowlatency-handoffs-v4–09.txt, June 2004.
11. [FMIP6DR], R. Koodli, "Fast Handovers for Mobile IPv6", IETF work in progress, Internet draft, draft-ietf-mipshop-fast-mipv6–03.txt, October 2004.
12. [CTP3374], J. Kempf et al., "Problem Description: Reasons for Performing Context Transfers Between Nodes in an IP Access Network", IETF, RFC 3374, September 2002.
13. [MIPNAI3846], F. Johansson and T. Johansson, "Mobile IPv4 Extensions for Carrying Network Access Identifiers", IETF, RFC 3846, June 2004.
14. [REGREQDR], E. Gustafsson et al., "Mobile IPv4 Regional Registration", IETF work in progress, Internet draft, draft-ietf-mip4-reg-tunnel-00.txt, November 2004.
15. [HMIP6DR], H. Soliman et al., "Hierarchical Mobile IPv6 Mobility Management (HMIPv6)", draft-ietf-mipshop-hmipv6-03.txt, October 2004.
16. [TEXTDR], M. Nakhjiri, "Time-efficient Context Transfer", Internet draft, draft-nakhjiri-seamoby-text-ct-03.txt, March 2003.
17. [RELCT], M. Nakhjiri, "A Time Efficient Context Transfer Method with Selective Reliability for Seamless IP Mobility", IEEE 58th, VTC proceedings, Orlando, October 2003.
18. [MOBTERM3753], J. Manner, and M. Kojo, "Mobility Related Terminology", IETF, RFC 3753, June 2004.

6

Remote Access Dial-In User Service (RADIUS)

In our treatment of authentication models in Chapter 2, we mentioned that gradually people realized that to handle client authentications in large networks it is more practical to have backend authentication servers that offload the network's front-end point of attachment. Later on these back-end authentication servers became full-blown AAA servers that use specific AAA protocols, not only handle authentication but also authorization and accounting as well. The most widespread AAA protocol today is Remote Access Dial-In User Service, RADIUS.

RADIUS was originally designed to serve the purpose of allowing a NAS to forward a dial-up user's request and its credentials to a backend server (three-party authentication model, described in Chapter 2). The Access-Request, Access-Challenge message structure in Radius attests to the fact that Radius was originally designed to accommodate PAP and CHAP. However, due to its extensible nature; RADIUS is able to support more complex EAP-authentication methods through support for EAP. Furthermore, RADIUS was later extended to provide authorization and accounting procedures.

In this chapter we intend to provide an overview of IETF RADIUS specifications. We will provide some highlights of current work in progress in extending RADIUS. In later chapters, such as Chapters 8 and 10, we will provide more examples of the usage of RADIUS in conjunction with Mobile IP and various EAP authentication procedures. We also report on some of the activities in the newly formed RADIUS extension working group [RADEX-TWEB] at the end of the chapter.

6.1 RADIUS Basics

RADIUS is a client–server mechanism, in which a Network Access Server (NAS) usually acts as a RADIUS client. It is, however, important to understand the distinction between a RADIUS client and an end client in a communications scenario. In the RADIUS context, the client is an entity that acts as a client in RADIUS messaging (a client–server protocol). As mentioned, NAS is the RADIUS client. The end user or the device that is authenticating to the network through the NAS is not the client in RADIUS discussions. We may call the end user or device: "the end client" or in the EAP discussions: "the supplicant" to make a clear distinction.

During authentication procedures, the RADIUS client is responsible for passing user information in the form of requests to the RADIUS server and waits for a response from the server. Depending on the policy, the NAS may only need a successful authentication or further authorization directives from the server to open its traffic ports to the client's traffic. As we will see in Chapter 10, the NAS may need to establish secure communications channels with the client before engaging in any communications with the end user. Furthermore, when accounting is required, the NAS is also responsible for collecting resource usage data and reporting back to the server.

The RADIUS server, on the other hand, is responsible for processing requests, authenticating the users, and returning the information necessary for client configuration to deliver the service to the user.

RADIUS specifications consist of several RFCs. The base RADIUS specification was revised several times (RFCs 2058, 2138) and is now included in RFC 2865 [RADIUS2865]. The base specification describes the process of authenticating a user to a server using PAP or CHAP, but does not describe support for accounting. The accounting procedures were standardized in a separate informational RFC document [RADACC2866]. RADIUS was later extended to perform several other functions such as authentication using EAP. Many of these functions are specified in further RFC specifications. We start the description of RADIUS by going through the protocol as described in the base RFC and later will go into details of the extensions.

6.2 RADIUS Messaging

The RADIUS message set is rather simple and consists of only eight messages, of which only the first four are specified in the base specifications. We will describe the function of these messages briefly. For now, here is a list:

- Access Request: This message is generated by the NAS (RADIUS client) towards the server to forward the request from or on behalf of a user. (NAS-->AS).
- Access Challenge: This message is sent from the RADIUS server to the RADIUS client (NAS) and is generally used to question the NAS or the user about something or perform some sort of negotiation.
- Access Accept: This message is sent from the RADIUS server to the NAS to indicate a successful completion of (and typically grant) the request.
- Access Reject: This message is sent by the server to indicate the rejection of a request.
- Accounting request: This message is sent from the client to the accounting server to convey accounting information regarding the service provided to the user.
- Accounting response: This message is sent by the server to the client to acknowledge that the accounting information sent by the client has been received and indicates the result of the performed accounting function by the server.
- Status-Server and Status-Client: These two messages are experimental.

It should be noted that newer RADIUS specifications such as [RAD3576] have defined a number of new RADIUS messages (codes). However, due to the large deployment base for RADIUS, the RADIUS community is very protective of the existing implementation and takes backward compatibility between those implementations and new features introduced to RADIUS very seriously. Many of these specifications, including [RAD3576], are categorized as an informational specification rather than standard ones due to the potential problems that they can cause with the existing implementations.

RADIUS can carry information regarding many different functions. The information is carried in the form of attributes. Each attribute can be a self-contained package including information of variable length and formed according to type, length, and value (TLV) format (Table 6.1). Typically, the main body of Access Request and Access Challenge messages are used to carry attributes from NAS to RADIUS server and vice versa. Since the attributes carry almost all information required to define the operation of the RADIUS, two different Access Request messages that carry different attributes perform different functions depending on the attributes they carry.

6.2.1 Message Format

RADIUS packet format is also very simple and consists of a header, including code, ID, length and an authenticator field, followed by the payload, which is a list of zero or more attributes (Table 6.2). The calculation of authenticator field is covered in detail later on and the attribute field is described in the following subsection. The end of attribute list is indicated by the RADIUS length field in the header.

Table 6.1 RADIUS attribute TLV format

Attribute type	Up to 1 octet (Max of 255 attributes) to describe the type of attribute
Attribute length	Length of the attribute in octets, indicating the entire length of the attribute payload, included type and length fields
Attribute value	Contains information that the attribute is supposed to carry

Table 6.2 Packet format for RADIUS messages: a header followed by a number of attributes

	RADIUS packet format	
Field name	Sub-field	Description
Header	Code	Identifies type of RADIUS packet (access request, response, etc.)
	ID	To match requests and responses
	Length	2 octet long showing the length of the entire message
	Authenticator	16 octet value that is calculated as described later
Attribute 1		First attribute in the packet
...		
Attribute N		Last attribute in the packet

6.2.2 RADIUS Extensibility

As mentioned earlier, the attributes are the main vehicle for carrying information over RADIUS messages. This means, the attributes also provide the opportunity to extend the functionality of RADIUS server to interact with many other entities and for many purposes. Also to support implementation-specific scenarios, vendor-specific attributes (VSA) can be defined, so that each vendor's NAS and RADIUS server can interact in a manner that is specific to the implementation defined by the vendor.

RADIUS base specification [RADIUS2865] specifies around 40 attributes, while several other attributes are defined by later RFCs, some of which are mentioned in later sections of this text. The list of attributes is rather extensive, so, we will refrain from providing a comprehensive list here. For such lists, the reader is referred to the IETF specifications, specifically [RADIUS2865]. These lists typically also include the capability of each RADIUS messages to carry the attribute. Table 6.3 shows some examples.

The IETF RADIUS extension working group [RADEXTWEB] is working actively on defining new attributes that extend RADIUS functionality in a way that preserves backward compatibility with the existing RADIUS deployments. Unfortunately due to the limited number of bits (8) in the type field of the attributes, the attribute space is limited to only 255 allowed attributes. This and the need for backward compatibility for the large RADIUS deployment base have caused the IETF RADIUS community to guard the attribute space tightly and strictly question any suggestions for standardization of new attributes. For more discussion, see the RADIUS issues section.

It is also theoretically possible to extend RADIUS functionality by creating new RADIUS messages. However, to preserve backward compatibility, the group is refraining from doing so.

6.2.3 Transport Reliability for RADIUS

RADIUS can be considered as an application layer protocol, running over a transport protocol. At the time of the RADIUS specification, UDP was deemed more appropriate than TCP, due to the fact that TCP session establishment is a time-consuming process involving state machines. As will be discussed later, the lack of reliability support in UDP causes some serious problems for RADIUS, especially for accounting functions.

IESG Note on reliability: It is interesting to note that the Internet Engineering Steering Group (IESG) which is the governing body of IETF, has issued a note on the front page of RADIUS RFC, stating that RADIUS suffers from performance degradation and packet loss in adverse network conditions and scalability and congestion controls should be better addressed in a successor AAA protocol.

Table 6.3 Example of RADIUS attributes and the quantity of each carried by various RADIUS messages

Attribute name	Attribute type	Access request	Access accept	Access reject	Access challenge
User name	1	0–1	0–1	0	0
User Password	2	0–1	0	0	0
CHAP Password	3	0–1	0	0	0
NAS IP address	4	0–1	0	0	0
NAS port	5	0–1	0	0	0

6.2.4 RADIUS and Security

Security protection in RADIUS is rather primitive. Two main functions are provided, one is attribute (mainly password) hiding and the other is authentication of certain messages. Both these functions are performed using MD5 hash functions and a secret that is shared between the RADIUS server and the RADIUS client (NAS). This secret is usually called the RADIUS shared secret. We discuss the details of message authentication first and then talk about attribute hiding functions.

6.2.4.1 RADIUS Message Integrity Protection

As seen in the description of the RADIUS message format, an authenticator field is added to all RADIUS messages. In the Access Request message, the field is called the request authenticator, while in the Access Challenge, Accept and Reject messages are called the response authenticator. However, the main difference is not in the terminology, but in the level of protection that each provides.

Request Authenticator
In Access Request messages, the request authenticator value does not include any cryptographic secret; it is merely a 16-octet (128) random number that is generated by the client and is added in the authenticator field. Since no secret keys are involved in production (by the client) or verification (by the server) of the request authenticator field, no valuable integrity protection is provided. All that is required from the request authenticator is that it should exhibit global and temporal uniqueness. Global uniqueness is required so that the same NAS can interact with geographically disparate RADIUS servers and temporal uniqueness is required to avoid replay attacks. Every time the client generates a new Access Request message with a new identifier (see message format), it generates a new authenticator. Note that retransmission of previous Access Requests include the same authenticator value as before. However, if the Access Request is regenerated as a result of the NAS receiving an Access Challenge from the server, a new ID and a new authenticator must be generated. In this case, NAS creates a CHAP challenge to the user, if the CHAP challenge value is shorter than 16 octets, it can be inserted in the authenticator field of the Access Request. Otherwise it is added in the CHAP challenge attribute.

The authenticator not only did not provide any integrity protection, but also guaranteeing temporal and global uniqueness was something that was beyond a primitive NAS with cheap and unsophisticated CPU. The required entropy behind such uniqueness simply could not be provided. Had the NAS identifier or IP address been included in generating the authenticator, at least the global uniqueness requirement could have been met to some degree.

The original design of RADIUS justified this "weak" security provisioning with two additional security functions. First, RADIUS servers check the source IP address of the received message, thinking that by checking that the message is coming from the right NAS, the message would not need additional protection. Furthermore, RADIUS was originally designed for remote authentication process (as the name implies). Since the user password is protected by a shared secret shared between the NAS and the RADIUS server, we would have another added security mechanism to protect us against security attacks. As we see later on, this thought process is not acceptable in this day and age (around 10 years later), when IP packets can easily be spoofed and password cracking tools are easily accessible.

Later on the community started demanding a specialized attribute called "Message Authenticator" attribute (type 80) to provide integrity protection for Access Request Messages. Message Authenticator attribute was first introduced in an informational specification [RADEXT2869]. When protecting the Access Request message with the Message Authenticator, the client calculates a hash (MD5) over the entire message (see Table 6.2) using the shared secret it shares with the RADIUS server and includes the result of the hash in the data field of the Message Authenticator attribute:

Message Authenticator value = MD5 (Code, ID, Length, Request Authenticator, Attributes)

When receiving an Access Request message that includes this authenticator, the RADIUS server calculates the authenticator, using the shared secret and if there is not a match, the server discards the message.

Even though the Message Authenticator can be used to protect RADIUS messages from the server (such as Access Accept), it is usually not used, since those message are protected natively as described below. RFC 2869 does not mandate the use of this authenticator in Access Request messages that carry User Password attribute and relies on the password hiding mechanism using the shared secret. We will explain the password hiding mechanism shortly. Finally, before we close the book on Request Authenticator, it is important to mention an exception where the Request Authenticator is calculated through a cryptographic process. That exception is RADIUS accounting. RFC 2866 [RADACC2866] demands that the Request Authenticator is inserted in the Authenticator field (see Table 6.2) of Accounting Request (code 4) messages and is calculated by a keyed MD5 hash of various message fields. The key is the RADIUS shared secret (between NAS and RADIUS server):

Authenticator value = MD5 (Code, ID, Length, 16 zero octets, request attributes, shared secret)

Response Authenticator

In contrast to the Access Request message, the messages from the RADIUS server to the client (Access Challenge, Access Accept and Access Reject) receive better integrity protection. The value of the authenticator field (called response authenticator) is a result of an MD5 hash of most of the message (except the authenticator field itself) as seen below:

Response Authenticator = MD5 (code, ID, Length, Request Authenticator, attributes, SS)

Where code, ID and length are fields in RADIUS packet itself (response ID matches the request ID), and Request Authenticator is taken from the corresponding Access Request message.

6.2.4.2 Attribute Hiding

As mentioned earlier, most of the functionality provided by RADIUS protocol is driven by the attributes that RADIUS messages carry. Security and privacy requirements dictate that some of these attributes must be protected from eavesdropping during transport. In RADIUS this is mostly provided by a feature called attribute hiding. Since RADIUS was initially designed to support PAP and CHAP-based authentication mechanisms (see Chapter 2 for a description of PAP and CHAP), support for hiding the attributes that carry passwords was added to basic RADIUS specifications. Attribute hiding performed for "User Password" is typically called password hiding. Password hiding is performed as follows:

- If the user password (UP) is less than 16 octets long, the client (NAS) generates a random Request Authenticator (RA) as described earlier and concatenates the shared secret (SS) that it (the NAS) shares with the RADIUS server with the RA.
- The NAS then calculates an MD5 hash of the concatenation and performs an XOR operation on the result and the user password.

$$B = MD5 (SS + RA) \qquad C = B \text{ XOR } UP$$

- The result of the operation above (C) is inserted in User-Password Attribute to be carried by the RADIUS message (typically Access Request packet).

When user password (UP) is longer than 16 octets, it is segmented into 16-octet chunks: P1, P2, P3,..., Pn, where Pn may be padded with zeros to make a 16-octet-long chunk. The following operations are then performed for password hiding:

$$B1 = MD5 (SS + RA) \qquad C1 = P1 \text{ XOR } B1$$
$$B2 = MD5 (SS \oplus C1) \qquad C2 = P2 \text{ XOR } B2$$

. .

. .

. .

$$Bn = MD5 (SS \oplus Cn-1) \qquad Cn = Pn \text{ XOR } Bn$$

The B values are sometimes referred to as key stream, since they are used in the XOR process, which is considered as encryption of the password chunks (Pi). The C values are the ones that are inserted in the User-Password Attribute (after concatenation of C1, C2, C3,..., Cn).

As elaborate as this may seem to a layperson, this password hiding procedure in RADIUS has many flaws. For one, the temporal and global uniqueness of the key stream depends on the uniqueness of the Request Authenticator and that has issues of its own, as discussed earlier. Due to lack of processing power, the NAS may not exhibit sufficient entropy to make the RA random enough, which means that the RA may repeat itself soon. This would mean the first block of key stream (B1) will repeat itself in two different messages. Second, password hiding relies on a shared secret that is used between the RADIUS client and RADIUS server to protect all the messaging between those two. This means this shared secret is used to protect the password from any user that is trying to authenticate to the RADIUS server through that same NAS. RADIUS specification is taking precaution here and recommends that the shared secret (that is itself considered a long-term secret) be at least as long as the passwords it is trying to protect. A known attack against this password hiding method is that an attacker submits an authentication request with a known password P and then collects the cipher text Ci from the attribute value that is produced. Knowing both P and Ci and using the XOR operations, the attacker can compute the key stream Bi. When the Request Authenticator repeats itself frequently, the attacker can start guessing other people's password (after calculating the first password chunk P1).

Later RADIUS specifications [RADTUN2868] providing support for users using tunneled dial-in methods (such as Layer 2 tunneling Protocol) introduce a salt (A) in the password hiding process to provide better uniqueness when creating the first block of key stream

$$B1 = MD5 \, (SS + RA + A) \qquad C1 = P1 \text{ XOR } B1$$

The salt, although providing better uniqueness, does not help the attacks much, since it is sent in the clear.

Finally, IPsec has been suggested to protect RADIUS messaging. The use of IPsec is described in the following subsection.

6.2.4.3 Security Vulnerabilities of RADIUS

As we have seen so far, whether it comes to authenticating a message or hiding an attribute, the only cryptographic secret available for RADIUS messaging is the shared secret configured between the NAS and the RADIUS server. The use of shared secrets as the basis for providing security functions within RADIUS has caused many vulnerabilities for RADIUS deployments. In the following we list some of the most important ones.

- *Static manually configured shared secrets*: No method for dynamic and automatic shared secret establishment is defined in the base RADIUS protocol. The pre-shared secrets are usually manually configured at the NAS. Due to the large number of NASes involved in the many networks, there have been a large number of cases, where the technician has simply configured a large number of NASes with the same shared secret to avoid the administrative burden. Furthermore, the shared secrets are long term (typically over the life the NAS) and the specifications define no methods to refresh the shared secrets.
- *Shared secret lookup*: To prevent spoofing, the RADIUS server uses the source IP address in the RADIUS UDP packet (rather than NAS IP address or ID attributes) to look the shared secret up. This is in part due to the need for support of hop-by-hop security when RADIUS proxies are implemented and in part due to the fact that the NAS ID (an NAI or MAC address) is only added as an attribute to the access request payload. This arrangement can potentially cause many problems in cases where the NAS IP address may change. For instance, a NAS may need to obtain its IP address dynamically through DHCP as may be the case for many WLAN hotspots. Managing WLAN hotspots with large number of access points without DHCP will cause administration problems.
- *Proxy chaining*: We will talk about RADIUS proxy chaining in more detail later on. For now, it is important to know that in deployments using RADIUS proxies between the NAS and the RADIUS server, the NAS only shares a secret with the first hop AAA proxy and not with the backend RADIUS server that is the ultimate destination. This means the trust between the NAS and the RADIUS server is only transitive, i.e. the NAS communicates with the RADIUS server based on a chain of trust rather than a direct trust relationship. If a proxy in the middle is rouge, security or fraud problems may arise.
- *Transport protection*: Attribute hiding provides selective application layer protection. It does provide any security protection (authentication or encryption) for RADIUS messages or the protocol layers (UDP, IP) that these messages are riding on. This means the IP address can easily be spoofed or other attributes could be changed.

6.2.4.4 RADIUS over IPsec

As a way of getting around the security vulnerabilities of the use of the shared secret for RADIUS, the community has started recommended using IPsec in some of the specifications for which security is extra important. One such specification is RFC 3579 [RADEAP3579] which defines the support of RADIUS for EAP. As discussed in Chapter 3, EAP provides a key management framework that allows the AAA server to assist the user and the NAS in establishing a secure communication channel before starting data traffic. As we saw there, the key material needed at the NAS needs to be transported from the AAA server to the NAS over the AAA protocol. To provide authentication and encryption support the use of IPsec is recommended for the protection of RADIUS messaging.

When non-null IPsec transforms are configured between the NAS and the RADIUS server, it is possible to skip the configuration of the RADIUS shared secret. The NAS and RADIUS server must assume that a zero length shared secret is configured in this case, specially for RADIUS servers that have no way of knowing whether the incoming traffic is protected through IPsec or not. Typically the RADIUS server applies an IPsec policy that accepts IPsec traffic but does not require IPsec-only traffic. An example of such policy would be "Accept IPsec, from any to me, destination port 1812", where UDP port 1812 is used for RADIUS authentication. This liberal policy is adequate since it is not fair to require support of IPsec at all RADIUS clients (which could be cheap access points). A typical IPsec policy at a NAS that supports IPsec would be "Initiate IPsec, from me to any destination port UDP 1812". This would cause the client to set up the IPsec SA prior to sending the IPsec traffic.

The discussion above brings us to the final point on use of IPsec and that is the required key management. It is recommended that Internet Key Exchange, IKE (see Chapter 4) is used to set the required IPsec SAs between the client and the server. However, remember that the shared secret may have not been configured at the NAS and this means the NAS may need to obtain certificates for IKE phase-1 authentication.

6.3 RADIUS Operation Examples

Now that we have covered the basics and messaging formats, let us go through some examples of RADIUS operation. We start this by going through the most classic applications of RADIUS that use RADIUS for password-based user authentication. After going through PAP and CHAP, we will go through the use of RADIUS for EAP. The discussion on interaction between EAP and RADIUS in this chapter will be rather brief. Instead we provide a lot of details on this interaction in Chapter 10, where the EAP-based authentication methods are discussed.

6.3.1 RADIUS Support for PAP

We discussed the use of Point to Point Protocol (PPP) in establishing dial-up connections for remote users in Chapter 2. The use of user passwords in the Password Authentication Protocol (PAP) was the most prominent way of performing user authentication. Initially the dial-up facilities had the ability to authenticate the users, but later on the arrangement was expanded so that the PPP users could authenticate to a backend authentication server. When RADIUS servers are used as backend server, the PAP is performed as follows:

- When a user tries to establish a PPP connection with the NAS and is configured to use PAP, the NAS prompts the user for user name and password.
- Upon receiving user name and password from the user, the NAS creates an Access Request message for the server as follows: The NAS creates a Request Authenticator (RA) and uses the RA and the shared secret it shares with the RADIUS server to hide the user password and produce the User-Password Attribute as described earlier. The NAS then waits for a response from the server. The NAS includes the RA in the Authenticator field of the Access Request message. A list of attributes for this message is shown in Table 6.4.
- If an NAS does not receive a response to its Access Request from the server within a pre-specified length of time, it can either retransmit the request to the same server, or try to send it to an alternate server (maybe in a round robin fashion). This is, however, not standardized.
- Upon reception of an Access Request, the RADIUS server runs a series of checks to see if all the conditions for Access Request are met. Among other things, the server checks to see if it has a shared secret with the client (NAS). It then retrieves the user password from the User-Password Attributes and checks to see if the user is within the database and the presented password matches what the server has.
- If the password check was successful, the RADIUS server sends an Access Accept message back to the NAS.
- If, on the other hand, any conditions for the authentication are not met, for instance, the user does not exist within the server's database, the password hash calculation does not succeed (due to wrong user password or wrong NAS shared secret), the server rejects the user's request by sending an Access Reject message to the NAS.

Credential error detection: One problem that the RADIUS security model has is revealed here: During PAP authentication, if the server's calculation of hash value fails, the server cannot tell whether the user password was wrong or the shared secret between the NAS and the RADIUS server was not correct. Ideally, the server should be able to independently check the validities of the client and user.

6.3.2 RADIUS Support for CHAP

We described Challenge Handshake Authentication Protocol (CHAP: RFC 1994) in Chapter 2. As we saw there, CHAP is one step above PAP as far as sophistication goes and could be more secure if done properly, since it avoids sending the password over the

Table 6.4 A list of attributes in the Access Request message for a PAP user

Attribute name	Description
User name	PAP ID for the user
User password	The MD5 hash of SS, RA and user password (as explained above)
NAS IP	IP address for NAS (this attribute could alternatively be an NAS ID)
NAS port	The port user connects to NAS, not the UDP port

communication channel. Instead of sending its password over the link, the user deploys the password in calculating the response to a challenge issued by the authenticator. Supporting CHAP was considered one of the fundamental functions of RADIUS as a AAA protocol for supporting dial-up applications. (Remember what the acronym RADIUS stood for?) We see that through the fact that an Access Challenge message is included in the RADIUS message set from early on to allow the server to send a challenge to the client for user authentication. However, for practical purposes nowadays the challenge is issued by the NAS rather than by the AAA server, while sending an Access Challenge message still indicates that the RADIUS server needs something from the user or the NAS to complete the authentication process.

The following describes the support of RADIUS for CHAP (shown in Figure 6.1).

- A user, that is configured to authenticate through CHAP, requests the NAS to connect to the network.
- The NAS creates a 16-octet challenge and sends it to the user through a CHAP challenge message. This message includes a CHAP ID as well.
- The user responds to the challenge using a response message including the same CHAP ID as used in the CHAP challenge message, a CHAP user name, and her response to the challenge. The response is calculated using a keyed MD5 hash, with the user's secret (known by RADIUS server and user) as the key:

$$MD5 \text{ (ID, secret, challenge)}$$

- The NAS then creates an Access Request. The NAS inserts the CHAP user name from the user in the User-Name attribute, and CHAP ID and CHAP password in the User-Password attribute and sends these attributes along with the Access Request to the RADIUS server, (Table 6.5).
- The RADIUS server looks up a password based on the User-Name and calculates a hash in the same way as the user did. However, if the CHAP-Challenge attribute is not present in the message, the server uses the value in the Request Authenticator field of the Access Request. The server compares the result of the hash with the value included in the CHAP-Password. If there is a match, the server sends back an Access Accept to the NAS. If not, the server sends an Access Reject.

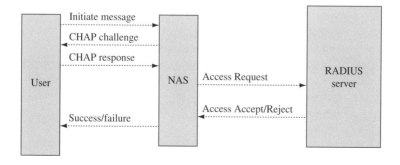

Figure 6.1 RADIUS messaging for CHAP as an authentication protocol

Table 6.5 Attributes sent along with the Access Request for CHAP authentication

Attribute name	Description
User name	CHAP user name
CHAP password	This attribute as others is in TLV (type length value form), however, the value includes both CHAP ID and the CHAP response of the user to the challenge presented by the, NAS, both in the CHAP response message to the NAS
CHAP challenge	The CHAP challenge that the NAS presented to the user (if this value is shorter than 16 octets, it can be placed in the request authenticator field)
NAS IP	IP address or identifier (do not say what type) for NAS
NAS port	The port the user connects to, not the UDP port

6.3.3 RADIUS Interaction with EAP

We discussed Extensible Authentication Protocol (EAP) at length in Chapters 2 and 3. As mentioned there, EAP is designed to provide support for a generic authentication protocol without requiring specific upgrades to the NAS. Since EAP relies on a backend server to understand and perform the actual authentication, RADIUS was extended to provide this support. We also mentioned in Chapter 3 that, when transiting between the NAS and the RADIUS server, EAP messages are carried over a AAA protocol. The RADIUS flexibility and extensibility provided by the concept of carrying information inside attributes allow RADIUS message even to carry messages from other protocols. The EAP-RADIUS framework allows the messages for EAP authentication schemes to be embedded inside RADIUS attributes and carried along RADIUS messages. The process in RADIUS is very simple (shown in Figure 6.2).

Messages from other protocols such as EAP can be embedded in the attributes and carried inside RADIUS Access Request and Access Challenge messages between NAS and server and vice versa:

- The requests from the RADIUS server to the user (EAP request messages) are carried inside RADIUS Access Challenge messages. The NAS decapsulates the EAP request from the Access Challenge and sends it over a layer-2 protocol to the user.
- The responses from the user to the RADIUS server are carried inside the EAP response to the NAS, which in turn encapsulates them inside Access Request messages sent to the server.

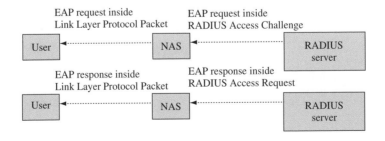

Figure 6.2 Embedding EAP authentication messaging inside RADIUS messages

Note that by encapsulation we mean that the EAP messages are included in a specific RADIUS EAP-Message attribute included in the RADIUS message. This attribute along with some others is specified in the RFC 3579 for RADIUS EAP support [RADEAP3579]. This specification specially requires support for the following RADIUS attributes:

- EAP-Message Attribute (type 79). This attribute encapsulates one fragment of the EAP message, which includes the PPP type, request-ID, length, and EAP-type fields.
- Message Authenticator Attribute (type 80). This attribute ensures the integrity of the message as was described in an earlier section of this chapter.

We will provide much more detail on EAP-based authentication methods in Chapter 10 and there we will see more on interaction of EAP and RADIUS.

6.3.4 RADIUS Accounting

The RADIUS base standards document [RADIUS2865] does not provide specifications for accounting support. Rather, RADIUS accounting is defined in a separate informational RFC [RADACC2866]. Accounting procedure is also based on a client–server model where the client (NAS) passes the user's accounting information to the RADIUS server, which hosts the RADIUS accounting machine.

RADIUS accounting uses two messages types: Accounting Request and Accounting Response, both of which are also transported over UDP. Accounting Requests are always sent from the RADIUS client towards the RADIUS server, while Accounting Responses are generated by the RADIUS server upon receiving and processing the Accounting Requests (Figure 6.3). However, as we will see later on, in roaming scenarios proxy servers may have a hand in the Accounting Request–Response exchange.

6.3.4.1 Basic Operation

The basic operation is shown in Figure 6.4. An NAS that is capable of supporting RADIUS accounting generates an Accounting Request "Start" at the start of operation and sends it to the RADIUS accounting server. This packet specifies, among other things, the type of service being delivered, and the user that the service is being delivered to. Upon reception of a valid accounting request, the server adds an accounting record to its log and acknowledges the request by generating an Accounting Response to indicate that the packet has been received.

At the end of service delivery the client will generate an Accounting Stop packet describing the type of service that was delivered and optionally statistics such as the actual session duration, disconnect reason, or number of input and output octets. It will send this to

Figure 6.3 Accounting messages in RADIUS

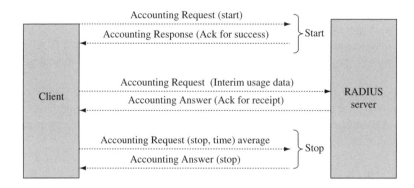

Figure 6.4 Message exchange sequence during a full accounting session

the RADIUS accounting server, which will send back an acknowledgement that the packet has been received. Of course, if the RADIUS accounting server is unable to successfully record, the accounting packet does not send an Accounting-Response acknowledgment to the client.

The RADIUS accounting specification introduces a number of new attributes, some of the more important of which are mentioned in Table 6.6. Note that as base RADIUS, RADIUS accounting is also an extensible protocol.

6.3.4.2 Security and Reliability of RADIUS Accounting

Think of your cellular phone bill statement when you think of RADIUS accounting. Do you like to see charges related to phone calls you never made or see entries that state your phone call was two and a half hours instead of 23 minutes? Now think of your operator for a second, even though most of us don't want to. Does the operator like to find out that people who have been making one-hour phone calls have received bills that have not listed the charges for these calls? Now you see why reliability and security for RADIUS accounting messaging are

Table 6.6 Examples of RADIUS Accounting Attributes

Attribute name	Description
Acct-Status-Type	When included in an accounting request, indicates whether the corresponding request marks the beginning of the user service (start) or the end (stop)
Acct-Input-Octets Acct-Output-Octets Acct-Input-Packets Acct-Output-Packets	These attributes are used to indicate the number of incoming and outgoing octets/packets transmitted through a specific client port during a session
Acct-Session-Time	Sent at the end of the session along with an accounting request to indicate how many seconds the user has received service for
Acct-Authentic	May be included in an accounting request to indicate how the user was authenticated, whether by RADIUS, the NAS itself, or another remote authentication protocol

so important. How could the operators prove their case without having reliable and provable records of the user's sessions?

Because RADIUS is transported by UDP, reliability in data transfer is not guaranteed. So it is possible that the accounting packets are lost in the network. Loss of RADIUS Accounting-Stops creates several problems as follows:

- Incomplete and incorrect billing information for metered services results in lost revenue or customer disputes that cannot be resolved.
- Some operator networks may implement a limit on the number of simultaneous sessions a user can run at any given time. If accounting packets carrying the stop indication for some previously ended sessions are not received by the server, the user will be out of luck when trying to start a new session.

RADIUS accounting RFC has taken an easy route for reliability and simply recommends that the client continues sending Accounting Requests until an acknowledgement is received. By including configurable parameters such as retransmission-intervals (back-off mechanisms) and the number of retransmissions in the NAS, and configuring those parameters more efficiently, some of those problems may be fixed. For instance, increasing the retry limit and overall time (timeout) allowed for retries improves the chances of a request getting through. Additionally, by waiting longer to retransmit, the NAS is not contributing as much to the congestion.

When it comes to security, RADIUS accounting is slightly better than the base RADIUS specification: both Accounting Requests and Accounting Responses are authenticated using an MD5 hash function and the shared secret configured between the client and the server. For the Accounting Requests, the hash function is performed as follows:

$$\text{Authenticator value} = \text{MD5 (Code, ID, Length, 16 zero octets, request attributes, shared secret)}$$

The 16-octet MD5 output is included in the Authenticator (called the Request Authenticator) field of the Accounting Request packet.

For Accounting Response packets, the value in the authenticator field (called the Response Authenticator) in the packet is calculated as follows:

$$\text{Authenticator value} = \text{MD5 (Code, ID, Length, Request Authenticator, shared secret)}$$

Where the Request Authenticator is extracted from the corresponding Accounting Request packet.

6.4 RADIUS Support for Roaming and Mobility

When it comes to support for mobility and multi-domain operation, RADIUS specification is very primitive. As we will see in Chapter 8, RADIUS specification provided by IETF does not provide any support for Mobile IP functionality, at least not at the time of writing. Most support for user mobility is in terms of support for roaming applications and is based on the work done by the "Roaming Operations" working group in IETF [ROAMOPSWEB], shortly named ROAMOPS. In the following subsection we will briefly show some of the highlights of their work.

6.4.1 RADIUS Support for Proxy Chaining

When it comes to support for roaming within RADIUS, one of the most relevant specifications that the ROAMOPS group generated is RFC 2607 [PROX2607], which defines the procedures for proxy chaining. For our discussion we can define proxy chaining as the procedures needed to forward the AAA packets between an NAS device and a home RADIUS server through a series of proxies when the user is roaming within a foreign domain.

The definition of a proxy in RADIUS is, on the other hand, very brief and its operation can only be understood through drawing inferences from the proxy chaining RFCs. We will see in Chapter 7 how Diameter takes great care in defining the terminology and the roles of various Diameter nodes, such as relay agents and proxies. Anyhow, [PROX2607] defines the RADIUS proxy as a node that can be employed to provide routing of authentication and accounting messages between a NAS and a RADIUS server. The proxy acts as a RADIUS server when dealing with a NAS, but acts as a RADIUS client when dealing with a RADIUS server.

6.4.1.1 Roaming Concepts

Roaming typically means that a user normally associated with a home operator is now somewhere else and requests service from a network that is operated and possibly owned by another operator. Obvious though it may be, people seem to forget that no roaming is possible unless a relationship exists between the two operators. Roaming relationships are based on roaming agreements and resemble trust relationships in a security framework in many ways:

- A roaming relationship can take the form of a roaming association between two peer-service providers/companies.
- A roaming relationship can exist between a service provider and a "roaming consortium".

Roaming relationship path: One common case for such relationships is shown by Figure 6.5. In this case, there is a series of proxies lined on the path between a local authentication proxy and the home authentication proxy. The set of relationships along this path is called a roaming relationship path. Each proxy acts as the server to the peer on one side and as a client to the peer on the other side. This means that each proxy shares a secret with the peers on either side.

Central proxy: As opposed to the serial configurations discussed in the previous case, in a large mesh of roaming partners (say, 100 partners), it was deemed to be more scalable, if a hierarchical forwarding model were deployed, in which a central proxy would route the requests to their destinations. The shared secret management would be more scalable, if each

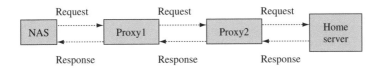

Figure 6.5 Proxy-chaining in RADIUS

partner only shared a unique secret with the central proxy (total of 100 secrets) as opposed to each partner pair sharing a secret (around 4,950 secrets in total). At the same time, fewer bilateral agreements would be required.

6.4.1.2 Proxy Chaining Operation

Figure 6.5 shows the concept of proxy chaining. The original AAA request is generated by the NAS and is sent towards Proxy1. Proxy1 examines the request and may forward the request as it stands to another proxy (Proxy2 in Figure 6.5). Proxy2 will then forward the request to the user's home server. Both Proxy1 and Proxy2 may find it necessary to modify the attributes in the packet after implementing some local policies. It is even possible that Proxy1 or Proxy2 might simply reject the request and send a AAA reject message to the previous proxy or to the NAS. The proxy may even decide to challenge the previous request sender. The proxies are allowed to hide some information in the packets, when needed. These functionalities were provided to compensate for the fact that RADIUS does not support capability negotiations between a NAS and a server that accommodate different feature sets.

6.4.1.3 Issues with Proxy Chaining

The fact that the RFC [PROX2607] is of informational type suggests that the implementation of procedures is not widely deployed in a standard manner. The roaming relationship path, i.e. the path to the next proxy towards the home server, is determined by the network access identifier (NAI). However, most of the RFCs do not specify the routing procedure along the roaming relationship path when using RADIUS. Diameter implements Realm routing tables and peer tables for this purpose.

Proxies can modify request and response messages without providing any notifications or signatures, unless some sort of authentication is required for the RADIUS message or the attributes. This not only means incompatibility between user requests and provided services, but also opens the doors for man-in-the-middle attacks by external parties and fraud, by involving parties without any possibility for detection by the end nodes. Even when the authentication is provided, it is based on the shared secrets between the neighboring proxies, which means the trust between the NAS and the home server only becomes transitive. Furthermore, end-to-end hiding of attributes cannot be supported this way. A large section in the RFC is devoted to describing the security vulnerability of RADIUS proxy chaining.

6.5 RADIUS Issues

With the recent increased attention to an integrated approach to network architecture design that considers many aspects of the network, such as mobility, QoS, call control and security, all at the same time, the role of AAA servers is becoming more and more central. This not only requires more interaction between the AAA servers and other entities in the network, but also puts more strain on the AAA servers and AAA protocols to sustain the new reliability, mobility and security demands. As we mentioned earlier, RADIUS suffers from a long list of security and reliability issues. Furthermore, RADIUS lacks support for IP mobility protocols. On top of all these, the number of allowed attribute types is limited to less than 255 and the length of the attribute value field is limited. This, along with the limited RADIUS message set, reduces the applicability of RADIUS as a AAA protocol of the future.

As we will see in the next chapter, Diameter as a successor protocol to RADIUS overcomes many of the RADIUS shortcomings. For that reason IETF decided to conclude the work of RADIUS standardization a long time ago [RADIUSWEB]. However, due to current economic downturn and the wide deployment base of RADIUS, the need was felt for a number of RADIUS enhancements before Diameter takes off as a widespread AAA protocol. This led to the creation of a new IETF working group, called RADIUS Extensions, or shortly named RADEXT [RADEXTWEB] for short. RADEXT has a very aggressive charter and has indicated early on that it does not intend to work on resolving RADIUS security or reliability issues. Nor does it seek to extend the current RADIUS message set. It seems that most of the current work revolves around standardizing new attributes that extend RADIUS functionality, while paying great attention to preserving backward compatibility with the existing deployment base. Compatibility with Diameter is also taken as seriously as possible. Among other things, the group has been examining interactions of RADIUS servers with SIP proxies for accounting and message authentication, with WLAN edge devices for support of WLAN link provisioning. Also the group is looking at revising specifications for network access identifiers (NAI) and applications that allow the users to prepay services.

6.6 Further Resources

The original IETF working group for RADIUS concluded its work back in July 2000, after producing and revising many standard track and informational RFC specifications. Some of these specifications have been mentioned in the text so far and some are listed below:

- RFC 2548 Microsoft Vendor-specific RADIUS Attributes (Informational)
- RFC 2809 Implementation of L2TP Compulsory Tunneling via RADIUS (Informational)
- RFC 2869 RADIUS Extensions (Informational)
- RFC 3576 Dynamic Authorization Extensions to RADIUS (Informational)

The reader is also strongly encouraged to follow the work of the RADEXT working group for more information on the progress of RADIUS specifications.

6.6.1 Commercial RADIUS Resources

Commercial RADIUS servers vary in price and capability as well as type of platform. Both software only and software/hardware combo platforms can be found. For instance, Funk Odyssey server, which includes 25 software licenses can run below $3000, while more sophisticated steel-belted servers from the same company are priced over $7000. RADIUS server providers are abundant, many of which support a wide range of network providers, such as ISPs, Wi-Fi, GPRS, and CDMA providers with a wide range of features, such as EAP/LEAP/MS-CHAP/accounting and interaction with a variety of mail and web servers and databases.

To name, but a few RADIUS server providers: Aradial, Bridgewater Systems, Cisco, Funk, IEA, and Interlink.

6.6.2 Free Open Source Material

The best-known open source implementation of RADIUS is FreeRADIUS software provided for Linux, Solaris and FreeBSD [FREERADIUS], and it enjoys popularity among a large range of users. It supports PAP, CHAP, and EAP with MD5, TLS, TTLS, and PEAP and interacts with a variety of open source and commercial databases such as open LDAP and Oracle. Since FreeRADIUS is really what it says it is (free!), the accompanying documentation can be very thin for beginners. However, a combination of the FAQs provided on the project website along with helpful guides in [HASSELL] could help such users on their way to RADIUS mastery.

It should be noted that FreeRADIUS only provides software for the RADIUS server and the user needs to provide her own client platform (most WLAN APs support RADIUS) or software. Fortunately, client software including user-friendly graphic user interfaces is provided for both Windows (NTRadPing) and Linux users (Radloginv4), courtesy of Mastersoft and IEA software, respectively.

6.7 References

1. [RADIUS2865], C. Rigney et al., "Remote Authentication Dial In User Service (RADIUS)", IETF Draft Standard, RFC 2865, June 2000.
2. [RADACC2866], C. Rigney, "RADIUS Accounting", IETF Informational RFC, RFC 2866, June 2000.
3. [RADEXT2869], C. Rigney, "RADIUS Extensions", IETF Information RFC, RFC 2869, June 2000.
4. [RADTUN2868], G. Zorn et al., "RADIUS Attributes for Tunnel Protocol Support", IETF Informational RFC, RFC 2868, June 2000.
5. [RADEAP3579], B. Aboba, and P. Calhoun, RADIUS (Remote Authentication Dial In User Service) Support for Extensible Authentication Protocol (EAP), RFC 3579, September 2003.
6. [ROAMOPSWEB], Roaming Operations Working group website URL, http://www.ietf.org/html.charters/OLD/roamops-charter.html, IETF WG, concluded January 2001.
7. [PROX2607], B. Aboba, and J. Vollbrecht, "Proxy Chaining and Policy Implementation in Roaming", IETF, RFC 2607, June 1999.
8. [RADEXTWEB], IETF RADIUS extension working group, Web URL, http://ietf.org/html.charters/radext-charter.html.
9. [RADIUSWEB], RADIUS working group website URL, http://www.ietf.org/html.charters/OLD/radius-charter.html, IETF, concluded July 2000.
10. [FREERADIUS], FreeRADIUS Server Project, http://www.freeradius.org/.
11. [HASSELL], J. Hassell, "RADIUS", O'Reilly, ISBN 0-596-00322-6, 2003.
12. [RAD3576], M. Chiba et al., "Dynamic Authorization Extensions to Remote Authentication Dial In User Service (RADIUS)", IETF, RFC 3576, July 2003.

7

Diameter: Twice the RADIUS?

7.1 Election for the Next AAA Protocol

Around the same time when the IETF RADIUS working group decided that it had concluded its work (first half of 2000), a new IETF working group called AAA working group had just started its work on finding the next big successor to RADIUS. Since, at the time several protocols were butting heads as contenders, the group decided to conduct a thorough comparison of the protocol proposals to make sure that the successful candidate was ready for prime time. The working group first created an RFC specification [AAAEVAL2989], defining a complete set of requirements for a protocol to serve as a AAA protocol. The group then assigned a team of experts to assess the suitability of each candidate proposal by evaluating how the proposal could meet these requirements.

The requirement defined in that RFC (2989) are similar to the requirements defined for support of network access server in RFC 3169 [NASCRIT3169]. We provided a list of the most important requirements mentioned in that RFC at the end of Chapter 1. However, RFC 2989 [AAAEVAL2989] provided a more complete set of requirements for overall evaluation of AAA protocols, that went beyond just supporting NASes. In summary, examples of these requirements are:

> Scalability, failover, mutual authentication between client and server, transmission level security, data object confidentiality, data object integrity, certificate transport, reliable AAA transport mechanisms, ability to run over IPv4, ability to run over IPv6, support for proxy and routing brokers, audibility, and ability to carry service-specific attributes.

The working group then asked supporters of each candidate protocol to submit proposals on their protocol in a way that best met these requirements. Four major contenders emerged: SNMP (Simple Network Management Protocol), RADIUS, COPS (Common Open Policy Service Protocol) and Diameter. However, since RADIUS seemed to need enhancements to be able to meet the requirements, it was enhanced into an upgraded protocol called RADIUS++

(or RADIUS v2). A list of the submitted documents, describing these proposals can be found in section 2 of RFC 3127.

Having those documents and the requirements in hand, the expert team studied the candidates mentioned earlier and presented its recommendations at the Pittsburg IETF meeting in August 2000 [PITTPROC] and published the results in RFC 3127. The interested reader is referred to the RFC 3127 [EVALRES3127] for the detailed version of this evaluation or to [PITTPROC] for the meeting notes. Here we only provide the final result of the evaluation:

- The SNMP-based proposal was deemed not to provide precise specifications for the use of SNMP for authentication and was hence was judged unacceptable, even though SNMP was deemed adequate for accounting.
- The proposal describing RADIUS++ was deemed too incomplete. Too much effort would be required to bring RADIUS++ to a level that it could meet requirements and the prediction was that the result was going to be very similar to Diameter, while RADIUS++ did not offer adequate backward compatibility with RADIUS.
- The COPS proposal was judged acceptable as a AAA protocol but the group felt such use would give COPS a multiple personality, since it is also used for QoS purposes. Using COPS as a AAA protocol would mean the firewalls would have to search inside each COPS packet to see if it was used for QoS signaling or AAA signaling and filter the packet accordingly. Not upgrading the firewalls could mean that an external party who otherwise should only be allowed to send AAA request messages, could set the QoS parameters for the routers inside the network. This was deemed to create difficult-to-accept implications for firewalls.
- The Diameter proposal was judged as acceptable but only slightly preferred to COPS, due to Diameter's better-prepared engineering and firewall-friendliness. However, requirement for support of SCTP as a transport protocol at firewalls posed a concern for the adoption of Diameter, since SCTP was considered to have a much smaller and slower deployment than TCP.

As the result of this evaluation, Diameter was declared the winner and the AAA working group decided to focus its efforts on further enhancement and standardization of Diameter.

7.1.1 The Web of Diameter Specifications

In the previous chapter we were complaining that, while RADIUS base specification defines support for many authentication mechanisms, it leaves the specification of accounting methods to another RFC. As shown in Figure 7.1, the relationships and dependencies between Diameter specification documents are much more complex, or less mildly put, peculiar to the eyes of uninitiated.

The Diameter base protocol was standardized (draft standard) as recently as September 2003. However, many of the important usages of the Diameter (applications) are defined in separate documents that often are still work in progress. In this section we intend to provide an overview of how the various specifications are tied to each other.

7.1.1.1 Diameter Base Specification

This specification [DIMETER3588] defines most of the Diameter basic building elements, such as a basic set of messages, attributes and the attribute structure. The base specification

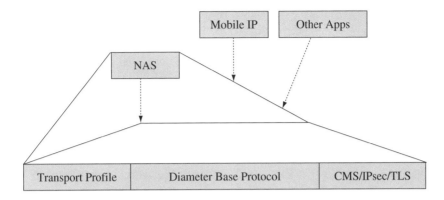

Figure 7.1 Interdependency of Diameter specifications

also goes to great lengths (compared to RADIUS) in defining inter-realm operations. For instance, it clearly defines the role of various types of Diameter agents in processing and routing of Diameter messages. Diameter base specification also spends a great deal of its bandwidth on explaining message transport concepts.

One thing that is new with Diameter is that the Diameter base specification defines the concept of applications. Diameter applications are services, protocols, and procedures that use the facilities provided by the Diameter servers, proxies and the Diameter protocol itself. A good example of a protocol using Diameter is Mobile IP. All Diameter applications must support the functionality specified in Diameter base specification (this explains why Diameter base protocol is at the bottom of the pyramid in Figure 7.1).

Possibly the most jaw-dropping surprise to a person coming from the RADIUS-planet is that dealing with NAS for authentication and authorization purposes is considered an application for Diameter. Despite all the volume and feel of completeness offered by Diameter base specification, no details on simple authentication and authorization procedures or even on interaction with NASes are provided in that specification. The interaction of Diameter with NAS for providing authentications is defined in a separate specification (NAS application). Also when it comes to authorization, the Diameter base protocol assumes that authorization request messages (including those for access authorization) are application specific and need to be defined by other application-specification documents. The Diameter base specification goes only as far as saying that the Diameter client is the entity issuing an authentication and/or authorization request on behalf of the user to the server, but does not offer any detail on the messaging procedure or the messages themselves. Oddly enough, Diameter base specification does define messages for terminating user sessions.

In contrast to RADIUS, Diameter base specification defines accounting methods and messages. The reason may be that it is required of all Diameter applications to support accounting procedures.

7.1.1.2 Security Specifications

The Diameter base specification also specifies end-to-end as well as hop-by-hop security mechanisms. Unfortunately transport layer security protocols, such as TLS or network layer security protocols, such as IPsec (see Chapter 4) would provide security for the entire Diameter message between the two nodes. However, as we explained in the previous chapter (see attribute hiding discussion in Chapter 6), often it is important to protect only specific data objects from some dubious intermediaries, while the message itself may have to be examined or processed by these intermediaries. Roaming scenarios, where Diameter signaling has to go through foreign-owned Diameter proxies, are perfect examples of this. To provide end-to-end data object security an additional security mechanism, called Cryptographic Message Syntax security (CMS) [CMS3370] module was suggested. Unfortunately, even though the specification describing the use of CMS with Diameter [DICMSDR] was pursued by the IETF AAA working group for a while and survived 4 revisions, it is no longer an active working group item. At the time of writing it is doubtful that this specification will be pursued by the group any longer.

7.1.1.3 Diameter Transport Profile

In the previous chapter, we discussed the issues with lack of reliability for RADIUS messaging due to its use of UDP. Diameter designers took reliability much more seriously and went as far as stating that Diameter specification consists of the base specification [DIMETER3588] and the AAA transport profile RFC 3539 document [AAATR3539]. This describes two of bottom blocks in the Diameter specification pyramid, shown in Figure 7.1. Oddly, the standards track RFC 3539 looks more like a recommendation document than a standards specification document. This document discusses issues of the interaction between AAA protocols and the transport layer and provides recommendations on how to deal with those issues. The only firm specification provided covers fail-over mechanisms. Still, there are several subtle inter-dependencies between Diameter base specification RFC and the AAA transport profile RFC that justify inclusion of the latter as a required Diameter specification:

- The Diameter base protocol specification defines the transport connection between Diameter nodes as a TCP or SCTP connection and puts great care on defining the transport layer concepts for Diameter. On the other hand, the transport profile document requires that Diameter agents and server must support both TCP and SCTP, while the client MUST support either TCP or SCTP.
- The transport profile document recommends that the AAA implementations make use of standards track and experimental techniques to deal with the transport issues described in that document.

The bottom line is that Diameter requires the use of reliable transport protocol. We will go through the Diameter transport concepts and requirements in more detail later on in this chapter.

So far, we have covered all the bottom blocks in the pyramid in Figure 7.1. All Diameter applications must be implemented together with the Diameter base protocol, Diameter transport profile and the proper security mechanisms defined for Diameter.

7.1.1.4 Diameter NAS Application

As mentioned earlier, the dealings of Diameter servers with NASes are considered as applications of the Diameter protocol and therefore are defined in a separate application specification document. However, due to the central nature of AAA servers, and the enforcing nature of the NASes as edge devices, almost all applications and services that deal with AAA servers need to do so through an NAS. That would mean the implementation of these applications must also support the Diameter NAS application specification. Thus Diameter NAS application procedures are fundamental for functionality of other Diameter applications. This should explain the awkward hump in the Diameter specification pyramid shown in Figure 7.1. Unfortunately all these intricate relations make understanding the lengthy and complicated Diameter specifications even more difficult.

7.1.2 Diameter Applications

As mentioned earlier, Diameter base specification only describes the support for accounting, while other protocols and services that use Diameter or Diameter servers are considered as applications for Diameter and are described in separate application-specific documents.

It can safely be assumed that not every Diameter node deployed in a Diameter infrastructure will support all the Diameter applications out there. Therefore, when two Diameter nodes intend to interact with each other on behalf of a Diameter application, each node needs to make sure that the other actually does support the said application. A Diameter feature called capability negotiation provides this assurance. We will provide more details on the messaging required for capability negotiation later. For now, suffice to say that to facilitate the capability negotiation, each Diameter application is assigned a standard unique application identifier by IANA, so that by passing the supported application IDs to the other party, each party can indicate what applications it supports. Since support for the Diameter base protocol is mandatory for all Diameter nodes, the Diameter base does not require its own application ID. Table 7.1 shows a list of application identifiers, defined in [DIMETER3588]:

In the following we provide a brief overview of some of these applications and leave the more detailed description for later.

- *Diameter accounting*: This is the only Diameter application whose specification is provided in the Diameter base protocol specification. We provide slightly more details on Diameter accounting later on in this chapter.
- *Diameter credit control application*: Due to the recently gained popularity of use of pre-paid cards for cellular network services, support of the pre-paid model is becoming an increasingly important practical issue for AAA infrastructures. Examples of services requested by end users are: messaging service, download service, and many other services controlled by

Table 7.1 Diameter application identifiers

Application	Application ID
Diameter common messages	0
NASREQ	1
Mobile IP	2
Diameter Base Accounting	3
Relay	0xffffffff
Diameter SIP application	To be determined by IANA

Session Initiation Protocol (SIP). The service provider must be able to determine the allowed coverage for each of the requested services prior to service initiation and cease the initiated service, once the applied credits are exhausted. The credit control application provides methods for real-time credit control for many services offered by the cellular networks to the end users. This application defines the interaction between the entity providing the service and the so-called credit-control server. The application specification is currently a work in progress [DIACCDR].

- *Diameter NAS application*: This application describes the details of the interaction of the Diameter servers with network access servers (NAS) for authentication and other procedures. These procedures are defined in [NASREQDR]. The NAS application would be the closest Diameter specification to RADIUS base specification, as RADIUS was also originally designed to provide authentication services to PPP users. The NASREQ application allows the NAS to support a variety of authentication mechanisms, such as PAP, CHAP and EAP. NASREQ supports RADIUS attributes as often as possible to provide backward compatibility with RADIUS. The NASREQ application is described in greater detail later on in this chapter.
- *Diameter Mobile IP*: While Mobile IP facilitates routing and transport of data for communications to and from a mobile host, Diameter, among other things, facilitates identity verification and authorization mechanisms for end hosts. The IETF Mobile IP community has also defined a number of security and key management mechanisms that are facilitated by AAA protocols, but has left the specification of the AAA mechanisms to support those mechanisms to the AAA community. While RADIUS does not provide any specific guidance for Mobile IP applications, Diameter Mobile IPv4 application specification [DIAMIPDR] closes the gap in a rather complete manner for Mobile IPv4. This application furthermore allows for mobility across administrative domains. We will come back to this in a brief segment of this chapter, but devote a lengthy part of Chapter 8 to a detailed description of Diameter Mobile IPv4 application.
- *Diameter EAP* is another Diameter application defining procedures for carrying EAP exchanges over Diameter messages between the NAS and Diameter servers. We will come back to this later in this chapter.

7.1.3 Diameter Node Types and their Roles

Before we venture into the detailed description of Diameter protocol, it is useful to go through the roles of various entities within a Diameter infrastructure. In order to make sure the reliability, security and routing behavior of Diameter functions are well specified and predictable, Diameter base specification categorizes the involved entities based on the roles they play in Diameter messaging:

- *Diameter Node*: A host process that implements the Diameter protocol.
- *Diameter Peer*: A diameter node that has a direct transport connection with another diameter node.
- *Diameter Client*: A device at the edge of the network that performs access control. Examples are NASes or Mobile IP foreign agents. Note, the end user, wishing to access the network is not the Diameter client, since the end user does not participate in Diameter signaling.

- *Diameter server*: A server is the device that handles AAA requests (authentication, authorization and accounting requests) for a particular realm. The server must support Diameter base protocol and additional Diameter applications utilized in the realm.
- *Diameter Agent*: The agent is a Diameter node that provides relay, proxy, redirect or translation services.
- Relay agents forward Diameter messages based on routing-related attributes with the message and special routing tables. Relay agents do not make any policy decisions and therefore do not examine any non-routing attributes within the messages. For that reason, these agents need to understand the semantics of routing-related attributes only. A relay never originates a message, but is capable of handling any Diameter application or message types.
- Proxy agents can be seen as relay agents that can also make policy decisions. They can track various states at the NAS device for resource provisioning purposes. Proxies typically do not respond to the client requests, but may originate Reject messages in cases where policies are violated.
- Redirect agents do not sit on the forwarding paths and do not alter any message attributes. So, they do not forward any messages, as relay agents do, but refer the client to a server by redirecting the messages according to a configuration. They are able to handle any message types but may be configured to only redirect certain message types.
- Translation agents perform protocol translations between Diameter and other AAA protocols, such as RADIUS. They are specially considered for backward compatibility with RADIUS.
- A broker is more of a business term than a technical one and therefore depending on the business model can take any of following roles: relay, proxy or redirect agent, but it does not act as a server.

7.2 Diameter Protocol

In this section, we will go through the basic building blocks of the Diameter base specification. We will describe the Diameter message format and the way different messages are distinguished and carry information around. We also go through the transport, reliability and security procedures of Diameter messaging.

7.2.1 Diameter Messages

As in RADIUS, Diameter messages are used for carrying various pieces of application or AAA information. The information, carried within a Diameter message, is typically called an attribute and is formed according to the attribute-value pair (AVP) format. For that reason Diameter attributes are frequently referred to as AVPs. The format for AVPs is described later on.

RADIUS is a server–client protocol, where requests are always issued by the client, while answers (challenge or accept/reject) are created by the server. In contrast, Diameter is a peer-to-peer protocol, which means both client and server can create either a request or an answer. Diameter messaging terminology does not include "message type", so one can say there are only two message types: requests and answers. Instead Diameter uses the concept of commands. The commands are distinguished from each through the use of a command code that specifies the type of function the Diameter message intends to perform. The action to be taken in conjunction with reception of each message is defined by the command code and the attributes included in the message. This will become clear as we go through a description of various commands and attributes.

7.2.1.1 Diameter Message Format

As shown in Table 7.2, Diameter messages consist of a standard header, and a number of AVPs. In Table 7.2 the header fields are shown in gray.

The header fields are as described in the following

- The Version field indicates the Diameter protocol version and is set to 1 for now.
- The Command flags field specifies 4 flags for now:
  - R flag (stands for Request) flag shows whether the message is a request or a response.
  - P flag (stands for Proxiable) shows if the message can be proxied, relayed or redirected or it must be locally processed.
  - E flag (stands for Error) to show if the message contains protocol or semantic errors.
  - T flag to show that a message can potentially be a retransmitted message after a link fail-over or is used to aid removal of duplicate messages.
- The command code value indicates the command associated with the message, such as "abort session request" or "accounting-answer", and so on. Every Diameter message must contain a command code so that the receiver can determine what action it needs to take for each message.
- Application ID identifies the specific application the message is used for, such as Mobile IP, accounting and so on.
- Hop-by-hop identifier field carries an identifier that is used to match request and responses over that hop. The sender of the request must ensure that the identifier is unique over the connection on that hop at any given time. The sender of a response must ensure that the identifier value is the same as that in the corresponding request.
- End-to-end identifier is an identifier used to detect duplicate messages. The identifier in a response message must match the identifier in the corresponding request message. The identifier must remain locally unique for at least 4 minutes. This identifier and the Origin-Host AVP (described later) are used together to detect message duplicates. Note duplicate request could cause duplicate responses but the duplications must not affect any states that were created by the original request.

7.2.1.2 Diameter Command Code (Message Types)

As mentioned earlier, the action to be taken upon reception of each message is defined by the command code in the Diameter message. As described previously, both request and answer

Table 7.2 Diameter message format and header

0 1 2 3 4 5 6 7	0 1 2 3 4 5 6 7 0 1 2 3 4 5 6 7 0 1 2 3 4 5 6 7
Version	Message length
Command flags	Command code
Application ID	
Hop-by-hop identifier	
End-to-end Identifier	
AVPs....	

Table 7.3 Common Diameter command codes

Command name(s)	Abbreviation	Code	Description
Capabilities-Exchange Request/Response	CER/CEA	257	Allows for discovery of a peer's identity and its capabilities, such as supported Diameter applications, security mechanisms, etc.
Disconnect-Peer-Request/Answer	DPR/DPA	282	Sent to a peer to inform sender's intentions to shut down the transport connection
Re-Auth-Request/Answer	RAR/RAA	258	Sent by a server to an access device, providing session service to user, to request the access device to re-authenticate and/or re-authorize the user
Session-Termination request/Answer	STR/STA	275	Sent by the access device that provided the service, on behalf of a user that is terminating a session, to the server that had authorized that session.
Accounting Request/Answer	ACR/ACA	271	Sent by a Diameter client to exchange accounting information with its peer.

messages corresponding to a command type use the same command code, but the request messages set the R flag within the message header. Typically an answer message includes a Result-Code AVP, indicating the result of the processing of the request at the receiver. Examples of some of the more relevant command codes that are defined in Diameter base specification are shown in Table 7.3.

Obviously, Table 7.3 does not include all the command codes defined in the Diameter base protocol specification. Furthermore, as mentioned earlier, the base specification leaves the definition of application-specific commands to other documents. For instance, authentication or authorization request/answer commands are defined in NASREQ specification. In the same way, other applications such as Mobile IP or credit control define additional command codes as needed for their operation. We will give examples of those commands when we describe each of these applications.

It is a requirement that each Diameter node must support all the command codes described in the Diameter base protocol, but only supports the command codes for the Diameter application it has advertised provide.

7.2.1.3 Attribute-Value Pair (AVP) Format

As mentioned earlier, most of the information to be carried by Diameter in the form of attributes are formatted as attribute-value pairs, commonly known as AVPs. AVPs can arbitrarily be added to messages, as long as the minimum required AVPs for a message are included and no AVP that is specifically excluded for inclusion in that message is added. The AVP format is shown in Table 7.4 with the AVP header in gray:

In the following we do not go through the description of every field, as that would be repeating the RFC specification. Instead we go through the fields important for our discussions in this chapter:

Table 7.4 Diameter AVP format

0 1 2 3 4 5 6 7 0 1 2 3 4 5 6 7 0 1 2 3 4 5 6 7 0 1 2 3 4 5 6 7	
AVP code	
Flags (VMPRRRRR)	AVP length
Vendor ID (optional)	
Attribute data	

- The AVP code identifies the type of information (attribute) included in the attribute data field of the AVP. AVP code values are standardized by IETF and controlled by Internet Assigned Number Authority (IANA) [IANAWEB]. New applications should try using the existing AVPs to the extent possible. Diameter saves AVP codes 0-255 for backward compatibility with attributes defined by RADIUS.
- Flags
  - "M" bit: The M bit within the flag fields of the AVP is called the Diameter "mandatory" bit and its purpose is to indicate whether a Diameter node requires its peer to support the attribute to process the message. If a Diameter node receives an AVP with an M bit set, but does not recognize the AVP or its value, it must reject the message carrying the AVP. The M bit is set according to the rules defined for that AVP.
  - "P" bit: The P bit indicates the need for encryption for end-to-end security. Diameter base protocol [DIMETER3588] specifies which AVPs must be protected by end-to-end security measures (encryption) if the message is to pass through a Diameter agent. If a message includes any of those AVPs, it must not be sent unless there is end-to-end security between the originator and recipient of the message.
  - RRRRR bits: Indicate existence of 5 reserved bits in the flags field.

7.2.1.4 Examples of Diameter Base Specification AVPs

Diameter base protocol and its applications define a large number of AVPs. Below, we go through a few Diameter base protocol AVPs that are required for the understanding of the discussions in this chapter:

- *Origin-Host AVP*: This attribute is added by the end point that originates a Diameter message and cannot be modified by relay agents. It must be present in all Diameter messages and is of type DiameterIdentity. It can resolve to more than one address, if the Diameter peer is using more than one address.
- *Origin-Realm AVP*: This attribute contains the realm of the Diameter message originator and must be present in all Diameter messages. Relay agents must not change this AVP.
- *Destination-Host AVP*: Destination-Host AVP is typically used to route a message to the user's home server or home domain and is used when the destination of the request is fixed, such as the following examples:
  - When the security mechanism uses a pre-established session key that is shared between the source and the final destination of the message.
  - When authentication exchanges that span multiple round trips are being performed.
  - For server-initiated messages (such as Abort session requests) that must reach a specific client.

An agent can forward a request to a host, described in the Destination-Host AVP, only if the host in question is included in its peer table (explained later on). This AVP must be present in all unsolicited agent-initiated messages and may be present in request messages but not in answer messages. An agent-initiated message is what it sounds like, a message that is initiated by that agent rather than one that is simply forwarded by the agent. Absence of this AVP will cause the message to be sent to any Diameter server supporting the application within the realm specified by the Destination-Realm AVP (below).

- *Destination-Realm AVP*: Contains the realm that the message is to be routed to. Again this attribute cannot be present in answer messages. This AVP is used to perform message routing decisions. Diameter clients insert the realm portion of the User-name AVP in their requests. Diameter servers, when sending requests, use the value of the Origin Realm AVP from a previous message received from the intended target host.
- *Routing AVPs*: Are used for routing purposes. Diameter agents need to process and change these AVPs and for that reason, these AVPs must not be protected by end-to-end security. We will discuss the routing of Diameter messages in the next section.
- *Result-Code AVP*: Indicates whether a particular request was completed successfully or not. All Diameter answer messages must include one Result-Code. The error codes, providing information on the kind of problem occurred during the processing of the request, are categorized in different classes, such as protocol errors, and transient versus permanent failures.

7.2.2 Diameter Transport and Routing Concepts

As mentioned earlier, Diameter designers defined the role of intermediary Diameter agents and proxies very carefully not to only allow for multi-domain operations but also to fix the reliability and transport issues experienced with RADIUS. In this section we will go through some of these concepts briefly.

7.2.2.1 Diameter Transport Concepts

Transport and reliability support is important part of Diameter design: the specification mandates the Diameter clients to support either TCP or SCTP, while requiring the agents and servers to support both TCP and SCTP. Due to the complexity of transport mechanisms such as fail-over, congestion control and multi-interface mechanisms, we do not go into the details of the transport requirements and recommendations for Diameter and refer the reader to the transport profile specification [AAATR3539]. Here we suffice with providing some of main concepts and their implications of Diameter messaging.

Diameter specifications define two important transport concepts, namely "session" and "connection". Understanding the distinctions made between these two concepts would help the reader to better understand the transport and routing design in Diameter specifications (Figure 7.2).

- Session is a logical concept at the application layer and is established end to end between a device and a server. A session is processed by end parties (device or server) and therefore is identified by a session ID AVP.
- Connection is, on the other hand, a transport level concept and is established between any two peers that send and receive Diameter messages.

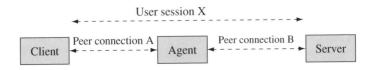

Figure 7.2 Connections versus sessions in Diameter

Note that the connection and the session may not have a direct one-to-one relationship. For instance, a user's authentication request creates a unique session, while the data for this session can be multiplexed with the data for many other sessions when carried over a single connection between each pair of peers on the path between the two end parties. On the other hand, the use of TCP or SCTP over each hop allows for detection and repair for transport failures at local level (hop by hop) at intermediary proxies. Lost packets can simply be retransmitted over that hop and avoid longer trips.

7.2.2.2 Diameter Routing Concepts

As we saw in the previous chapter, RADIUS base specifications do not provide much guidance on the operation of RADIUS proxies in multi-domain environments and roaming scenarios. The process of routing for RADIUS packets is not always clear from the RADIUS specifications. Sometimes specifications made by other standards groups (such as the roaming operation) need to be consulted. Diameter takes a long leap from that standpoint and defines the routing procedures at the Diameter nodes very clearly. We went through the definition of various Diameter nodes and agents earlier. Here we will first go through some of the major routing-related concepts and tools as defined in Diameter base specification to help the reader understand how Diameter nodes use these tools to forward Diameter messages.

- *Peer table*: The Diameter peer table is used in message forwarding and is referenced by the realm routing table (described below). Each Diameter node keeps a peer table including entries on each of its peers. Each entry in the peer table includes information on the identity of the peer, useful state information, whether the peer entry is statically configured or dynamically discovered and the expiration time for the entry for this peer in the table (if the peer was dynamically discovered).
- *Realm-based Routing table*: Diameter agents consult their realm-based routing tables to forward the message towards the destination or next appropriate "AAA hop" that may reside in other realms. The table contains the following fields:
  - Realm name field: This is similar to the network prefix in IP routing and is the primary key for table look-up, however, some application only require longest match rather than exact match.
  - Application identifier fields: This is similar to host identifier in IP routing, in the sense that it is the secondary key for table look-up. The reason is that the same entry may have different destinations depending on the type of application in the session. Note that the target server must have already advertised support for that application.
  - Location action: This field indicates whether a message is to be processed locally, relayed to next hop server, or redirected the sender.

- Server identifier: The identifier for one or more servers to which the message is to be routed. Note that these servers must be the peers of the current Diameter node (they must exist in the current node's peer table).
- Static or Dynamic and expiration time fields: these fields have similar meaning as those in the peer table.

7.2.2.3 Diameter Message Routing and Forwarding

Unless a Diameter node is a final destination for a Diameter request and must process the message locally, the node needs to forward the request by consulting its Diameter peer table. This table includes all the peers with which a node can communicate directly.

Diameter messages that may be forwarded by other agents must include the target realm information inside the Destination-Realm AVP and the AVPs that specify the type of application in use. Example of these AVPs are Auth-Application-Id, Acct-Application-Id or Vendor-Specific-Application-Id. Diameter requests are routed towards their final destination using a combination of Destination-Realm and Destination-Host AVP (sounds like combination of network prefix and host identifier in an IP address?).

In the interest of keeping this text short, we will not go through the details of Diameter message routing and forwarding in this text and we refer the reader to the rather detailed description of generation, forwarding (including proxying or redirecting) and processing of request and answer messages in the Diameter base specification [DIMETER3588]. As usual we provide a few important notes to help the reader in understanding the concepts:

- Requests that cannot be proxied and must be handled locally must carry an indication that shows they are for local consumption. The most obvious indication is of course when the request includes a Destination-Host AVP that contains the local host's identity. Other less obvious indications are when the Destination-Host AVP is not present, but a Destination-Realm AVP is present and includes a realm at which the server is configured to be local and the Diameter application in question is also supported. Another indication is when the request contains neither of the Destination-Realm or Destination-Host AVPs.
- Requests that need to be sent to a specific realm, but can be processed by any server within that realm, must contain only Destination-Realm and not the Destination-Host AVP.
- Requests that need to be sent to specific home server in a given realm must contain both Destination-Host and Destination-Realm AVPs.

7.2.3 Capability Negotiations

Diameter base specification defines a capability exchange procedure, which allows the peers to become aware of each other's capability in supporting various functionalities and applications. The capability exchange is also used to negotiate towards a common set of functionalities for the Diameter application, when both nodes support a specific application.

The two peers must carry out a capability negotiation exchange prior to bringing up the connection. The negotiation consists of a peer sending a capabilities-exchange request (CER), to which the other peer responds with a capabilities exchange answer (CEA).

When sending the capability exchange related message, each peer indicates its support for each given application by advertising the relevant application identifiers. For example, applications supporting the NASREQ-application include a value of 1 in the Auth-application-ID or Acct-Application-ID AVP of the capabilities-exchange request and answer commands. Once the receiving peer receives a capability-exchange-request message from a peer, it examines the applications supported by the peer and if it finds some common applications that the receiver itself supports, it will cache the information on applications supported by the peer along with the identity of the peer. This optimizes the future interactions of the two peers as well, since each peer will avoid sending commands related to the applications that the other peer does not support. If the receiver does not find any applications common between itself and the sending peer, the receiver returns a capabilities-exchange-answer with a Result-Code AVP set to DIAMETER-NO-COMMON-APPLICATION and disconnects the transport layer connection.

The capability exchange mechanism also serves as a vehicle ensuring the security of the Diameter session. When the receiver does not have any security mechanisms in common with the sender of the capability-exchange request, the receiver returns a capabilities-exchange answer with the result code AVP of DIAMETER-NO-COMMON-SECURITY and can close the connection.

Since the Diameter session may have to be routed through multiple nodes and agents until it reaches the destination peer, the Diameter session must be routed only through those nodes that have advertised the capability to support the applications required by the session.

Whether or not a node is allowed to receive and respond to capability-exchange requests from unknown peers is a matter of network policy and the node can discard any pending transactions with such unknown peers in case of a transport failure.

7.2.4 Diameter Security Requirements

RADIUS's traditional reliance on hop-by-hop security based on shared secrets has created many problems for more modern applications of AAA protocols. RADIUS does not provide integrity protection in a symmetric manner (different mechanisms for access requests and responses are suggested as we saw in Chapter 6). This issue with RADIUS security procedures is especially severe when attribute hiding is important or the network needs to deal with mobile clients that have non-static IP addresses. Some specifications have tried to provide guidance on the use of IPsec for RADIUS. Still, the details on the use of IPsec for providing security services for specific RADIUS applications are not clear.

Once again, Diameter steps up and tries to accommodate for the shortcomings presented by RADIUS.

- Diameter starts by mandating support for IPsec for both Diameter clients (NAS) and servers. Diameter further mandates support of TLS by Diameter servers, while leaving the support of TLS for clients such as NAS and Mobility agents, such as FA, optional.
- One reason for easing the requirement on TLS support for NAS and edge devices is to relax the need for PKI certificate support by these devices, so IPsec can primarily be used for edge traffic or intra-domain traffic. On the other hand, it is recommended that inter-domain traffic is protected by TLS.
- Even though IPsec or TLS can provide security protection over a connection (see above), end-to-end security protection of Diameter messages may be required. One example is when the confidentiality of an attribute needs to be protected from these intermediaries. As opposed to RADIUS support for hop-by-hop security, Diameter base specification also encourages the use of specific CMS extensions [DICMSDR] for this purpose.

More guidance on using IPsec and TLS for Diameter is provided in the following subsection.

7.2.4.1 Use of IPsec or TLS for Diameter

Diameter requires transmission level security (through IPsec or TLS) over each connection. Note that a connection may only extend between two Diameter intermediaries (e.g. a client to a relay, an agent to another agent, and so on). Diameter base specification assumes that Diameter messages are secured by using either IPsec or TLS. When IPsec is not used, only the capability exchanges between two Diameter nodes can be done without TLS support (more details on this in the next subsection). The two peers indicate the support for TLS to each other through the use of Inband-Security-ID AVP (with a value of TLS). The peers need to start the TLS handshake following the capability exchange (CER/CEA) messaging. If the TLS handshake fails at this point, the connection between the two peers must be closed.

All Diameter implementations must support the IPSec transport mode with encryption and authentication algorithms to provide per-packet authentication and confidentiality. They must also support IPSec anti-replay mechanisms. Furthermore, Diameter mandates the use of Internet Key Exchange (IKE) for peer authentication, key management and negotiation of IPsec security associations in all Diameter implementations. However, Diameter relaxes the requirements on support of all IKE authentications (see Chapter 4) and only requires the use of pre-shared secrets for IKE authentication, while leaving the support of certificate-based authentication as a choice for implementers. One issue that arises with use of certificates with IPsec and IKE is that, since the use of port identifiers is prohibited in IKE Phase 1, it is not possible to uniquely configure root certificate authorities (CAs) for each application individually. This implies a flaw in the use of certificates for IPsec-based security provisioning of Diameter messaging: the same policy must be used for all applications. The reason is that, since the authentication occurs only within Phase 1 of IKE between the client and the server and it is usually not possible to define separate trust or authorization schemes for each application during IPSec SA establishment, which happens later during phase 2.

When using TLS, the Diameter node that initiates the connection acts as the TLS client, while the diameter node (other peer) that accepts the connection acts as the TLS server. Diameter peers, implementing TLS to secure their connections, need to mutually authenticate each other as part of TLS session establishment. To support the TLS mutual authentication, the peer acting as the TLS server must be able to request a certificate from the peer acting as the TLS client and the TLS client must be able supply the certificate. Again, the issue that arises with use of certificates is that both peers need to trust the root CA that has issued these certificates. Even though TLS is much more flexible than IPsec with configuring the root CAs, it may still be possible that different CAs are used to generate certificates for different Diameter usages.

Finally, the specifications recommend that a Diameter peer implements the same security mechanism (IPsec or TLS) across all its peer-to-peer connections to avoid inconsistency, redundancy or even inadequacy in security provisioning.

7.2.4.2 Path Authorization: Impact of Security on Authorization and Accounting

As part of the transmission level security at each connection, not only are the two Diameter peers on each side of the connection ("AAA hop") required to authenticate each other, but also they need to authorize both the connection and the session. For instance, the mere fact that a peer has been successfully authenticated does not mean that it is authorized to act as a Diameter server supporting the applications it is advertising. Therefore the following is also required:

- Authorization of functionality: Before initiating a connection, a Diameter peer must check that its peers are authorized to act in their roles. Before bringing up the connection, authorization checks are performed at each connection along the path. This includes Diameter capability negotiations (CER/CEA) to determine what applications are supported by each server. This is to make sure that Diameter messages relating to a session are routed along a path that only includes authorized nodes that have advertised support for the Diameter application required by that session.
- The home server, prior to authorizing a session, must check to make sure that the route traversed by the request is accepted, i.e. the request has not gone through untrusted realms. This is accomplished by checking the physical route by examining Route-Record AVPs. A DIAMETER_AUTHORIZATION_REJECTED error message is issued if the traversed route is not acceptable.
- Accounting messaging may also be tied to the aforementioned authorization. A home server may want to create a policy to accept only accounting requests for sessions for which specific authorization responses have been issued by the server.
- Local Diameter agents, proxying for Diameter messaging, also need to check the routing records included in the authorization responses coming from a home server. This is to make sure that the route taken by the message is an acceptable one. By forwarding the authorization response further down the path, a local agent is accepting the financial risk involved in authorizing the session. The same responsibility is involved in creating an accounting request corresponding to an authorization response.

7.3 Details of Diameter Applications

As mentioned before, most of the Diameter applications are defined in separate specification documents, except the Diameter accounting application that is described within the Diameter base specification [DIMETER3588]. We provided a high level overview of several Diameter applications earlier. In this section, we will go through several of these applications in more detail.

7.3.1 Accounting Message Exchange Example

In contrast to RADIUS, accounting methods are specified within the Diameter base protocol specification. Even though the messaging for accounting in the Diameter base specification is rather brief, the IETF AAA working group has spent some energy on providing more guidance on accounting management and its attributes by producing two other RFC specifications ([ACCMGM2975] and [ACCATT2924]). The highlights of accounting management ([ACCMGM2975]) documents were given in the first chapter of this book. The reader is referred to [ACCMGM2975] for more information on accounting attributes for a variety of AAA protocols.

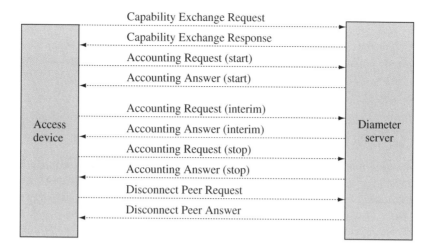

Figure 7.3 An example of Diameter accounting application exchange

Figure 7.3 shows a sample message exchange for Diameter accounting application. The names of the messages are self-explanatory. The capability exchange messaging is included for completeness to make sure that both nodes support the accounting application.

7.3.2 Diameter-Based Authentication, NASREQ

As mentioned many times so far, modern networks rely on three-party authentication mechanisms, where a network access server (NAS) at the edge of a network interacts with a backend AAA server through a AAA protocol to accomplish access control. We also saw that NAS-AAA server interactions to accomplish access control and authentication were central to the design of RADIUS and its messaging, since RADIUS was designed as a client–server protocol.

In contrast to the RADIUS specification, Diameter is a peer-to-peer protocol with more than just authentication and accounting in mind. The Diameter NAS application specification, describing the interaction between the NAS and the Diameter server is not fully standardized yet and is still in Internet draft form [NASREQDR]. As we will see in Chapter 10, modern authentications are mostly EAP-based. This means Diameter EAP application [DIAEAPDR] must also be implemented along with Diameter NAS application to perform EAP-authentication with a Diameter server through a NAS. We will go through Diameter EAP application in more detail later on in this chapter. Here we provide more details on the use of NAS and its interactions with the Diameter server.

Another valuable element included in Diameter NAS application specification is consideration relating to RADIUS. Diameter was designed to eventually serve as a successor to RADIUS, and yet the designers were completely aware of the currently large and persistent deployment base for RADIUS. Co-existence with RADIUS was considered especially important for NASes. This along with the historical reasons (NAS was probably designed among the first Diameter applications) led to the NAS application specification spending a great deal of its bandwidth to defining the roles and requirements for the so-called translation agents. The addition of material on RADIUS interactions to the Diameter NAS specification makes this specification rather unique compared to other Diameter applications, such as Mobile IP, which only define the functionality of the application in question. Unfortunately in the eyes of a person uninitiated to Diameter, this is another obscure point with Diameter: Diameter-RADIUS translations are explained in an application document rather than in the base documentation. We will provide more detail on the operation of these agents later on. For now let us focus on the description of messaging for NAS application.

7.3.2.1 Commands Introduced by NASREQ

We will provide more detail on the NAS operation shortly, but need to acquaint ourselves with the most important commands used in the NAS specification first. The NAS application defines a number of commands that mostly relate to authentication and authorization. Our guess is that since the Diameter base protocol specifies accounting messages, the NAS application focuses on authentication and authorization commands. We like to think that this guess is confirmed by the fact that NAS specification defines a AA-request and a AA-answer command that deal with the first two "A"s of AAA: authentication and authorization.

- *AA-request (AAR) command (code 265)* is sent by a NAS to request authentication and/or authorization for a given user. In the request messages, the "R" bit within the command flags field is set to indicate a request. All requests must include information to uniquely identify the source of the call. This could include user-name, NAS port identifier, and so on. For authentication requests, the user-name and related authentication AVPs need to be present. For authorization requests, information on the called and caller stations is included. The specification includes a detailed list of AVPs related to authorization, authentication and other AVPs carried by this command. Among important AVPs carried by AAR are NAS-ID, NAS-IP address, NAS-port, Auth-request type, user-password, CHAP-Auth, CHAP-challenge. We will describe some of these AVPs further below.
- *AA answer (AAA) command (code 265)* is sent in response to AA-request message and can include related authorization AVPs, if the authorization was requested and processed successfully. This message may carry a text message for the user through a Reply message AVP. The NAS will prompt the user's client software with the text message. In the answer messages, the "R" bit within the command flags field is cleared.
- *Re-Auth-Request (RAR) command (code 258)* is already defined by the Diameter base protocol specification, but is mentioned here for completeness of our discussion on NAS. The NAS application specification allows the server to initiate a re-authentication and/or re-authorization for a session.
- *Re-Auth-Answer (RAA) command (code 258)* is also defined by Diameter base protocol and is sent in response to Re-Auth-Request and must include Result-Code AVP.

7.3.2.2 NASREQ AVPs

Diameter reserves the AVP codes 0-255 for backward compatibility with RADIUS. The NAS application assigns AVP code values 363-366 and 400-406 from the code space defined in Diameter base specification. The values 363-366 and 406 are assigned to support accounting mechanisms involving a NAS, while the values 400, 401 are for NAS-Filter-rule and tunneling. Values 403-405 are assigned to CHAP algorithm, CHAP ID and CHAP response AVPs. Note that CHAP-challenge AVP is the same as CHAP-challenge attribute (60), defined by RADIUS. Also, it should be noted the code value intervals mentioned above are not exclusive to NAS application. The Mobile IP application uses some of those code values, but there is no usage overlap between NAS and Mobile IP applications.

Aside from the newly introduced AVP codes, the NAS document uses a large number of AVPs already defined in the Diameter base protocol, but groups these AVPs into several groups, of which the most important ones are NAS session AVPs, NAS authentication AVPs, NAS authorization AVPs, NAS accounting AVPs, NAS tunneling AVPs and other AVPs for authorization of use of various network types, such as IP, framed-versus non-framed links, IPX.

To provide some examples on NAS application AVPs, we describe some of the AVPs related to PAP and CHAP authentications in more detail and refer the reader to the draft [NASREQDR] for a more complete list of the AVPs.

- *User-Password AVP*: carries the user's authentication password for password-based authentication mechanisms or the input that the user provides in multi-round authentication mechanisms. The password may be a permanent user password or a one-time password. Due to the sensitivity of permanent passwords, Diameter messages carrying these passwords must be protected by end-to-end security mechanisms such as CMS, when going over untrusted proxy environments. In such cases the hop-by-hop protection provided by IPsec or TLS is not enough.
- *Password-Retry AVP*: When an authentication failure occurs, the server sends an AA answer message indicating the failure and at the same time can include a Password-Retry AVP indicating the number of authentication attempts a user can pursue.
- *CHAP-Auth AVP* is used to carry information necessary for performing a challenge-based PPP authentication. CHAP-Auth AVP is a grouped AVP, meaning that it can include other AVPs, such as CHAP-Algorithm AVP, CHAP-Ident AVP, and CHAP response AVPs. CHAP-Algorithm AVP includes an identifier that indicates what algorithm should be used to compute a response to the challenge, while CHAP-Ident AVP includes an identifier that is used in calculation of the response. So far MD5 is standardized as a CHAP algorithm. CHAP response AVP includes the authentication data calculated by the user in response to the challenge.
- *CHAP-Challenge AVP* includes the challenge that was sent by the NAS to the CHAP peer. Note that this AVP is not defined as part of the group within CHAP-Auth AVP. If the CHAP-Auth AVP is included in a message, the CHAP-challenge AVP must also be included.

7.3.2.3 Diameter NAS Messaging

An example of Diameter server-NAS message exchange is shown in Figure 7.4. It should be noted that the messaging between the user and the NAS is not considered as Diameter messaging and is only included for completeness.

As we can see, the concept is rather simple:

- Upon arrival of a new call, service request or re-authentication request from a user, the NAS creates a Diameter AA-request message (AAR), including user identity, user authentication information and other call-related information and sends the message to a server.
- The server processes the request and returns a Diameter AA-answer (AAA), including authorization information or a Result-Code AVP, indicating a failure. Alternatively, the

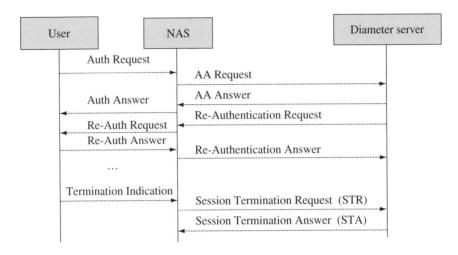

Figure 7.4 Messaging exchange for Diameter NAS application

server may indicate that further rounds of exchange are required to complete the authentication process. Note that, Diameter also allows authorization-only requests, which do not include any authentication-related data. Enabling this feature may, however, involve serious security risks, if not designed properly. Furthermore, since the corresponding functionality does not exist in RADIUS, traversal of such requests through Diameter-RADIUS translation agents (explained later on in this chapter) may lead to a failure in completion of the client–server exchange regarding authorization-only requests.

- When the authentication or authorization exchange is completed successfully, the NAS application starts a session context. If accounting is to be implemented, an accounting message is sent to record the start of the new session. Depending on the network policies, failure in establishing accounting procedures for the session may result in failure to establish the session. The Diameter server informs the NAS about the maximum time allowed before the next re-authorization or re-authentication by using an Authorization-lifetime AVP.
- The session starts and the NAS application starts the session context. If accounting is required, the application needs to send accounting-related messages. The NAS may attempt re-authentication or re-authorization before the end of the authorization life time. Although the servers typically implement a grace period (Auth-Grace-period AVP) before it releases all the state information regarding the user, the server may issue unsolicited re-authorization, re-authentication requests prior to the expiration of the Authorization-lifetime to avoid a service termination.
- When the user indicates to the NAS that it needs to terminate an existing session, the NAS issues a session-termination-request (STR) to the Diameter server. When accounting procedures are implemented, termination of session needs also to be communicated to the server.
- Upon receiving and processing a valid session-termination request, the Diameter server responds with a session termination answer (STA), including a Result-Code AVP. When the session is terminated, the server will release all the resources that are tied to the session ID for the terminated session. Also all the proxy servers on the proxy chain will release any relating resources.

7.3.3 Diameter Mobile IP Application

When discussing Mobile IP registration, we discussed the use of AAA servers in the authentication of Mobile IP signaling messaging to the AAA server, FA and HA. However, we also mentioned that Mobile IP specifications only discuss the authentication of registration request and responses and key distribution from the AAA server to the mobile device. They do not describe how the same key material is distributed to the Mobile IP agents. Those specifications simply state that the AAA servers-mobility agents (FA and HA) interactions are handled by AAA protocols. IETF AAA WG has been working on providing specifications to use Diameter to carry out the initial authentication and key distribution required for Mobile IP signaling for quite some time [DIAMIPDR]. The specification not only provides details on authentication, but also authorization and accounting as well. For instance, it not only helps specify how the FA authenticates a mobile node that is visiting the foreign network, but also how the FA informs the mobile node's home administrative domain about the resources that the MN consumes in the foreign network. It should, however, be noted that this specification only applies to Mobile IPv4 and not Mobile IPv6.

We will go through Mobile IP-AAA interactions in detail in Chapter 8 and refer the reader to that chapter for the section describing Diameter Mobile IPv4 application.

7.3.4 Diameter EAP Support

As mentioned in Chapters 2 and 3, EAP provides a framework for providing authentication and access control services over many different types of access media. We also mentioned that EAP signaling between the NAS and the AAA server is carried over a AAA protocol. We briefly described encapsulation of EAP messages inside RADIUS and the RADIUS attributes used for that purpose in Chapter 6. In this section we do the same for Diameter. The details of Diameter support for EAP are provided in the Diameter EAP application specification [DIAEAPDR].

As one can imagine, since the Diameter EAP application has to do with authentication and carrying authentication-related messages between NAS and AAA servers, it is closely tied to the Diameter NASREQ application. This is to the point, that for a Diameter agent to support the EAP application, it must also support the NASREQ application. A Diameter agent may advertise its support of EAP, so that the client and NAS know whether they can engage in an EAP exchange. When a server cannot support EAP, it should allow the NAS to negotiate other authentication mechanisms such as PAP or CHAP with end user/device.

The model for carrying EAP over Diameter is very similar to what was shown for RADIUS-EAP interaction in the previous chapter:

- EAP messages from the NAS towards the AAA server are carried inside Diameter-EAP-request messages. The EAP message itself is carried as an EAP-payload-AVP.
- EAP messages from the AAA server to the NAS are carried inside the Diameter-EAP-answer messages. The EAP message itself is carried as an EAP-payload-AVP. The Result-Code AVP may be used to indicate additional information to the NAS, for instance, at the start of an EAP conversation, upon receiving the first EAP-request message from the NAS, the Diameter server may send a DIAMETER-MULTI-ROUND-AUTH inside the Result-Code AVP to indicate to the NAS that it is expecting further EAP-request messages regarding the ongoing authentication procedure.

Although often the EAP server is collocated with the Diameter servers, since EAP and Diameter are still separate mechanisms, their interaction requires some subtle considerations:

- The EAP is spoken between the EAP peer (end user/device) and the EAP server, rather than with the NAS. On the other hand, the NAS speaks only Diameter and may not understand any of the EAP of messaging, except the EAP success and EAP failure messages and acts only as a pass-though for the rest of EAP messaging.
- The result of the authentication process is typically conveyed directly to the EAP peer through the EAP success or failure messages, while the result of the access authorization process is conveyed to the NAS through Diameter Result-Code AVP in the Diameter answer. If the two processes have conflicting outcomes, e.g. an EAP success is sent to the peer along with the Diameter message to the NAS carrying a failure Result-Code AVP, the peer may believe it has the right to access the network, while the NAS cannot grant any access to the peer. The conflict may happen if there are Diameter proxies along the way that make their own authorization decisions.
- Many newer access networks today strive to protect the user's identity by allowing the user not to send its actual identity over the initial and unprotected EAP requests. Other times only the EAP messaging and not the Diameter messaging may carry the user identity. A NAS that does not have the capability to understand EAP signaling may not be informed about the peer's true identity. Care must be taken so that at some point some sort of peer identity is conveyed to the NAS, so that the NAS can authorize the peer for access and account for her resource usage.
- Often sensitive key information may need to be carried through EAP messaging. Diameter standards still do not provide specifications on end-to-end security protection (Diameter CMS was never standardized) and only provide guidance on IPsec or TLS usage for provisioning of hop-by-hop security. For that reason, care must be taken to protect the confidentiality of sensitive keys in transit.

7.4 Diameter Versus RADIUS: A Factor 2?

Now that we have completed our humble attempt at a description of Diameter, we devote this section to a comparison of RADIUS and Diameter as well as the co-existence of both during migration phases.

7.4.1 Advantages of Diameter over RADIUS

In the following subsection we list a number of improvements that Diameter as a AAA protocol offers over RADIUS.

7.4.1.1 Fail-Over

Fail-over is defined as the process of forwarding all the pending requests with an agent to another agent, once a transport failure with the first agent is detected. For this to be possible, it is, however, required that the nodes have agreed on failure support by setting up a flag in their Diameter messaging.

RADIUS does not define a standard fail-over mechanism, and as a result, fail-over behavior can differ between RADIUS implementations. Diameter, on the other hand, is more resilient towards transport failures and provides a well-defined fail-over behavior. Diameter supports application layer acknowledgements and specific watchdog mechanisms to detect lack of activity. Diameter fail-over mechanisms are defined in [AAATR3539]. A pending message queue for every peer is maintained at a Diameter node. Upon receiving a response, the corresponding request is removed from the queue.

7.4.1.2 Server-initiated Messages

Support of server-initiated messages is only optional in RADIUS [RAD3576] and this makes it difficult to implement features such as unsolicited disconnects or re-authentication/re-authorizations on demand across a heterogeneous deployment. Support for server-initiated messages is mandatory in Diameter.

7.4.1.3 Reliable Transport

Using UDP as a transport and lack of retransmission specifications in RADIUS makes reliability an issue for use of RADIUS for accounting: Packet loss may translate directly into revenue loss. Diameter runs over reliable transport mechanisms (TCP, SCTP) as described earlier.

7.4.1.4 Capability Negotiation

The client and server have no way of indicating their support of various attributes to each other and RADIUS does not support error messages. This means performing capability discovery and negotiation to a mutually agreeable service may be very difficult with RADIUS. Diameter includes support for error handling, capability negotiation, as well as ways to indicate support of attribute-value pairs (through the mandatory flag).

7.4.1.5 Security and Audibility Issues

As we saw in the previous chapter, RADIUS defines an application-layer integrity protection (message authentication) scheme that is only required for Access Response packets. The authentication is based on shared secrets, but the trust is only established between neighbor hops and not end to end. Malicious proxies between client and server can modify attributes or even packet headers without being detected. Per packet confidentiality is not supported, only attributes can be hidden. RADIUS accounting protocol has replay protection issues. Support of IPSec is not required in RADIUS. Use of IPSec is only defined when RADIUS is used with IPv6. Even then, the use of IKE limits the usefulness of IPSec for various applications. Also not having the possibility of setting up a certificate hierarchy makes usage of RADIUS in roaming applications, where inter-domain AAA trust relationships are required, difficult.

On the other hand, Diameter defines both transmission-level security and end-to-end security and requires mandatory support of IPSec and optional TLS support at the clients. However, data object security is not mandatory in Diameter.

7.4.1.6 Diameter Support for Agents and Inter-Domain Roaming

RADIUS does not provide support for agents and proxies clearly. Since the expected behavior is not defined, it varies between implementations and interoperability problems may arise. Although the concept of RADIUS proxy chaining via intermediate servers is defined, due to lack explicit support for proxies and data object- and transmission-level security, RADIUS-based roaming is vulnerable to attacks and fraud, and as a result, it may cause problems for wide-scale deployment.

Diameter defines the role of agents and proxies and their behavior explicitly. By providing explicit support for inter-domain roaming, message routing and transmission-layer security, Diameter addresses the RADIUS limitations.

7.4.1.7 Peer Discovery and Configuration

RADIUS implementations typically require that the name or address of servers or clients be manually configured, along with the corresponding shared secrets. This results in a large administrative burden, and creates the temptation to reuse the RADIUS shared secret for many clients (NAS) and that can result in major security vulnerabilities.

Through DNS, Diameter enables dynamic discovery of peers. Derivation of dynamic session keys is enabled via transmission-level security.

7.4.1.8 Backward Compatibility with RADIUS

While Diameter does not share a common message format with RADIUS, considerable effort has been expended in enabling backward compatibility with RADIUS, so that the two protocols may be deployed in the same network. It is, however, expected that translations will need to take place through gateways enabling communication between legacy RADIUS devices and Diameter agents. We will talk about this shortly.

7.4.2 Issues with the Use of Diameter

The completeness (and complexity) of Diameter protocol specifications as well as the early 2000 economic downturn have worked against Diameter. At the time of writing, the Diameter base protocol as an RFC is about one year old, so there are very few commercial Diameter servers around (Interlink, HP-UX). Deployment has been slow since not many vendors have stepped forward with Diameter offerings. Some vendors claim that Diameter will be used by only IPv6 users.

Diameter requirements of support of either TCP or SCTP at the clients and support of both TCP and SCTP at the agents and servers put greater than ever burdens on communications links and networks. The lightweight characteristics of UDP are now replaced with complicated TCP/SCTP session set-ups and implementation issues.

There is a large RADIUS deployment base out there and unless a proper migration plan that includes deployment of translation agents and co-existence of RADIUS and Diameter, migration to Diameter will not be simple.

7.4.3 Diameter-RADIUS Interactions (Translation Agents)

As mentioned earlier, Diameter and RADIUS may need to co-exist within the boundaries of the same operator administration for a long migration period. In the process of designing Diameter, a great amount of effort was focused on providing facilities for RADIUS–Diameter co-existence. An example of such efforts was to ensure that RADIUS attribute space is included as-is inside the Diameter attribute space to eliminate the need for attribute conversion. However, Diameter creates a superset of RADIUS attributes and messages and for that reason co-existence of Diameter and RADIUS requires specialized effort.

Due to its focus on authentication and authorization, Diameter NAS application is the one Diameter specification with the most similarity to RADIUS. For those reasons, the Diameter NAS application is the first specification that describes interoperability between Diameter and RADIUS implementations. This interoperability is envisioned through an architecture consisting of distinct RADIUS and Diameter systems meeting each other through a translation agent at their boundaries.

Since Diameter functionality is superior to that of RADIUS and there are many differences between RADIUS and Diameter, there can be many variants and implementations of translation agents in proprietary non-IETF manners. Also due to the simultaneous standardizations of RADIUS and Diameter, various RADIUS messages may be handled differently by different translation agents along the process, while none of those translation agents can be assumed to have access to complete and accurate session state information.

The Diameter NAS application describes many requirements and conversion procedures for RADIUS–Diameter translation agents. We do not go into the exact details of how the translation of a message, an attribute or AVP or a service such as security is performed in every instance. Instead we provide some highlights to give the reader some idea of the issues to be considered in the translation:

- RADIUS security mechanisms are hop by hop, while Diameter may apply end-to-end security mechanism. The Diameter agent will have to decrypt the RADIUS messages and attributes and secure the information in a Diameter-specific manner as explained earlier. For instance, when the translation agent receives a RADIUS message including a User-Password Attribute encrypted with a RADIUS shared secret over the link, the agent must decrypt the password from the RADIUS link and forward the password information inside a Diameter message that is protected by Diameter security mechanisms. The authenticator value in RADIUS messages including message-authenticator attribute (defined in [RADEXT2869]) must be verified by the translation agent, but not included back in the Diameter message created by the agent.
- RADIUS supports neither peer-to-peer architecture nor server-initiated messages, while Diameter defines a large number of command codes that can be used in both request and response messages in a peer-to-peer manner. When negotiations involve multi-round message exchanges, RADIUS only offers Access request for client to server and access challenge for server to client messages. The translation agent must create access challenge messages or access request messages based on the Diameter commands.
- RADIUS servers are assumed to be stateless, while Diameter nodes maintain state and, as mentioned earlier, Diameter agents may have a distorted picture of the overall end-to-end session.

- Diameter AVPs are defined in fully qualified domain name (FQDN) format, while RADIUS attributes are not. The translation agent must change the information format according to the system to which the message is being forwarded. A prime example is converting RADIUS NAS IP address attribute into Diameter Origin-Host AVP in FQDN format.
- Diameter supports grouped AVPs. When the translation agent receives RADIUS attribute/s that will have to be part of a grouped AVP, the agent must extract the related RADIUS attributes to build the Diameter grouped AVP. The prime example is handling of CHAP exchanges. The CHAP-password attribute of RADIUS includes the response as the attribute data, but includes CHAP-ID in the "header" of the attribute. In contrast, Diameter defines a grouped CHAP-Auth AVP that includes CHAP-response and CHAP-Ident as sub-AVPs. This conversion needs to be done by the translation agent.

In summary, the translation agents act as gateways responsible for interoperability between RADIUS and Diameter. The problem is: Diameter does not specify the details of the operation of these agents. Therefore, it cannot be assumed that implementing Diameter specifications alone will lead to ready backward compatibility with RADIUS in a plug and play manner.

7.5 Further Resources

For more information on the standardization of Diameter in IETF, the reader is referred to the website for the IETF AAA working group
http://ietf.org/html.charters/aaa-charter.html

For information on open source implementations of Diameter, the following website is a good starting point:
http://www.opendiameter.org/

One of the major third generation cellular network standards (3GPP) is using Diameter for AAA procedures. The standardization is performed partly at IETF in documents such as [DIA3G3589] and partly in 3GPP, such as [3GPPDIA]. This would mean most Diameter activities can be traced by following the activities around 3GPP cellular networks.

7.6 References

1. [AAAEVAL2989], B. Aboba et al., "Criteria for Evaluating AAA Protocols for Network Access", IETF, RFC 2989, November 2000.
2. [NASCRIT3169], M. Beadles, D. Mitton, "Criteria for Evaluating Network Access Server Protocols", Internet Engineering Task Force, RFC 3169, September 2001.
3. [PITTPROC], IETF 48th Proceedings website, AAA Working Group meeting minutes, http://www.ietf.org/proceedings/00jul/00july-58.htm, Pittsburg, July 31–August 4, 2000.
4. [EVALRES3127], D. Mitton et al., "Authentication, Authorization and Accounting Protocol Evaluation", IETF, RFC 3127, June 2001.
5. [DIAMETER3588], P. Calhoun, "Diameter Base Protocol", RFC 3588, IETF, September 2003.
6. [NASREQDR], P. Calhoun et al., "Diameter Network Access Server Application", Internet draft, IETF work in progress, draft-ietf-aaa-diameter-nasreq-16.txt, June 2004.
7. [AAATR3539], B. Aboba, J. Wood, "Authentication, Authorization and Accounting (AAA) Transport Profile", IETF, RFC 3539, June 2003.
8. [DICMSDR], P. Calhoun, W. Bulley, S. Farrell, "Diameter CMS Security Application", Internet draft, IETF work in progress, draft-ietf-aaa-diameter-cms-sec-04.txt, March 2002.

9. [CMS3370], R. Housley, "Cryptographic Message Syntax (CMS) Algorithms", RFC 3370, IETF, August 2002.
10. [ACCMGM2975], B. Aboba et al., "Introduction to Accounting Management", IETF, RFC 2975, October 2000.
11. [ACCATT2924], N. Brownlee, A. Blount, "Accounting Attributes and Record Format", IETF, RFC 2924, September 2000.
12. [IANAWEB], Internet Assigned Number Authority, website URL, http://www.iana.org/
13. [DIACCDR], H. Hakala et al., "Diameter Credit Control Application", IETF work in progress, Internet draft, draft-ietf-aaa-diameter-cc-06.txt, August 2004.
14. [DIAMIPDR], P. Calhoun et al., "Diameter Mobile IPv4 Application", IETF work in progress, draft-ietf-aaa-diameter-mobileip-20.txt, August 2004.
15. [DIAEAPDR], P. Eronen et al., "Diameter Extensible Authentication Protocol (EAP) Application", IETF work in progress, Internet draft, draft-ietf-aaa-eap-07.txt, June 2004.
16. [RAD3576], M. Chiba et al., "Dynamic Authorization Extensions to RADIUS", IETF, RFC 3576, July 2003.
17. [RADEXT2869], C. Rigney et al., "RADIUS Extensions", IETF, RFC 2869, June 2000.
18. [DIA3G3589], J. Loughney, "Diameter Command Codes for Third Generation Partnership Project (3GPP) Release 5", IETF, RFC 3589, September 2003.
19. [3GPPDIA], "Diameter Applications, 3G Specific Codes and Identifiers", 3rd Generation Partnership Project, June 2004.

8

AAA and Security for Mobile IP

In Chapter 5 we described Mobile IP as one of the most prominent methods for providing mobility for IP network users. We also explained some of the security measures provided for Mobile IP control signaling, such as authentication of Mobile IP transactions between the mobile node and Mobile IP agents and between Mobile IP agents themselves. We described how authentication extensions can be calculated and added to the registration messages to provide integrity protection (authentication) for these messages. However, as mentioned there, calculation of those authentication extensions requires pre-established trust relationships (e.g. security associations including shared secrets) between the mobile node and Mobile IP agents. Unfortunately, Mobile IP base specification [MIP3344] does not provide any details on how these SAs are established. The implication is that, using the base specification alone, if by time of Mobile IP registration the mobile node has not yet established any security associations with its HA or the current FA, it cannot calculate any of those authentication extensions. In this chapter we focus on solving the problem of establishing the security associations required for Mobile IP signaling.

From an administrative point of view, when dealing with large networks serving many roaming mobile nodes, it is not scalable to manually configure security associations between a mobile node and all foreign agents or even between all FAs and HAs beforehand. This may not even be feasible, since the mobile node may not know what HA it is assigned to it at time of booting up. Hence, mechanisms should be provided to set up these security associations dynamically and when required, which is typically the time of Mobile IP registration.

In recent years, many wireless system designers, such as those working in the WLAN 802.11 community, have started using AAA infrastructures for dynamic and real-time generation of security association and distribution of session keys for sessions between end nodes and their network point of attachment. It is safe to assume that the AAA server of the service provider or the enterprise with which the mobile node has been initiated, has access to mobile nodes and its user's identity and authentication credentials or at least has the functionality to understand and verify this data when necessary. Such a trust relationship allows the AAA server to authenticate the node, its user and its messaging.

The idea here is when a Mobile IP agent receives a registration request from a mobile node, with which it does not have a trust relationship, it can request the AAA server not only to authenticate the MN and its message, but also to generate the necessary keying material to establish trust relationships for the protection of future Mobile IP transactions.

Using the AAA infrastructure brings advantages beyond providing authentication and key management services. Authentication is only part of the security battle; the use of newly acquired IP addresses must be authorized by the network authorities. When the AAA server has access to the user's service profile, it can make decisions regarding authorization of the user for using network resources as well. Looking at Mobile IP registration process from that angle, we realize that Mobile IP registration request is really an authorization request. Through the registration, the mobile node asks the HA to grant it the right to have two IP addresses and to use the HA resources required for support of traffic routing for those two IP addresses.

Another advantage of using AAA methods is that, while Mobile IP provides the routing mechanisms required for the mobile node's traffic as the node roams across different networks, the AAA infrastructure can make sure that no business or trust agreements between these networks are violated and the security and integrity of these networks are not threatened as a result of this roaming. Using the AAA infrastructure, the foreign network can simply refer to the client's home domain to verify its legitimacy, provide the required services, collect the accounting data for the provided services and bill the user's home domain accordingly, without having to deal with the client directly.

Even though by now we think we have made a good attempt to describe the rationale behind a joint Mobile IP-AAA signaling procedure, we must admit that understanding the IETF specifications for this procedure is a fairly time-consuming and involved task. Part of the specification is defined by the IETF Mobile IP working group, while the other parts of the specification are defined by the AAA community. To make matters worse, on each side only one camp has shown interest in supporting this interaction: On Mobile IP side only Mobile IPv4 has taken the time to define this interaction, and on the AAA side only Diameter is working on providing specification for a Mobile IP application. At the time of writing, Mobile IPv6 has just started producing a statement for the problem [MIP6BOOT] and may or may not adapt a similar process to that for Mobile IPv4. On the AAA side, RADIUS has currently no support for Mobile IP operation, even though some standard bodies, such as 3GPP2 (3rd generation partnership project) have developed vendor-specific RADIUS attributes to provide such support. Continuing with our complaints, we note that even the specifications for required Mobile IP functionality and extensions are spread over multiple Mobile IP documents [MIP3344], [MIPCHAL3012], [3012bis], and [MIPKEYS3957]. Hence we devote an entire chapter to describing what we call Mobile IP-AAA signaling and its variants:

- How Mobile IP registration signaling is extended to outsource the task of mobile node authentication to the AAA servers.
- How Mobile IP registration signaling is extended to request the AAA server to help the mobile node and Mobile IP agents with the generation of security associations and keys for the protection of Mobile IP control signaling.
- How Diameter protocol and servers interact with Mobile IP agents to handle the authentication and key generation requests in a scalable and dynamic fashion.
- How the 3GPP2 community is using RADIUS to achieve similar functionality.

8.1 Architecture and Trust Model

Mobile IP-AAA signaling is about establishing trust relationships needed for Mobile IP signaling based on relationships provided by the AAA infrastructure. Thus, it is useful to go through the assumed trust model. The trust model includes the architecture elements involved in the signaling and shows what trust relationships exist prior to the start of Mobile IP-AAA signaling and what trust models are generated as a result of this signaling. As one can imagine, the trust model depends on the mobility pattern of the mobile node, the network topology and the administration policies of the networks the model is trying to connect with. It is therefore important to revisit the trust model for every scenario.

Figure 8.1 shows the Mobile IP-AAA signaling trust model for a very generic scenario, where the mobile node attempts to connect with a foreign network that belongs to an administrative domain separate from the administrative domain to which the mobile node and its home network (and HA) belong. The administrative domain for the foreign network is served by a local AAA server (LAAA or AAAL), while the mobile's home network is served by the so-called home AAA server (HAAA or AAAH). In the model shown in Figure 8.1, the foreign network deploys a foreign agent (FA) to assist the visiting mobile nodes with CoA acquisitions and Mobile IP signaling. Note that FA is a Mobile IPv4-only concept that does not exist for Mobile IPv6. Furthermore, even some networks implementing Mobile IPv4 may not deploy FAs, which means the mobile node must use a co-located care of address (CcoA) and register directly with the HA. In such cases the model may be significantly simpler. Another simpler case is when the FA belongs to the same administrative domain as the HA and hence is served by the HAAA. On the other hand, to keep the description of the protocols from becoming too complicated, we assume that the two networks have compatible administration and security policies. Also for this discussion we do not get into how routing of AAA messages between two different administration domains takes place in general. We simply assume that the local AAA server in the foreign network has a way of determining how to contact the home AAA server for the mobile node.

Figure 8.1 shows a variety of security associations (SAs). Before going into the description of each sort, we would like to point out that in this discussion security association does not necessarily refer to IPsec security associations. By SA, we simply mean a security context that exists between the two entities and defines the keys, transforms, and algorithms that are used to perform the security functions related to the Mobile IP-AAA signaling procedure.

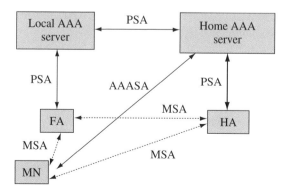

Figure 8.1 Trust model for AAA-Mobile IP registration

We will discuss these functions in more detail later on in this chapter. On the other hand, as we will see shortly, depending on the network infrastructure design, there are cases where an SA may actually refer to an IPsec SA.

8.1.1 Timing Characteristics of Security Associations

8.1.1.1 Pre-established SAs (PSA)

These are the SAs that ensure the security of the operation of the AAA infrastructure when it comes to supporting the Mobile IP-AAA interaction. Thus, these SAs must exist prior to the start of Mobile IP registration. To discriminate between pre-established SA and other SAs, the PSA are shown with solid arrows in Figure 8.1. An example of a PSA is one that exists between the LAAA and the HAAA. This PSA ensures that both the LAAA and HAAA can trust the integrity of the messages coming from the other peer and can communicate sensitive key material in a secure fashion. As discussed below, this SA can also be an IPsec SA and may either be a direct SA or one established through a third party broker. One implication of assuming a PSA between LAAA and HAAA is that the domains that both the LAAA server and the HAAA server belong to must have a trust relationship with each other, either directly or through brokers.

Mobile IP Agent-HAAA PSA: It is assumed that both the FA and the HA have a permanent trust relationship (PSA) with the AAA server within the domain they belong to. These PSAs are indicated as FA-LAAA PSA and HA-HAAA PSA in Figure 8.1. As we will see shortly, the HAAA-HA and HAAA-FA links may have to carry AAA messages that includes actual keys for the HA and FA, respectively. In such cases the AAA messaging should be protected by a strong end-to-end encryption mechanism such as IPsec ESP or TLS record layer (as shown in Figure 8.2), especially when AAA proxy agents are present between the HAAA and Mobile IP agents. Note that implementation of TLS requires the existence of certificates and PKI management functions. Another issue with TLS is that since the key transfer is in only one direction (HAAA to the agent), the HAAA-HA or HAAA-FA links may only be required to provide encryption in one direction, while providing only integrity protection in the other direction. TLS usually cannot accommodate this sort of asymmetric requirement.

The IPsec (and IKE) have their own set of issues, for instance, IKE and IPsec mechanisms place great emphasis on the exact IP interfaces that are used by Mobile IP agents and HAAA. This means careful configurations are required when HAAA server are multi-homed (the HAAA server may own multiple IP interfaces).

RADIUS/Diameter message carrying keys	
TLS record protocol (optional)	IPsec ESP
TCP	
IP	IP

Figure 8.2 Protecting AAA signaling between HAAA and Mobile IP agents when encryption is required

8.1.1.2 Mobility Security Associations (MSA)

The term mobility security association is introduced by IETF Mobile IP community [MIPKEYS3957]. The MSA refers to a security context (SA) that the MN establishes with a Mobile IP agent. The MSA describes what keys and algorithms the MN and the Mobile IP agent should use to provide the corresponding security protection. When these MSAs have been established, the MN and the Mobile IP agents (FA and HA) are able to authenticate the Mobile IP control signaling (registration requests and replies).

In contrast to the PSAs, the MSAs are those SAs that do not exist prior to Mobile IP registration and are generated as a result of Mobile IP-AAA signaling (with the help from the AAA infrastructure). This means establishment of these MSAs involves generation and distribution of keying material during Mobile IP-AAA signaling as well. It should be noted, these MSAs are not generated in a way that pleases the die-hard cryptographer fans. As opposed to methods such as IKE that use public key cryptography (Diffie–Hellman) for key exchange, Mobile IP-AAA signaling uses hash functions and nonces in getting the keys to MN (as we will see in this chapter). However, the goal of these MSAs is not to provide secure links for the interaction of the MN to the network. The MSAs and the corresponding keys are only suitable for what they are intended for, i.e. authentication of Mobile IP registration request and response messages. Given that this only accounts for a small portion of data that is exchanged rather infrequently, the chances of security attacks trying to break these keys are low. The designers of these procedures realize this and do mention that these MSAs are not intended to provide protection for volumes of traffic between the mobile node and the mobility agents.

Another point to note is that the MSAs are unidirectional SAs. In other words, the MSA protecting the signaling from MN to HA (MN-to-HA MSA) is different from the MSA protecting the signaling in the other direction (HA-to-MN MSA), even though in the majority of the cases both MN and HA use the same key to calculate the authenticator field within the authentication extensions. Each MSA comprises a security parameter index (explained in detail below), a shared key and an algorithm type and other information, such as lifetime, and so on.

One final but important note: Once the MSA are established, Mobile IPv4 registration does not need to invoke the AAA infrastructure as long as the MSAs (and the keys) have not expired.

8.1.1.3 AAASA

AAASA is the pre-established SA that exists between the mobile node (MN) and its home AAA server. Even though the AAASA is actually a special case of a PSA, due to its fundamental importance to Mobile IP-AAA signaling, it deserves its own description here. The reason is that, this SA is the most fundamental SA for Mobile IP-AAA signaling, since it allows the Mobile IP agents to consult the AAA server when no prior trust relationships exist to support Mobile IP signaling (MSAs described earlier). When the mobile node shares an SA with the AAA server, the AAA server can verify the MN's credentials and help both MN and the Mobile IP agents to create the needed trust relationships.

We expect that when each node is initiated into a network service provider or an enterprise network owner, it is given an identity and a set of credentials that are recognizable by the AAA server within the domain. The credentials should be such that they are sufficient to perform a set of security functions such as authentication or key negotiation.

For our Mobile IP-AAA signaling discussion, we assume that the credentials given to the mobile node include a key called the AAA key or at least can be used to derive the AAA key.

The AAA key is a key shared between the MN and HAAA and is the fundamental component of the AAASA. Mobile IP specifications assume that the AAA key is a symmetric key and leave the administration specifics of AAA key configuration to the implementers. When the credentials given to the mobile node are public key based (certificates), a symmetric key can be derived based on the certificate. We will describe how the Mobile IP specifications use the symmetric AAA key to provide authentication and key exchange.

8.1.1.4 Lifetimes

Now that we have described the temporal distinctions between these SAs, we should mention a commonality between all of these SAs as well: Almost all the SAs in the model should be created for a limited period of time to ensure robustness against dictionary or off-line attacks. These means the keys have limited lifetime and must be refreshed prior to expiration.

8.1.1.5 Security Parameter Index (SPI)

As mentioned earlier, many of the security associations (SAs) in the Mobile IP-AAA specifications, such as [MIPKEYS3957] are not real IPsec SAs, but using a terminology similar to that of IPsec is rather convenient when it comes to describing the security relationships and operations. In particular, the Mobile IP-AAA signaling specifications use the concept of security parameter index (SPI) to point to various SAs. However, when reading these specifications, keeping track of various SPIs becomes a daunting task. For that reason, we provide some guidelines on nomenclature of the SPIs early on. The reader can move on to the next sub-section only to read the highlights of the signaling process, if she so wishes.

We use the notation that Diameter Mobile IP application specification [DIAMIP] provided to describe the SPIs for MSAs and expand it to provide a general guideline for SPIs naming and assignment. As mentioned earlier, the mobility security associations (MSAs) are unidirectional, i.e. X-to-Y MSA is different from Y-to-X MSA. The X-to-Y MSA is used to describe the security context needed for the protection of messages from X to Y. The receiving party (Y) uses the X-to-Y SPI to locate the security context needed to verify an authentication extension provided by the sending party (X). Hence, the following general rule applies:

1. Y assigns an X-to-Y SPI and sends it to X.
2. X uses the X-to-Y MSA as the security context for calculating the X-Y authentication extension to be sent to Y and includes the X-to-Y SPI value in the extension.
3. Y uses X-to-Y SPI value to locate the security context it has with X and verifies the authenticator within X-Y authentication extension.

For instance, the HA uses the MN-to-HA SPI to locate the security context needed to verify the Mobile-home authentication extension, calculated by the MN. Since MN-to-HA SPI is an internal HA index, the HA assigns this SPI and send it to the MN in advance (when the MSA is being established).

Diameter Mobile IP application defines the notion of an MN-AAA SPI as the security parameter index (SPI) that HAAA will use to locate the AAASA and AAA key required to verify the authentication material provided by the MN.

8.1.2 Key Delivery Mechanisms

As mentioned earlier, Mobile IP-AAA accomplishes two goals:

1. Allows the Mobile IP agents to outsource the authentication of the mobile node's initial Mobile IP registration requests to the AAA server.
2. Uses the AAA server in generating the key materials needed for the creation of MSAs between the mobile node and its Mobile IP agents.

We mentioned earlier that, even though the MSAs between the MN and the Mobile IP agents are unidirectional, the MN and the Mobile IP agents dealing with the MN use the same (symmetric) key in both directions. Unfortunately, the designers of Mobile IP key management procedure [MIPKEYS3957] only specified the methods by which the keying material are delivered to the MN, but left the specification of the key delivery process to Mobile IP agents to AAA designers. This, although inconvenient, is understandable since the Mobile IP is really about the signaling between the MN and its Mobile IP agents (HA and FA), while any interaction between these agents and the AAA servers is really not within the realm of Mobile IP design. The plus side of including the Mobile IP agents inside the AAA infrastructure is that the AAAH server can use a AAA protocol (such as Diameter) in conjunction with the PSAs that exist within the AAA infrastructure to deliver the MSA key material to the Mobile IP agents. This, however, further complicates the key management for Mobile IP signaling and is the source of confusion for most people encountering the Mobile IP-AAA specifications for the first time.

- *Keys to Mobile IP agents*: The Mobile IP agents are part of the AAA infrastructure. We do not want to go as far as saying these agents are AAA clients, since requiring all the AAA client functionality from the mobility agents may be too much. The HA and FAs should support accounting or authorization signaling. They simply need to be able to interact with their own AAA server and have pre-established SAs (PSAs) with the server. These PSA may even support end-to-end IPsec or TLS channels and hence the AAA server can deliver the keys to these agents through a AAA protocol either in the form of encrypted attributes or over a completely encrypted channel. Note that delivery of keys to Mobile IP agents is not defined in Mobile IP specifications and is a function that has to be supported by the AAA protocol. As we will see later in this chapter, currently only Diameter provides support for this procedure, while RADIUS at the moment has no IETF specification for this functionality.
- *Nonces to mobile node*: On the other hand, the key material for the MN is delivered through the Mobile IP agents, using extended Mobile IP signaling. However, we know that the MN initially does not have any trust relationship with these agents. In fact, the MN needs to receive these key materials to create the MSAs with the agents. This means the MN cannot trust any data it receives from these Mobile IP agents, let alone the key materials that are the very source of trust. Furthermore, the MN cannot verify whether the Mobile IP agents have actually received the key materials that they are presenting to the MN from the AAA server. Finally, the keys must be delivered to the MN in a secure manner. For all these reasons, the AAA server cannot send the full MSA keys to the MN, since the keys traverse Mobile IP agents that are not trusted by the MN. Instead the AAA server sends random nonces that are cryptographically related to the actual MSA keys to the Mobile IP agents. The MN receives these nonces from the agents through Mobile IP signaling and

then, based on the AAA key that the MN shares with the HAAA, the MN can calculate these keys (the same keys as the AAA has sent to Mobile IP agents through AAA infrastructure) using a hash function and some other information in a manner similar to the following formula:

$$MSAkey = HMAC\_SHA1(AAA\_key, \{key\_material \parallel IP\_address\})$$

Note that the nonces will still travel from the AAA server to the Mobile IP agents through the AAA server but they do not need the same level of protection as the actual keys. This way, both the MN and the Mobile IP agents create the keys independently even though they initially do not trust each other.

After our previous discussion of key delivery process, understanding Figure 8.3 should be straightforward. The aim of Figure 8.3 is to emphasize the difference between the ways key materials are sent to MN and the protocol over which the key materials are sent to the Mobile IP agent.

One important point, Figure 8.3 (especially the lightning ray) is trying to make is that the Mobile IP-AAA interaction assumes Mobile IP signaling is used between the mobile node and the Mobile IP agents, while AAA protocols are used between the Mobile IP agents and the AAA servers. This implies the minimal required functionality for conversion or at least encapsulation of Mobile IP messages into AAA messages and vice versa. We recommend that this functionality is incorporated within the Mobile IP agents rather than within AAA servers, since burdening the AAA server with Mobile IP software not only makes the AAA server busier and complicated, but also will be a source of problem for networks deploying legacy AAA servers.

8.1.3 Overview of Use of Mobile IP-AAA in Key Generation

Our experience shows that once the trust model for Mobile IP-AAA signaling and the roles that each of Mobile IP and AAA protocols play in key distribution are understood, designing the rest of the interaction is much more straightforward. In the following sections we will describe the syntax and semantics of the messages and their extensions and revisit the entire process again. For simplicity (since this is an overview, after all), we assume that no FA is present and the MN sends its registration request directly to HA. Figure 8.4 shows an overview of the process. The notation X-Y-SA {message} denotes protection of a message with an SA between entities X and Y.

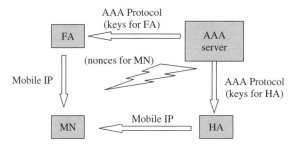

Figure 8.3 Distribution of keys from AAA server to the Mobile IP agents and mobile node is performed by different protocols

AAA and Security for Mobile IP

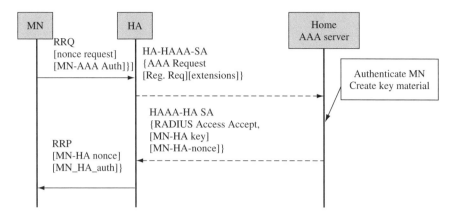

Figure 8.4 Overview of MSA key generation using Mobile IP and AAA protocols
Note: Solid lines indicate the use of Mobile IP, while dotted lines indicate the use of a AAA protocol.

The process is as follows:

- The mobile node sends a Mobile IP registration request (RRQ) towards the HA. When the MN does not have any trust relationship (MSA) with the HA, the MN cannot authenticate to the HA therefore must authenticate to the HAAA instead using an MN-AAA authentication extension. This extension also includes material to be used by the HAAA to perform MN authentication. The MN also requests a nonce from the HAAA. This request is made in the form of a nonce request extension to the Mobile IP RRQ. The exact format of these extensions is explained shortly.
- The HA captures the RRQ, but since it does not trust the MN, the HA creates a AAA request message, including the necessary information from the RRQ to help the AAA server with the authentication and key generation process. The exact format of the AAA request message depends on the AAA protocol that is used. We will discuss this in more detail later on when describing Diameter Mobile IP application.
- The HAAA verifies the MN-AAA authentication based on the information provided in the RRQ. If successful, the AAA server creates the nonces and related keys for the MN and HA, respectively, and creates a AAA response for the HA. The response message includes the keys for the HA and the nonces for the MN in separate attributes that may be protected differently.
- The HA extracts its own keys and the nonces (for MN) from the message from the AAA server, creates the MSAs for communications with the MN and processes the Mobile IP registration request from the MN. Upon successful registration, the HA creates a Mobile IP registration reply and includes the nonces in nonce reply extensions added to the reply. The HA authenticates the registration reply by calculating a Mobile-Home authentication extension (using the MSAs it just created).
- The MN captures the nonces from the nonce reply extension, creates the keys for MSA with the HA (using its AAA key) and verifies the Mobile-Home authentication extension. If there is a match, the MN trusts the registration reply and the HA from now on.

The rest of this chapter provides signaling and protocol details of the procedure that we have just described. Since the procedure is the result of an interaction between the Mobile IP and

AAA protocols, therefore, Mobile IP and AAA protocols needed to be extended to support this interaction. We devote the rest of this chapter to describing the extensions to Mobile IP and to AAA protocols. At this point, of the two main AAA protocols, only Diameter provides support for this interaction through IETF standards, while RADIUS extensions have not quite found their way into IETF yet. It should also be noted that most other extensions described here are the result of work in progress in IETF and are not completely stable yet.

8.2 Mobile IPv4 Extensions for Interaction with AAA

Now that we have shown the highlights of the Mobile IP-AAA interaction, we can go into more detail on the extensions that are being added to the Mobile IPv4 registration messages to support the process.

When discussing Mobile IPv4 registration in Chapter 5, we described the three Mobile IPv4 authentication extensions as specified by RFC 3344 [MIP3344]:

- Mobile-Home Authentication extension
- Mobile-Foreign Authentication extension
- Foreign-Home Authentication extension.

We also explained the Mobile-IP agent advertisement challenge extension and Mobile-Foreign challenge extension that are exchanged between the FA and the MN to provide replay protection for the foreign network during registration request. As we mentioned, these challenge extensions are described by RFC 3012 [MIPCHAL3012] that is currently being revised into a new RFC [3012bis].

The RFC 3012 also provided an interesting extension called generalized Mobile IP Authentication extension, for use of a third party verification infrastructure (that we guess is a general term for AAA infrastructure!) to help the FAs and HA to verify the mobile node's credentials. When the verification infrastructure is a AAA infrastructure, this extension is further specified as the MN-AAA Authentication extension. Adding the MN-AAA authentication extension to the registration request is a powerful concept, since it provides a way of creating all the trust relationships needed for secure operation of the mobility framework from a single trust relationship between the mobile node and its AAA server (AAASA). When the mobile node adds this extension to its registration request, the mobility agents simply contact the HAAA server for the security services that we described earlier, without worrying about their integrity being threatened. The syntax of this extension is explained in the following.

8.2.1 MN-AAA Authentication Extension

RFC 3012 [MIPCHAL3012] and its ongoing revision [3012bis] specify the so-called generalized Mobile IP authentication extension to allow the MN to present its credentials to an entity that understands these credentials. Since the main goal of that specification was to protect the integrity of a foreign network from malicious use of Mobile IP registration signaling, the designers tried to keep the trust model rather generic by assuming that this entity could be part of a generic third party verification infrastructure. Hence, the specification introduced a generalized Mobile IP authentication extension to allow the MN to authenticate to the generic verification infrastructure (Table 8.1).

Table 8.1 Generalized Mobile IP authentication extension

Field	Description
Type	36
Subtype	Type of entity to which MN authenticates, AAA server = 1
Length	4 (for SPI) plus length of authenticator
SPI	Security parameter index identifying SA with the entity
Authenticator	Data = preceding MIP data and extensions ‖ type, subtype, length, SPI Authenticator value = HMAC-MD5 (data, data length, key, key length)

It can be noted that by defining two fields of type and subtype, the generalized authentication extension in RFC3012 almost redefines the guidelines for Mobile IP extensions. While the type field defines the type for the extension in parallel to those defined in base Mobile IP specification (RFC 3344), i.e. the subtype defines the type of the verification infrastructure used by MN and mobility agents. For instance, when the verification infrastructure/entity is a AAA infrastructure/server, the subtype value is equal to 1 and the extension can be called MN-AAA authentication extension.

The reason for introducing the subtype field in the extension, we guess, was to be more frugal with the limited type space for extensions. RFC 3344 had already occupied a unnecessarily large chunk of type space by reserving the types 32, 33 and 34 for the Mobile-Home, Mobile-Foreign and Foreign-Home authentication extensions (see Chapter 5). RFC 3012 uses type number 36 for a generic extension that along with the subtype field might possibly have been used by RFC 3344. However, due to backward compatibility and the need to keep the already deployed implementations of Mobile IP, the original extensions are kept in place.

Now let us go back to the most important Mobile IP extension in support of Mobile IP-AAA interaction, i.e. MN-AAA Authentication extension. The MN uses the AAA key that it shares with the HAAA (AAASA) to calculate the authenticator within the MN-AAA authentication extension:

authenticator = HMAC_MD5(Data, Data length, key, key length)

As shown above, any other message fields that require integrity protection can be included in the calculation of the authenticator field as "data". All other extensions added to the registration request, including those requesting key material (see below) would be added to "data" prior to the MN-AAA authentication extension, so that they also can be protected by the authenticator field within this extension. Also it should be noted that at the time of writing most designers realize that SHA1 is providing better security strength than MD5 as a hash algorithm when comes to providing message authentication and, as we can see in the next subsection, the Mobile IP key management specifications are now mandating the use of SHA1 as a hash algorithm. Still at this point the RFC 3012 and its revision require only MD5 for calculation of MN-AAA authentication extension. We believe that [3012bis] should be revised to mandate SHA1 for this purpose; otherwise this would cause confusion for the implementers. However, to be faithful in our citations of IETF standards (the version of the [3012bis] stated in the references), we stated that MD5 is used in the calculation of MN-AAA authentication extension.

It should be noted that due to the limitations imposed by some AAA protocols and servers (such as RADIUS), the "data" field included in the MN-AAA Authentication

8.2.2 Key Generation Extensions (IETF Work in Progress)

In the previous subsection, we described the process by which the mobility agents can mediate in putting the MN in contact with the HAAA server. Here, we describe the extensions that allow the MN to request and receive the keying material from the HAAA server. In the same manner as the MN-AAA authentication, these extensions are also still under development in IETF [MIPKEYS3957]. But we feel that even if all the details are not set in stone yet, this is the way Mobile IPv4 and AAA protocols will interact and understanding the syntax and semantics will provide the reader with enough background to follow the progress of the standards.

The main principles of the key generation process through Mobile IP-AAA signaling were described in the treatment of the trust model earlier in this chapter. As we mentioned there, the MN creates the keys and MSAs with Mobile IP agents (HA and FA) based on the nonces it receives from the HAAA and the key it shares with the HAAA (AAA key). To request the nonces for each Mobile IP agent (FA and HA) from the HAAA server, the mobile node adds a nonce request extension for each nonce to its registration request. The formal name for a nonce request for MN-HA MSA for instance is Generalized MN-HA key generation nonce request extension, whose syntax is shown in Table 8.2.

So far, only one subtype (subtype 1) has been defined, which describes how the MN can create the MSA with an HA by using a transform indicated by the AAA SPI. For nonce requests, this fields has of course zero length.

The HAAA creates a nonce for the MN-HA MSA, and according to [MIPKEYS3957] inserts the nonce into the "key generation nonce" field of the subtype data within a "Generalized MN-HA key generation nonce reply extension" (Table 8.3). This extension is added to the Mobile IP registration reply sent back to the MN.

The data for the subtype is defined in Table 8.4.

In the same manner, to create MSAs with the FA, the MN requests nonces for the MN-FA-MSAs from the HAAA server. The format for the Generalized MN-FA key generation nonce request extension and the Generalized MN-FA key generation nonce reply extension is very similar to those described for MN-HA nonces.

Table 8.2 Generalized MN-HA key generation nonce request extension

Field	Description
Type	(TBD) Not assigned by IANA at this point
Subtype	=1, number assigned to identify the entity, used to generate the MN-HA key
Length	Length of the extension=length of subtype data plus 4 bytes for SPI
Mobile Node SPI	Security parameters index that the MN assigns for MSA created (HA_MN MSA) for use with the registration key (HA-MN key). The HA must later on include this SPI in MN-HA authentication extensions for messages from HA to MN.
MN-HA generation nonce request subtype data	Data needed to carry out the creation of the registration key on behalf of the MN. In MN-HA key generation nonce request, this field has zero length.

Table 8.3 Generalized MN-HA key generation nonce reply extension

Field	Description
Type	Not assigned at this point
Subtype	Number assigned to identify the way the subtype data in this extension is used to obtain the registration key (MN-HA key)
Length	Length of the extension = length of subtype data plus 4 bytes for SPI
MN-HA generation nonce request subtype data	Encoded copy of the keying material, possibly along with some other information needed for the creation of the MSA

Table 8.4 "MN-HA key generation nonce from AAA" subtype data

Field	Description
Lifetime	The duration of time (in seconds) for which the key material can be used to create the key.
AAA SPI	The SPI that the MN must use to determine the transform to use for creating the MN-HA MSA. For instance, the MN uses this SPI to locate the AAA key.
HA SPI	The SPI for the MN-to-HA MSA that the MN creates based on the nonce.
Algorithm identifier	An identifier selected from an authentication algorithm table to indicate the exact algorithm to use for computation of MN-HA authentication extension.
Key generation nonce	A 128 bit random number serving as nonce.

8.2.3 Keys to Mobile IP Agents?

If we look closely, the key generation extensions defined for Mobile IP only go halfway, i.e. if only Mobile IP registration extensions were implemented, only the MN would get the keys to be shared with the Mobile IP agents (HA and FA). The Mobile IP specifications leave one "small" detail out, and that is how does the AAA server get the same keys to the HA and FA? The reason is maybe understandable: that part of key distribution is done through AAA signaling (see Figure 8.3) and it is beyond the jurisdiction of the Mobile IP standardization working groups to design standard procedures for AAA infrastructure. Now that we have the readers' attention, we can reveal that so far we have not shown the details of how the Mobile IP agents and the AAA servers speak to each other. In the next section we will describe what is needed from the AAA protocols to support the interaction with Mobile IP.

8.3 AAA Extensions for Interaction with Mobile IP

As mentioned previously, the AAA infrastructure and the AAA protocol also need to support the interaction with the Mobile IP agents:

- Since the Mobile IP agents use AAA protocol to interact with AAA servers, they will be considered as AAA entities. The AAA functionality for Mobile IP agents needs to be defined clearly.

- The AAA protocol needs to specify the attributes and their semantics for carrying Mobile IP-related information between the AAA entities. The AAA protocol also needs to specify in what AAA messages these attributes must be carried.
- Functionality required for support of the Mobile IP authentication and key generation extensions at the AAA servers must be defined.

Of the two most prominent AAA protocols, only Diameter provides detailed specification for support of Mobile IP. At the time of writing no IETF standards for RADIUS interaction with Mobile IP exist, even though various communities have designed their own specifications for this interaction, as explained at the end of this chapter the authors have attempted to start an IETF standardization very recently [RADIUS MIPDR].

8.3.1 Diameter Mobile IPv4 Application

Diameter considers Mobile IPv4 as an application of its services and provides detailed specification for support of Mobile IP-AAA signaling [DIAMIP].

- Diameter Mobile IP application defines new commands and command codes for interaction between Mobile IPv4 agents and the AAA server and defines these interactions for a variety of mobility and trust scenarios, depending on the way the mobile node registers. Aside from the support for the mobile node's initial registration signaling, Diameter even describes the procedure to support Mobile IP handovers.
- Diameter Mobile IP application defines specific attribute value pairs (AVPs) to transport Mobile IP-related material between the AAA servers and Mobile IP agents (FA and HA).
- Furthermore, Diameter supports dynamic assignment of home address, and home agent for the mobile nodes, when needed. Diameter even provides support for provisioning of Mobile IP home agents within foreign networks. This is desirable in roaming scenarios, since when the deployed HA is located in the home network far away from MN's current location, long round trip delays are introduced due to the triangle routing through the HA. Diameter also provides guidance on usage of Network address identifier (NAI) for the MN and routing of the Diameter messages towards the HAAA.
- Finally, the Diameter Mobile IP specification [DIAMIP] provides the best one-stop-shop as reading material on the Mobile IP key generation support procedures and the detailed definition of the MSAs and the SPIs.

In the following, we provide an overview of the Diameter Mobile IP application as described in [DIAMIP]. It is important to note that this specification is still work in progress and not finalized at the time of writing.

8.3.1.1 Diameter Model for Mobile IP Support

As we explained in Section 8.1.3, Mobile IP-AAA interaction assumes Mobile IP agents can act as a translation/conversion point for Mobile IP and AAA signaling. The Diameter Mobile IP application follows this model and assumes Mobile IP agents are able to create, send, receive and understand the Diameter messaging and AVPs defined in Diameter Mobile IP application.

As we explained in our treatment of trust and architecture model, Mobile IPv4-AAA signaling for a mobile node connecting to a foreign network is influenced by whether or not the foreign network deploys foreign agents (FA):

- FA-based CoA: When the foreign network has deployed foreign agents and the MN is using a FA-based care of address (CoA), the MN must register through an FA. The FA would then forward the registration request to the HA directly. When the necessary MSAs for support of Mobile IP authentication do not exist, the FA must forward the request to the AAA server.
- Co-located CoA: When no FAs are deployed, the MN sends its registration request directly to the HA. When the MN does not share an SA with the HA, the HA can then send the request to AAA server to authenticate the MN prior to processing the registration request.

As we can see, the Diameter Mobile IP interaction must be designed based on the mode of CoA registration. The path taken by Diameter signaling depends on whether FAs are deployed and the administrative domain to which both FA and MN belong. When foreign administrative domains are involved, the FA must first contact its local AAA server (LAAA), which in turn contacts the home AAA server (HAAA) for the MN and the trust model for Diameter Mobile IP interaction is more complicated (Figure 8.1). To keep this discussion simple, we assume the mobile only moves within the boundaries of its home domain, i.e. a case when the FA is also served by the AAAH and up do not consider the LAAA-HAAA interaction here. The mechanics of the process are still the same, but that assumption makes the figures and the understanding of the problem a bit simpler. The reader is referred to [DIAMIP] for a treatment of that model.

Figure 8.5 shows how a Diameter server interacts with the Mobile IP agents in each of the two Mobile IP modes explained above. As we can see in Figure 8.5. Diameter Mobile IPv4 application has defined new message types (commands and command codes) for this interaction. In the following we provide a brief description of each of the new Diameter command codes. We will go through the messaging procedure afterwards.

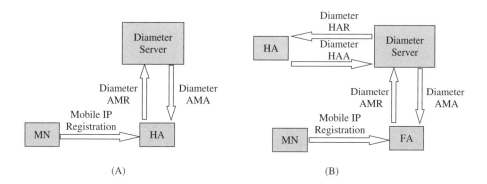

Figure 8.5 Interaction between Diameter server and Mobile IP agents (A) Co-located mobile nodes (B) Mobile nodes using a foreign agent

Diameter defines four new commands for interaction with Mobile IP agents:

- *AA-Mobile node Request (AMR)*: The AMR command is sent from the Mobile IP agent, receiving the Mobile IP registration request, to the Diameter server (in this case HAAA). Through this message, the Mobile IP agent forwards the MN's request for authentication and key generation to the HAAA. As we can see in Figure 8.5, for co-located MNs the AMR is created and sent by the HA (Figure 8.5(A)), whereas for MN using an FA, the AMR is created by the FA. The Mobile IP agent includes necessary information related to MN's registration, authentication and key generation inside specific Diameter AVPs that are explained later on. The Mobile IP agent is assumed to have the minimum necessary information required for routing of the Diameter message towards the HAAA.
- *AA-Mobile node Answer (AMA)*: The AMA command is sent from the Diameter server, acting as the HAAA for the MN, towards the Mobile IP agent that had sent the AMR. The HAAA server includes the result of processing of the AMR inside a Result-Code AVP. When processing is successful and key materials were requested, the AMR includes AVPs that carry information to be used for generation of MSAs at MN and at the Mobile IP agent receiving the AMR. For FA-based MNs (Figure 8.5(A)), the Diameter server has already contacted the HA for the MN and received a registration reply. Hence, for these nodes the AMA sent to the FA already includes the registration reply as an AVP (detailed in the following subsection). For co-located MNs (Figure 8.5(B)), the HA processes the Mobile IP registration request first after it receives the AMA from the Diameter server. Regardless of the case, when the Mobile IP agent receives the AMA from the Diameter server, the agent sends the result of Mobile IP-AAA signaling in the form of a Mobile IP registration reply message with related extensions to the MN.
- *Home agent MIP-Request (HMR)*: The HMR command is sent from a Diameter server to HA to request the HA to process the Mobile IP registration request. As shown in Figure 8.5(B), this is applicable only to the cases when the MN registers through an FA rather than directly with the HA. Obviously this Diameter command needs to include Mobile IP registration request as an AVP, so that the request can be conveyed to the HA. The command includes key material for HA and nonces for MN for MSAs with HA and FA as well. However, for obvious reasons the key materials for the FA are not included in this message.
- *Home agent MIP-Answer (HMA)*: The HMA command is sent back from the HA to the Diameter server that has sent the HMR to the HA.

We will describe the messaging procedure for Diameter Mobile IP signaling shortly. However, it is useful to go through the details of the new AVPs that are defined for support of Mobile IP within Diameter.

8.3.1.2 New Diameter AVPs for Mobile IP Support

Diameter Mobile IP application [DIAMIP] defines a large number of AVPs for support of Mobile IPv4 authentication and key distribution. Table 8.5 shows a number of AVPs defined in the application, while the full list of AVPs is provided in [DIAMIP]. In order to provide clarity, we have grouped the AVPs into two tables to help the understanding of the use of these AVPs. A brief description of each AVP is provided in Table 8.5. This description should be enough to gain an understanding of the purpose of most of these AVPs. We will describe the use of many of these AVPs when we go through the details of

Table 8.5(a) Diameter AVPs defined for support of Mobile IP-AAA authentication

Attribute name	AVP type	Description	
MIP-reg-request	320	Carries the Mobile IP registration request sent by the MN to the FA.	
MIP-reg-reply	321	Carries the Mobile IP registration reply sent by the HA to the FA	
MIP-MN-AAA-Auth	322	Carries ancillary data to assist the AAA server with verification of authentication data in the registration request. Note that the authenticator value itself is not included here. This AVP is of a grouped type, which means it can carry a number of other AVPs: • MIP-MN-AAA SPI • MIP-Auth-Input-Data-Length: The length of the data portion of the registration request to be used as input when calculating the authenticator. • MIP-Authenticator length: The length of the authenticator to be validated by the AAA server. • MIP-Authenticator offset: The offset into the registration request data, where the authenticator value (to be verified) can be found.	
MIP-MN-address	333	Carries mobile node's home IP address	
MIP-HA-address	334	Carries home agent's IP address	
MIP-Feature-Vector	338	Carries flags with values set by the FA or by the LAAA operating the FA domain. This AVP is included by the FA in message to the LAAA, some example values provided	1 MN home address requested 2 HoA allocable only in Home Realm 4 HA requested 8 Foreign HA is available 16 MN-HA key requested 32 MN-FA key requested 64 FA-HA key requested 128 HA in foreign network 256 Co-located mobile node
MIP-MN-AAA-SPI	341	Indicates the MSA, which the HAAA must use to validate the authenticator value included by the MN in the registration request.	
MIP-FA-Challenge	344	Contains the challenge advertised by the FA to the MN. This AVP must be present in the AMR message if the MN has used RADIUS-style MN-AAA computation algorithm.	

Table 8.5(b) Key distribution AVPs

Attribute name	AVP type	Description
MIP-FA-to-HA-SPI	318	Assigned by HA, used by FA in calculations of FA-HA-auth. extension
MIP-FA-to-MN-SPI	319	Assigned by MN, used by FA in calculations of MN-FA-auth. extension
MIP-HA-to-FA-SPI	323	Assigned by FA, used by MN in calculations of HA-FA-auth. extension

(*continued overleaf*)

Table 8.5(b) (*continued*)

Attribute name	AVP type	Description
MIP-MN-to-FA-MSA	325	A grouped AVP, from the AAA server to HA inside HAR message, including • MIP-MN-to-FA-SPI (allocated by FA) • MIP-algorithm type, • MIP-nonce (to be used for generation of MN-FA session key)
MP-FA-to-MN-MSA	326	A grouped AVP, from the AAA server to FA inside AMA message, including • MIP-FA-to-MN-SPI (allocated by MN) • MIP-algorithm type, • MIP-session key (with FA)
MIP-FA-to-HA-MSA	328	A grouped AVP, sent to FA in an AMA message, including • MIP-FA-to-HA-SPI (allocated by HA) • MIP-algorithm type, • MIP-session key (shared by HA and FA)
MIP-HA-to-FA-MSA	329	A grouped AVP, sent to HA in an HAR message, including • MIP-HA-to-FA-SPI (allocated by FA) • MIP-algorithm type, • MIP-session key (shared by HA and FA)
MIP-MN-to-HA-MSA	331	A grouped AVP, sent from AAA either inside HAR to HA for FA CoA MNs, or inside AMR to CcoA MNs message, including • MIP-MN-to-HA-SPI (allocated by HA) • MIP-algorithm type, • MIP-nonce • MIP-replay mode
MIP-HA-to-MN-MSA	332	Similar to MN-to-HA MSA, including • MIP-HA-to-MN-SPI (allocated by MN) • MIP-algorithm type, • MIP-session key • MIP-replay mode
MIP-nonce	335	This AVP contains the nonce selected by the HAAA and sent to the MN for creation of the associated security association based on the methods defined by Mobile IP key management procedures.
MIP-Session-Key	343	Contains the session key selected by the HAAA and sent to the Mobile IP agent for the associated security association.
MIP-Algorithm-Type	345	Contains an identifier for the algorithm (selected by the HAAA) to be used for calculation of authentication extensions. Value = 2 for HMAC-SHA1.
MIP-replay-mode	346	Contains an identifier for the replay mode (selected by the HAAA) that the HA requires of an authenticating MN. Values: 1 for no replay protection, 2 for timestamps, 3 for nonces
MIP-MSA-lifetime	367	Represents the period of time (in seconds) over which the session key or nonce is valid.

Diameter Mobile IP messaging and procedures. In the following, we provide a few more notes for understanding some of these AVPs:

- The authenticator value within the MN-AAA-Authentication extension of Mobile IP registration request is not included in the Diameter MIP-MN-AAA-Auth AVP. Instead this AVP is built as a grouped AVP (see Table 8.5) consisting of shorter AVPs including only the ancillary data to help the AAA server to calculate its own copy of the authenticator value. To help the AAA server in verifying this copy with the copy calculated by the MN (and included in the registration request) an MN_Authenticator_offset AVP is included in the messaging.
- The MIP-Feature-Vector AVP provides a powerful way of conveying a large amount of scenario-specific information to the AAA server. This vector consists of a number of flags, each indicating a different request or piece of information to the AAA server. For instance, when bit 16 is set, it means the MN asks the AAA server to create keying material for an MN-HA MSA.
- Since the AAA server sends the nonces to the MN, and keys to the Mobile IP agents, the AVPs carrying information for MSAs between the MN and its Mobile IP agents are different and contain different information. For instance the MIP-MN-to-HA-MSA includes the nonces, while the MIP-HA-to-MN contains the actual session key.
- Diameter supports a number of advanced Mobile IP functionalities to increase the flexibility and scalability of mobility management for roaming environments. For instance, Diameter allows the home IP address of a mobile node and/or its home agent to be assigned by a AAA server. Furthermore, to reduce the traffic due to triangle routing, Diameter even allows an HA to be allocated to the MN in the foreign network. We do not go through the details of these procedures, but this note serves to explain some of the AVPs listed in the Tables 8.5(a) and (b).

8.3.1.3 Diameter Mobile IP Messaging Overview

As mentioned earlier, the exact messaging flow depends on the trust model and Mobile IP deployment scenario. In this section we briefly go through the messaging procedure for the each of the two Mobile IP scenarios described earlier, namely, when Mobile IP FAs are deployed and when they are not.

Procedure for Mobile Nodes Using an FA
In this case the MN registers through an FA. The Diameter Mobile IP operation is as follows (shown in Figure 8.6):

- The MN sends a registration request to the FA and since it does not have any MSAs with FA or HA, it includes a generalized MN-HA key generation nonce request extension and a generalized MN-FA key generation nonce request extension with its registration request. To comply with the challenge response mechanism (if implemented by the FA), the MN includes a Mobile-Foreign Challenge extension, including the challenge from the latest agent advertisement or registration reply to the message. Finally, the MN authenticates the entire message by using its AAA key with the HAAA to calculate the authenticator within the MN-AAA-Authentication extension and includes the extension to the registration request.

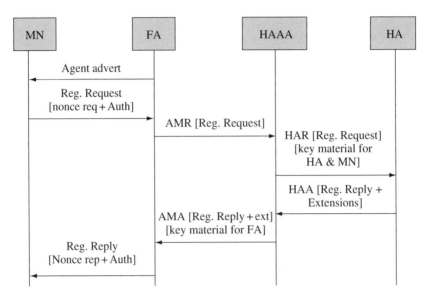

Figure 8.6 Diameter Mobile IP messages for MNs registering through a Mobile IP FA

- Once the FA receives a registration request message from a mobile node, the FA creates an AA-Mobile-Node-Request (AMR) towards the HAAA server for the MN. Since the HAAA is not be required to understand Mobile IP messages, the FA extracts any information, pertinent to HAAA, from the registration request and forms an appropriate AVP, accordingly. Examples are MIP-registration request AVP and MIP-Feature Vector AVP. The latter AVP indicates the types of advanced Mobile IP features and the keying materials that are requested from the HAAA. The FA also extracts information from the authentication extension and includes a MIP-MN-AAA-Auth AVP, so that the HAAA can authenticate the MN and its message. When the FA implements challenge response mechanisms, the FA also includes the challenge information in a MIP-FA-challenge AVP within the AMR.
- The HAAA processes the AMR, extracting the AVPs related to authentication and registration key materials. If the authentication is successful, the HAAA creates the requested key material and creates a Home-Agent-MIP-Request (HAR) message for the HA, so that the HA can process the Mobile IP registration request itself. For that reason, the HMR includes the MIP-registration request AVP as well as any other AVPs that carry the keys for the HA and nonces for MN (including nonces for MN-FA-MSA). The HAR message may include information in MIP-MN-address AVP and MIP-HA-address AVP, to provide these addresses to the HA (depending on types of requests made to the HAAA through the MIP-Feature-Vector AVP in the previous AMR).
- The HA receives the HAR and extracts the Mobile IP registration message. The HA may assign a home address and create the binding for the MN or just simply use the binding based on the home address provided to the HA (either by the MN through the registration request, or by the HAAA). The HA uses the key material from the Diameter server to create the MSAs with the MN, and with the FA (if required). To create the MSA with the MN, the HA extracts the session key from the MIP-HA-to-MN-MSA AVP and keeps a copy, while it copies the nonces for the MN from the MIP-MN-to-HA-MSA AVP and MIP-MN-to-FA-MSA

AVP into a generalized-MN-HA key generation nonce reply extension and a generalized-MN-FA key generation nonce reply extension, respectively. Finally the HA creates a Mobile-Home authentication extension for the registration reply, using the keys for the MSA with the MN to authenticate the entire message (including the nonce reply extensions). The HA creates a Home-agent-MIP-answer (HMA) message (including a MIP-registration reply AVP) for the HAAA.
- The HAAA server forwards the registration request along with the AVPs carrying keying material for MSAs that the FA needs to create with the MN and HA inside an AA-mobile-node-answer (AMA) to the FA. The HAAA also sends the nonces for an MN-to-FA MSA inside an AVP (with the same name) to the FA.
- The FA extracts the registration reply as well as the AVPs including the key materials from the Diameter AMA. The FA creates the necessary MSAs with the MN and the HA and authenticates the registration reply using the MSA it just created with the MN and appends a Mobile-Foreign authentication extension to the registration request. When a challenge/response is implemented, the FA may include a challenge for MN to use with the upcoming registration request in the registration reply.

Procedure in Absence of FAs
In this scenario, the MN registers its collocated CoA directly with the HA (Figure 8.7). The procedure is as follows:

- The MN forms a registration request including the extensions for authentication to AAA and nonce requests for an MSA with the HA (see previous case for details). The MN sends the registration request to the HA. No challenge-related extensions are required in this case.
- The HA builds an AMR message based on the registration request in a similar manner as described for the FA in the previous case and sends the AMR to the HAAA server.
- After processing the authentication and creating the key material, the HAAA sends an AMA back including the session keys for the HA and the nonces for the MN, within the MIP-HA-to-MN-MSA and MIP-MN-to-HA-MSA AVPs, respectively.
- The HA processes the registration request and creates the MSA with the MN. The HA creates a registration reply and authenticates the reply using the newly created MSA.

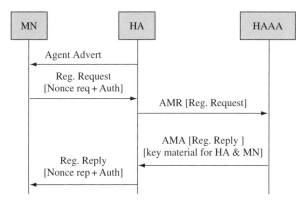

Figure 8.7 Diameter Mobile IP messages for MNs registering directly with a Mobile IP HA

8.3.2 Radius and Mobile IP Interaction: A CDMA2000 Example

Cellular standard development organizations (SDO), especially the Third Generation Partnership Project 2 (3GPP2) tend to borrow many Internet-related standards from IETF in their network architecture and protocol development efforts. However, 3GPP2 has restricted itself to only using IETF specifications that are in RFC format and are rather stable and accessible. This is to avoid deploying unstable specifications that may cause interoperability issues between 3GPP2 operators. For this reason Internet drafts cannot be deployed for 3GPP2 implementations. 3GPP2 has implemented both Mobile IP and AAA mechanisms. However, when the mobility standards were being developed, Diameter base specification [DIA3588] was still merely an Internet draft, so the 3GPP2 community decided to use RADIUS for its AAA infrastructure.

As opposed to Diameter, IETF RADIUS specifications do not provide any guidance on how a RADIUS infrastructure is supposed to interact with Mobile IP agents. Furthermore, the Mobile IP key management specifications are in Internet draft format, even at the time of writing. For these reasons, the 3GPP2 designers had to work around the lack of IETF specifications for both RADIUS–Mobile IP interaction and Mobile IP key management by developing these procedures on its own. However, it is easy to see that the specification is greatly inspired by the work in IETF [CDMAMIP].

We provide a brief description of the CDMA2000 procedure for Mobile IP-AAA interaction as the closest real-life example we can find for RADIUS–Mobile IP interaction. Before diving into the interaction, it is useful to know a few more details about the way CDMA2000 systems handle data traffic.

CDMA2000, although a highly sophisticated technology, still follows the circuit switched mentality, where the network treats and handles voice and data traffic separately. CDMA2000 networks deploy a specific entity called the Packet Data Service Node (PDSN), which handles the data traffic and in the case of mobile nodes deploying Mobile IP, it acts as a Mobile IP foreign agent. The PDSN also supports RADIUS.

As opposed to voice connections, for data connection, the end node must establish a point-to-point (PPP) connection (see Chapter 1) with the PDSN. When the user is not mobile and does not deploy Mobile IP, the PPP link establishment also includes a PAP or CHAP authentication. However, when the user is mobile, a Mobile IP-AAA registration-authentication procedure is performed instead of the simple CHAP or PAP procedure used for stationary users.

8.3.2.1 Mobile IP Support Within CDMA2000

As mentioned in Chapter 5, while authentication of Mobile IP registration messages between MN and HA is mandatory, Mobile IP RFCs [MIP3344] does not mandate authentication of messages between MN and FA. As opposed to IETF that only concerns itself with communications on a single layer of the TCP/IP stack, CDMA2000 standards design across multiple layers. For instance, the CDMA2000 designers can rely on the fact that some sort of authentication exists between an MN and the access link it is using and hence a Mobile IP type authentication between MN and PDSN, acting as FA (at layer 3) is not required. For Mobile IP-AAA signaling this means one less MSA (the MSA between MN and FA) and shared key to worry about. For the remaining MSAs within the Mobile IP-AAA trust model, i.e. the FA-HA and MN-HA MSAs, the CDMA2000 designers could not rely on IETF Mobile IP key management Internet drafts, since no RFCs existed. Hence the nonce request and response mechanisms described in Section 8.2.2 could not be deployed either:

- Since both FA (PDSN) and HA are part of the wired network, it opted to use a direct IKE dialog between the FA and HA to establish an IPsec security association between those two. The FA and HA may exchange certificates for IKE authentication. In the absence of certificates, the shared secret for IKE authentication can be requested from the RADIUS server. In the following we describe each of these alternatives briefly:
    - Public key X.509 certificates: When the PDSN receives the RRQ from the HA and the PDSN does not share an SA with the HA, the PDSN verifies to see if it has certificates that belong to HA, or the HA owns any certificates. The problem with this method is that it requires the PDSN and HA to have certificate-processing capabilities.
    - When PKIs and certificates are not available, dynamic IKE pre-shared secret distribution through the home RADIUS server can be used. The process for establishment of this pre-shared secret, called the "K" key is described shortly.
    - A statically configured IKE pre-shared secret is loaded into both HA and PDSN. Our guess is this option may not be desirable, if scalability and system administration burden are a concern.
- Unfortunately, the specification is not very clear about how the key for MN-HA MSA is requested or generated. It is mentioned that if the MN-HA key is being transferred from the RADIUS server to the HA, it must be encrypted with RSA-MD5. However, the specification omits the details on how this key is delivered to the MN. The fact that the standards allow the same secret to be used for the calculation of the MN-AAA authentication extension and the MN-HA authentication extension implies that many systems may use the AAA key (the key the MN shares with the AAA server) as the MN-HA key, as well. Not entirely a robust trust model!
- The CDMA2000 model handles the challenge response mechanism slightly differently from what was described earlier. A CHAP-based mechanism is used instead as discussed briefly later on.

8.3.2.2 RADIUS Support, or Not!

The RADIUS–Mobile IP interaction for CDMA2000 network uses many of the conventional RADIUS attributes, such as User-Name and User-Password to carry information to RADIUS server [CDMAMIP]. However, a fair amount of Mobile IP information must also be carried by the AAA protocol (as we saw in our Diameter Mobile IP discussion), but IETF RADIUS specifications do not provide support for Mobile IP. This means there are no standardized attributes (with standardized type numbers) to carry Mobile IP information within RADIUS framework.

The lack of IETF standardized attributes forced CDMA2000 designers to define proprietary attributes using the RADIUS vendor-specific attribute (VSA) format. Vendor-specific attributes all have attribute type 26/X, with 26 to indicate they are vendor specific and X to indicate the sub-type, i.e. the information they carry. Table 8.6 shows a subset of the VSAs as defined in [CDMAMIP] as an example to help understand the RADIUS–Mobile IP interaction. Many of the attributes shown in Table 8.6 are described in the following text and therefore, we will not go through their details in this section.

Table 8.6 RADIUS vendor-specific attributes defined by CDMA2000 for Mobile IP support

Attribute name	Attribute type	Description
IKE pre-shared Secret Request	26/1	Sent from PDSN to the AAA within a RADIUS access request to request a pre-shared key for IKE
Security level	26/2	From AAA server to PDSN or to HA
Pre-shared secret	26/3	Sent from AAA server to PDSN
'S' key	26/54	From AAA server to HA to send the HA-AAA shared key ('S' key)
'S' Request	26/55	From HA to AAA server to request the HA-AAA shared key ('S' key)
KeyID	26/8	
MN-AAA removal indication	26/81	From AAA server to PDSN, to indicate to PDSN to remove challenge and MN-AAA extension before forwarding the registration request to the HA
MN-HA SPI	26/57	From HA to AAA server to request an MN-HA shared key
MN-HA shared key	26/58	From AAA server to the HA to send the MN-HA shared key

RADIUS Server's Role in Distribution of Pre-Shared Secret for IKE Between CDMA2000 PDSN and HA

As mentioned earlier, depending on the network security policy, the PDSN and the HA may need to engage in an IKE procedure to establish an IPsec secure channel (security association). However, the authentication exchange within the IKE requires that a pre-shared key exists between the PDSN and the HA. CDMA2000 allows the pre-shared key, called the "K" key, to be requested from the RADIUS server. Based on this request, the RADIUS server generates the key and delivers it to these two entities.

To request the pre-shared key for IKE from the home RADIUS server, the PDSN sends a RADIUS Access-Request message to the server. This request includes an IKE pre-shared Secret Request attribute. The Home RADIUS server distributes a key identifier and pre-shared secret for IKE to the PDSN. The RADIUS server needs to distribute the same key to the HA, but to handle the key transport security, the home RADIUS server and the HA are required to share a security association (a shared secret "S"):

$$K = HMAC\_MD5(Home\_RADIUS\_IP\_addr \mid FA\_IP\_address \mid timestamp, "S")$$

This, however, means that the HA must retrieve the "S" key from the RADIUS server in advance. To do this, the HA sends a RADIUS Access Request message to the RADIUS server and adds a "S" Request attribute to the message. The RADIUS server responds by sending a RADIUS Access Accept including the "S" key inside an "S" key attribute. The specification does not go into the details of the "S" key it protected during this transfer and simply says that an RSA-MD5 method needs to be used to encrypt the "S" key and its lifetime.

The PDSN, on the other hand, simply receives the "K" key through the pre-shared secret attribute included in the RADIUS access accept message from the RADIUS server to the PDSN.

Support for SPIs and Challenge/Response

Some AAA servers only admit a single security association. This means they are not able to use the SPIs to find the security association that would be required to verify the submitted Mobile IP authentication extensions. In such cases a specific SPI number, defined by RFC 3102 as CHAP-SPI, is reserved and used by the entity that calculates the authentication extension (such as MN). We refer the reader to [MIPCHAL3012] and [CDMAMIP] for more details and suffice to say that this SPI indicates to the receivers (such as a AAA server), what method the receiver must use to verify the authentication extension.

8.3.2.3 CDMA2000 Messaging Procedure

Now that we have gone through the labor of understanding what was required of proprietary RADIUS servers and Mobile IP agents to support CDMA2000 specifications, we will go through the messaging process to put it all together (Figure 8.8):

- Immediately following the completion of the PPP link negotiation, or as a response to the MN's agent solicitation, the PDSN starts sending agent advertisements to announce that it can act as a Mobile IP FA. The PDSN includes an MN-FA challenge extension in its agent advertisements to deliver a challenge to MN.
- Once the PDSN receives the registration request from the MN, it forwards the registration request inside a RADIUS Access Request message to the RADIUS server for authorization. If the registration request includes a specific address for the home agent, the PDSN needs to verify whether an SA exists with the HA or not. If not, it needs to establish the SA, possibly by asking the RADIUS server for help. The PDSN may communicate the FA challenge information to the home RADIUS server in a RADIUS Access Request

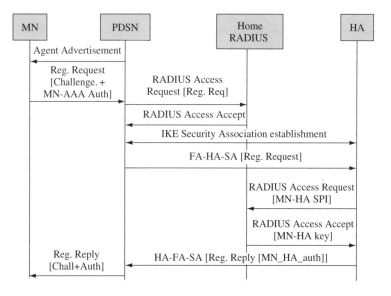

Figure 8.8 RADIUS–Mobile IP interaction for CDMA2000

message, depending on what SPI the MN has included in the calculation of the MN-AAA authentication extension.
- If the authentication succeeds, the home RADIUS server sends a RADIUS Access Accept message back to the PDSN. Otherwise, the home RADIUS server sends a RADIUS Access reject message back to the PDSN. The Access Accept message may include a RADIUS attribute "MN-AAA removal indication" to indicate to the PDSN that the PDSN does not need to include an MN-FA challenge extension and MN-AAA authentication extension in the registration request that it forwards towards the HA.
- The PDSN forwards the registration request towards the HA either as received from the MN or after removing some of the extensions based on the directive by the RADIUS server (as described above). The PDSN may be required to have a security association with the HA for forwarding the registration request. If a shared key is configured between the PDSN and HA, the PDSN will use the shared key in the calculation of the Mobile IP Foreign-Home authentication extension included in the registration request. Note that when the home network authorizes the IPsec service, the PDSN will not forward the registration request to the HA unless an IPsec SA is established between the PDSN and HA. If an IPsec SA does not exist, but IPsec is authorized, the PDSN establishes an IPsec SA with the HA using IKE and one of the authentication options that were explained previously.
- When the HA receives the registration request from the PDSN, it authenticates the registration request using the MN-HA shared key. For initial registration requests, when no MN-HA shared key yet exists, the HA needs to retrieve the keys from the home RADIUS server. To retrieve the MN-HA shared key from the home RADIUS server, the HA sends a RADIUS Access request message to the home RADIUS server and includes a user name, user password (the password is configured between the HA and AAA server) and an MN-HA SPI to indicate that MN-HA keys are required.
- The home RADIUS server responds with a RADIUS Access Accept message including the MN-HA shared key encrypted with RSA MD5 back to the HA.
- The HA saves the MN-HA shared key and then calculates an MN-HA authentication extension (according to RFC 2002) based on the newly received MN-HA shared key. For later registrations the HA uses the MN-HA shared key to verify the MN-HA authentication extension included in the registration request by the MN.
- The HA sends a registration reply, including the MN-HA authentication extension for authentication to MN to the PDSN and uses the IPsec channel between the PDSN and HA.

8.4 Conclusion and Further Resources

In this chapter we provided an overview of the important aspects of a system integrating Mobile IP and a AAA infrastructure to provide secure roaming. Unfortunately this interaction is rather complicated and the standards are not quite stable yet, but we tried our best to provide the right amount of detail without diving too deep into every technical detail. In fact, we must confess that the level of complexity involved in Mobile IP-AAA interaction was the main reason we decided to write this book: to help engineers not familiar with AAA protocols to understand the procedure without having to first plow through lengthy and frequently changing IETF documentations. The fact that Diameter itself is almost non-existent in the real world and there are no IETF RADIUS–Mobile IP documentations does not make the implementation of Mobile IP-AAA signaling any easier. There are almost no vendors that at this point claim support for Mobile IP-AAA signaling, since there are very few commercial networks that deploy this interaction. From what the authors found out, the very few vendors that advertise mobility functionality for their AAA server are those that support parts of the CDMA2000 way of implementing Mobile IP-RADIUS interaction. However, the authors are hoping to initiate the standardization process for RADIUS–Mobile IP interaction in IETF very soon [RADIUS MIPDR].

8.5 References

1. [MIPCHAL3012], C. Perkins, P. Calhoun, "Mobile IPv4 Challenge/Response Extensions", RFC 3012, November 2000.
2. [3102bis], C. Perkins et al., "Mobile IPv4 Challenge/Response Extensions (revised)", IETF work in progress, draft-ietf-mipv4-rfc3012bis-02.txt, June 2004.
3. [MIPKEYS3957], C. Perkins, P. Calhoun, "AAA Registration Keys for Mobile IPv4", IETF, RFC 3957 March 2005.
4. [DIAMIP], P. Calhoun et al., "Diameter Mobile IPv4 Application", IETF work in progress, draft-ietf-aaa-diameter-mobileip-20.txt, August 2004.
5. [CDMAMIP], 3GPP2, "CDMA2000 Wireless IP Network Standards: Simple IP and Mobile IP Access Service", 3GPP2 X.S 0011-002C, V 1.0 August 2003.
6. [MIP6BOOT], A. Patel et al., "Problem Statement for Bootstrapping Mobile IPv6", IETF work in progress, draft-ietf-mip6-bootstrap-ps-00, July 2004.
7. [MIP3344], C. Perkins, "IP Mobility Support for Ipv4", RFC 3344, August 2002.
8. [RADIUS MIPDR], M. Nakhjiri et al., "RADIUS mobile IPv4 extensions", IETF, Internet draft, draft-nakhjiri-radius-vnip4-00.txt, February 2005.
9. [DIA3588], P. Calhoun, J. Loughney, E. Guttman, G. Zorn and J. Arkko, "Diameter base protocol", IETF, RFC 3588, Sept. 2003.

9

PKI: Public Key Infrastructure: Fundamentals and Support for IPsec and Mobility

We have discussed the topics of authentication and key management and the related services in relatively great length in earlier chapters (especially Chapters 2 and 3) and pointed out the strength of public key cryptography in providing these services on many occasions. As noted earlier, public key cryptography involves a key pair of which one has to be kept very private (private key) and the other that should be widely publicized (public key). The reasons for this are very simple:

- When sending a signed message, the sender (signer) uses the private key to produce the signature, while the recipient uses the public key of the sender (signer) to verify the signature.
- When encrypting a message, the sender of the message uses the public key of the recipient to encrypt the message. Only the recipient, that is also the sole owner of the private key, is able to decrypt the message.

One might think keeping the private key private must be the hardest part in management of public keys. While this is true, keeping the public key public is not an easy task either. How do we make sure that the recipient of a signed document has the public key of the signer, or the sender of a secret document has the public key of the recipient? One way the engineers have dealt with this is to produce public key certificates. The certificate includes the private–public key pair owner's identity and public key. By sending this certificate to the other party (we call the other party, the certificate-using party), the sender makes sure that the receiver can have the sender's public key when needed.

Producing and managing certificates is an involved task, since it will provide the fundamental source of trust for all the entities that perform security functions based on certificates. There are many requirements that must be met when managing certificates and many functions that need to be performed to meet these requirements. Putting the burden of all these functions on certificate authorities (CAs) has been shown to be not a scalable and robust approach. To offload the busy CAs, many of these functions are outsourced to entities outside a CA. To ensure the integrity of the process, public key infrastructures that define the relationship between these entities are put in place. This chapter discusses the design and management of a public key infrastructure (PKI) that is used to support public key certificates for authentication and encryption using public key cryptography. However, before going on, we would like to point out that a PKI simply provides a way of assigning various entities in the network certificates and managing these certificates. It may also provide methods for creating the public and private key pairs. But it is wrong to assume that by having a PKI or buying one, the job of authentication or key management design is done. This is similar to thinking that by simply spending tens of thousands of dollars on building a modern kitchen, anybody without cooking skills or knowledge of food ingredients would be able to prepare elaborate pasta dishes. Apologizing for our ignorance of geological events, even filling the freezer with frozen dinners would not help a cave man (or a slightly more sophisticated cave woman). Assuming he or she is a post Ice Age person, used to warm food, he or she would first need to figure out the buttons on microwave oven. In the same manner, the certificates provided by the PKI must be worked into the authentication or key management schemes. The PKI does not replace such schemes. We provided examples of this by describing the use of certificates in authentication (Chapter 2) and key management (TLS). We will also talk briefly about the use of certificates in IKE.

9.1 Public Key Infrastructures: Concepts and Elements

9.1.1 Certificates

Digital certificates are used to present information about the user's identity and public key for various purposes. However, a trusted authority needs to guarantee the information in the certificate, especially the public key data, so that such information can be trusted to be free of tampering. Without an assurance from a trusted and known authority, the certificate is almost the same as a business card. Any person can choose any title with an affiliated company, phone number and other information and print a business card on her own. There is almost nothing on the business card (except possibly the company's logo) that can be guaranteed. On the other hand, in the case of a digital certificate, a trusted certificate authority (CA) issues the certificate on behalf of the user and signs the certificate with its own (the CA's) private key. This way the CA vouches that the user, whose identity and public key are included in the certificate, does in fact hold the corresponding private key. A user that has been granted the certificate can from now on use the certificate as a proof of identity and as a way to inform its peers about its own public key.

Of course, the format in which the certificate is produced and presented plays an important role for the interoperability of various elements of the security architectures. Table 9.1 shows the format called X.509 for certificates [X5093280].

Table 9.1 X.509 Version 3 format for digital certificates (shaded fields are protected by the issuer's signature)

Field name	Description
Version	Version number for X.509 certificate. 3 for V3 (especially when the certificate includes extensions).
Serial number	The certificate serial number assigned by the issuer of the certificate (typically the CA).
Signature	This is the signature algorithm identifier not the actual signature. This field is a copy of the algorithm identifier below, but it is protected by the issuer signature.
Issuer	The distinguished name of the issuer of the certificate.
Validity	Indicates the validity period of the certificate. This includes both the start of validity and the expiry date for the certificate.
Subject name	The distinguished name of the owner of the certificate (which is also the holder of the private key corresponding to the public key shown by the certificate).
Subject Public key info	The public key of the owner of the certificate.
Optional issuer Unique ID Optional subject Unique ID	These two optional fields contain unique identifiers to handle some issues relating to reuse of subject names and issuer names. Their use is not widespread.
Extensions	This field was a new field for V3 certificates (compared to V2). The extension is formatted in a standardized manner, including extension identifier, value, and flags. Examples of information carried by extensions are policy information, usage for the public key within certificate (key encryption or digital certificate).
Signature algorithm	An algorithm identifier, indicating the algorithm the issuer of the certificate has used to sign the certificate. This field is not protected, but it is repeated in the protected part.
Signature value	The actual value of the digital signature made by the issuer to vouch for the validity of the certificate.

9.1.2 Certificate Management Concepts

There are many details that need consideration when issuing and using certificates for users or devices. For example, the CA needs to carefully examine the user's identity and legitimacy before registering the user and issuing a certificate to the user. Also, when an entity receives a certificate from a peer, it needs to first examine the authenticity of the certificate by verifying the signature added by the CA (see X.509 certificate format above). However, this can only be done if the receiving entity has a copy of the public key of the CA available. The system also needs to make sure the certificates used by various users and entities are valid. This is partly accomplished by maintaining a secure and tamper-proof record of the validity start time (and eventual termination time in case of revocation) of the certificates. If different certificate types are generated, the type of certificate should also be included in the record to support any type-based policy dispute resolutions. This of course implies that policies and procedures themselves must also be kept as secure as possible from tampering.

As the use of certificates has become more widespread and complex, the CAs and their certificate directories have evolved into full-blown PKIs, in which many of the CA's original responsibilities are now delegated to other entities. We will describe these entities and their interactions later on. However, in the following we will first go through some of the certifications issues without getting into the specific PKI elements that are responsible for each specific task. For simplicity we refer to any PKI element as "the PKI". Also in this subsection, unless specifically noted, by "user" we do not refer to a human user. Any end entity (user or device) capable of these interactions and using a certificate is called the user. When necessary, we will make clearer distinctions between human users and devices.

- *Registration of users*: The process by which the user makes itself known to the PKI and qualifies for a certificate from the PKI. The PKI must authenticate the user and if the user already possesses a private key, must also assess the user's proof of possession of this private key (as explained below). The authentication means the PKI verifies that the user actually possesses the identity that she is requesting to be included as certificate subject name (the identity the user claims when presenting the certificate). This verification must be performed with a pre-determined level of assurance. Determination of this level of assurance for authentication is a matter of network policy, which in turn depends on the intended usage environment and application for the certificate. Also if other information is to be included in the certificate, the CA must make sure of the correctness of this information during the registration process. Authentication may or may not be the first act in the certification process as explained below.
- *Initialization and key generation*: At this stage a public key pair is generated for the end entity or the user. Depending on the capability of the end entity, the key pair may have been generated at the end entity, or by the CA. Entities other than the CA may also generate the key pair. The location of key pair generation and its storage have a significant impact on the security and integrity of the PKI and its certificates. This creates many alternatives in handling the order in which the certification activities take place. For instance, when the end entity generates its own key pair, it can also generate a certificate request and send it to the PKI prior to authenticating to the PKI. The CA must then ensure that the user actually possesses the private key corresponding to the public key presented to the CA. This is referred to as proof of possession. This may be done by the client including a digital signature or through a challenge response interaction between the PKI and the client.
- *Certification*: After completion of public key pair generation and proof of possession, the CA issues a certificate for the user's public key and either returns the certificate to the user or posts it in a repository accessible through the external or internal websites. Depending on the operation policy, the certificate is either returned to the user directly or through another PKI entity called the registration authority (explained later). If the user has not yet been authenticated, it must be authenticated before receiving the certificate. When a registration authority is deployed, this entity would perform an authentication of entity before handing out the certificate.
- *Certificate revocations and CRLs*: Certificates have a validity period but may be rendered invalid due to reasons other than lifetime expiration, for instance, due to compromise of

the user's private keys or change of associations between the user/device and its domain CA. The former can happen when a certificate is stored in a device and the device gets stolen or lost. The latter can happen when the user leaves a company or organization or when the device is no longer used for service. At any rate, when the certificate is deemed not usable, the PKI (usually the CA) revokes the certificate and must make sure that the entities (certificate-using entities) do not accept any revoked certificates as valid. In the X.509 infrastructure, a CA periodically issues a data structure called the certification revocation list (CRL). The CRL is time stamped and includes the serial number for all revoked certificates. This implies that the CRLs can become long over time.

- *Certificate status checking*: When a certificate-using entity is verifying a certificate, it not only needs to check the certificate signature and validity period of the certificate from the certificate itself, but it also needs to acquire a recent CRL, to check that the certificate serial number is not on that CRL. The required level of CRL timeliness depends on the local policy, since choosing the interval between the CRL publications presents a design trade-off between security and network cost. The cost is not only imposed on the PKI (the CA must churn out more CRLs), but also on the network. The CRL must be downloaded by the certificate-using system and this involves unnecessary consumption of network bandwidth. Instead of downloading the entire CRL, the entities that need to verify the validity of the certificate, can outsource the certificate status checking to the CA or specific CRL servers using request/response procedures or full-blown certificate status checking protocols, such as "Online Certificate Status Protocol (OCSP)" [OCSP2560], which claim to reduce the latency in status checking and do not require CRL downloads.
- *Certification path*: As mentioned earlier, in order to vouch for the authenticity of the certificate, a CA must sign the certificate with its own private key. However, several implicit assumptions are made here:
  - The certificate-using entity, i.e. the entity that receives a certificate from or on behalf of a peer entity, needs to be able to trust and identify the CA that has signed the certificate.
  - Since the public key of the CA is not included in the certificate (X.509 only includes an identifier for the CA and its signature algorithm), the certificate-using entity must have access to the public key of the CA to verify its signature.

  If the certificate-using entity, i.e. the entity that receives a certificate from a peer, does not trust or know the CA, a certification path has to be built. Building a certification path basically means the signature by the first CA must vouched by signatures from further CAs, until one ultimately reaches a CA that the certificate-using entity trusts. This is CA is called a trust anchor CA and is either a root CA at the top of a hierarchy or a CA that is a peer to one of the CAs in the chain. Typically this is the CA that also signs the certificates for the certificate-using entity itself. The path that starts from the CA that signed the certificate to the CA that the receiving entity trusts is called the certificate path.
- *Certification hierarchy*: As seen, the certification path can be based on a hierarchy leading to a root CA or a cross-certification between peer CAs that mutually trust each other. Privacy Enhanced Mail (PEM) standards for using X.509 certificates for email define a very rigid certification hierarchy with well-defined authority levels and name subordination rules (called X.509 naming tree). A certification authority at each hierarchy level can only issue certificates for those lower-level authorities, whose names are subordinate to that CA in the naming tree. This way, a CA in an organization can only certify entities in that organization's name tree. X.509 V3 allows the certification path to start with either a top

level CA or a CA in the user's own domain (does not necessary have to be a top-down hierarchy). More details on certification paths can be found in [X5093280].

- *Certificate chain*: When a certification path is built and the CAs sign each other's signature, they will add their own certificate including their name and public key as a way of guaranteeing their signature as well as providing the user with the public keys that are required to verify these signatures. The ultimate message includes a chain certificates from each of the involved CAs. A certificate chain describes the certification path that leads to the trusted CA. The end entity examines each of the signatures on the chain until it recognizes a CA that it trusts. Another usage of a certificate chain is when the user receives a certificate signed by a CA but cannot obtain the public key of the CA to verify the signature on the first certificate. If a CA certificate that includes a CA's public key is included with the original message, the receiving entity can easily check the validity of the certificate. Due to the relatively large size of each certificate, certificate chains increase the amount of signaling. When a choice is available, the naming trees and trust domains should be chosen in a way that limits the need for use of certificate chains. Loading devices with certificates from a few known CAs may also spare the user from having to use long certificate chains.
- *Cross-certification*: If two CAs are not part of the same hierarchy, but establish a trust agreement, each can vouch for certificates signed by the other, by issuing a certificate for the other CA. It has been suggested that they can go as far as issuing certificates for the users within the other CA domain. To do this each CA needs to include a signature key so that the second CA can issue certificates on behalf of the first CA. This can of course have major security implications.
- *CA's certificate profile*: Just as a gold versus a premium credit card indicates the bank's level of trust in the credit card's holder, the certificate policy may guide the certificate-using entity to the level of trust with which she can use the public key included in the certificate, or the type of security function the certificate can be used for (producing signatures for authentication or encryption of data), or whether the certificate is issued to a user or a device or a server. A CA specifies the type of information that will go into the certificates and the contents of the CRLs. The CA must not only protect the integrity of its definition of the profile, but also make sure that all the certificates and CRLs issued conform to the specified profile. The CA must also make sure that the certificate requests comply with the appropriate certificate policy. For more information on certificates and CRL profiles, the reader is referred to [X5093280].
- *Certificate policies*: Together with certificate profiles, certificate policies open the door for customizing the PKI and certification process for their own operational needs. We mentioned some examples for profiling certificates. Certificate policies can define how to treat certificates belonging to each profile and can provide a lot of flexibility to bring the cost of certification process down. As we will see later, they can even boost the efficiency or even security of the overall operation of the network. In initial PKI designs, each CA used to only issue certificates under a single policy. This not only required deploying different CAs for different applications (different policies), but also it required users to track which policy was implemented for each certificate and thereby to match policies with user names and directories. Developing certificate policy extensions accompanying the certificate resolved these problems. The user can simply and readily find out under which policy the current certificate is issued.
- *Cryptographic modules*: The integrity of the entire PKI relies on the trust put upon the CA and its ability to protect its private key from disclosure. This private key must be protected

both when in use and when in storage. Typically CAs implement the cryptographic algorithms using a cryptographic module. Hardware implementations of the cryptographic modules are safer since they tend to keep the private key of the CA out of the host system's memory and operating system software. This is especially a plus when the CA is located at a place with weak physical security.

- *Archives*: The CA must be able to maintain archives on certificates even after their expiration or revocation. The PKI must also be able to attest that a certificate that is now expired was valid at the time it was used for certification. It may also be required that the archive maintains information on if and why the certificate was revoked. This is a very important enabler for forensic studies by law enforcement agencies.
- *Key pair recovery*: As an option, the keying material for the client user can be backed up at the CA or a backup server. Example of the keying material can be only a password giving access to the keys, or in some cases, the private key that corresponds to the public key in the client's certificate. One can argue about the benefit of public key cryptography, when big brother is holding a copy. On the other hand, in law enforcement cases, where forensics play an important part, or in mission-critical scenarios where robustness is an important requirement, key recovery becomes an important feature. If a user loses its private key (a disk drive becomes damaged) or forgets a password, PIN or a hardware token, the key pair recovery mechanism provides for recovery of decryption keys from an authorized key backup facility.
- *Key pair update*: If the certificate expires or is revoked (due to employee terminations, compromise of private key, and so on), a new public key pair and certificate must be generated.

As we can see, the certification process involves many complicated and computing-intensive functions. In order to ease the burden put on the CAs, many of these functions are outsourced to other entities. However, to ensure the integrity of the process, public key infrastructures that define the relationship between these entities are put in place. In the following subsection we describe various elements of a PKI.

9.1.3 PKI Elements

The Public Key Infrastructure (PKI) as defined by IETF PKIX working group (PKI for X.509) consists of the following elements [X5093280] as shown in Figure 9.1:

- *End entity*: An entity, bound with a certificate generated by the CA in the PKI. The end entity can be an end user or a server, a router or even a PKI entity dealing with the CA.
- *Certificate-using entity*: In PKIX terminology, the entity, which uses another entity's certificate for the purpose of public key or identity verification, is called the certificate-using entity. Note that this entity is not the certificate holder, but an entity that has a trust relationship with the certificate holder based on the certificate it receives from the certificate holder.
- *Certificate Authority (CA)*: The entity in a public key infrastructure (PKI) that is responsible for issuing and typically revoking end entity certificates (users or devices), publishing the certificates and CRLs, and controlling compliance with policies. For scalability reasons, the CA may offload some of its administrative duties to a registration authority (see below) and repositories.
- *Registration Authority (RA)*: This is an optional element in a PKI. When they exist, RAs are delegated those of CA's original responsibilities that have to do with verification of

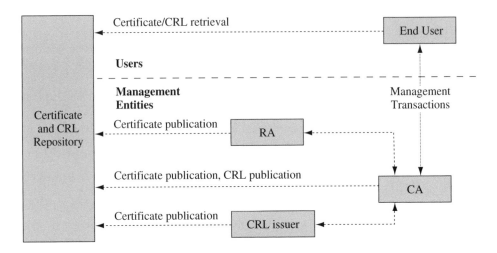

Figure 9.1 PKI entity and management functions
Note: Not all the shown elements exist at every PKI implementation.

certificate contents and certification requests. More concrete examples of responsibilities of RA include one or more of the following: user authentication and registration, public and private key pair generation and generation and submission (to the CA) of certificate requests on behalf of users. There is, however, one CA responsibility that cannot be delegated to an RA: the RA can never issue a certificate. Implementation of an RA has the benefit of retaining control over the registration process within the organization, even in cases when the CA is outsourced to an external agency.

- *Repository*: A collection of systems that are in charge of storage and distribution certificates and certificate revocation lists (CRL). For storage, many types of directory and directory access methods, such as LDAP, HTTP and FTP are used.
- *CRL issuer*: Again this can be an optional PKI entity, to which the CA delegates the responsibility of publishing the certificate revocation lists. This may be referred to as a CRL server. The CRL server may use either of the push or pull methods to entities that require the latest CRLs. The push method may be useful for network servers, such as VPN gateways (or access control servers) that need to frequently examine certificates from users requesting service. If the VPN gateway receives the latest CRLs as they arrive, it does not ask for them every time it receives a certificate from an end user. The pull method is useful for entities with sporadic needs for CRLs, such as an entity that has received a certificate from a server and needs to verify the validity of the certificate. The CRL server may also support certificate status checking procedures such as Online Certificate Status Protocol (OCSP), so that CRL server itself, instead of the certificate-using devices, checks the certificate against the CRL based on the request from those entities. This feature is useful when timeliness and bandwidth-efficiency are required of the status-checking process.

9.1.4 PKI Management Basic Functions

The categorization of a function as a PKI management function can vary greatly from one protocol to another. Many variants of PKI management protocols have been proposed with varying degrees of functionality richness and complexity. One feels that the most basic functionality to be expected from a PKI is to issue and revoke certificates, and therefore the most basic management protocol needs to support the transactions that involve certificate requests and revocation requests. Therefore we discuss the transactions that cover these basic functionalities first and then go into more complicated and rare features.

9.1.4.1 Basic PKI Transactions

In the following we describe the basic transactions involved with issuing and revoking certificates:

- *Certificate requests*: Certificate requests are either for a subject name that already owns a certificate from the CA or for a subject that does not own a currently valid certificate from the CA. Therefore, the reason for requesting a new certificate may either be that the subject requests a new certificate for the same key pair or that it requests a certificate for a different key pair. The former is called an initial certificate request, while the latter is called a basic certificate request. A third kind of certificate request can be originated from a different CA. Either a subordinate CA in the same hierarchy or a peer CA can make such a request.
- *Certificate response*: For now, we only consider certificate requests that have not been rejected. Hence, certificate response will define the act of responding to the certificate request, either by sending a certificate response message including the requested certificate, or returning the certificate to the user through some other means.
- *Certificate revocation requests*: Revocation request can come from either the subject of the certificate or another entity.
- *Certificate revocation response*: Typically this is not in the form of an actual message. The effect of a successful certificate revocation process may not be seen until the next issued CRL.

The complexity and richness of the management protocol dictate which of these messages are implemented and how they are implemented. For instance, a certificate request transaction may include messaging for a request, a response and a confirmation message or simply messaging for only the request. The request message itself may be issued by the entity itself, an end entity outside the PKI (such as a VPN server) or an RA within the PKI. The request can be received by different entities as well as described later on.

9.1.4.2 Enrollment and Authentication

The enrollment process is usually highly dependent on the administration policy and is by and large impacted by a trade-off between security and the acceptable PKI administration overhead. The details of the authentication and proof of possession also depend on the type of certificate request. For instance, when a client requests a certificate, if the client is already known to the CA, such as a client that is requesting a new certificate, while holding another valid certificate (basic certificate request), then the client can authenticate itself to the CA using the valid certificate along with the certificate request. On the other hand, when an unknown client requests a certificate for the first time (initial certificate request), the CA must have a way of verifying the identity of the client. This means the CA must use a separate channel to verify the client's identity (authentication) and other claimed information. This is typically done in an out-of-bound manner which depends on whether a CA deals with the client directly, or through a registration authority:

- When the CA deals directly with the client, according to the so-called two-party mode, the out-of-bound model can be email or regular mail. For instance, if the user has claimed to possess a certain email address, the CA may generate a random value called authenticator and send it to the user's email address. The user returns the authenticator in a response email and thereby proves that it owns that email address. As we can see, this method has limited applicability, since proof of possession of an email address is hardly the same as proof of possession of an actual identity (such as what is indicated in a driver's license for a person or an actual hard-coded MAC address for a device). In the case above, the issued certificate can only be used to sign or encrypt data transmitted through email and not through any other means. A more sophisticated implementation may require the user to possess a cryptographic module (with embedded private key) that creates a response to the authenticator from the CA. Since certificates often include more information than just a form of identity and public key, even possession of the cryptographic module does not guarantee the correctness of all the information claimed for certification. This means the two-party model is inherently flawed in that it cannot provide a secure means of verification of the information presented by the client.
- The limitation of the two-party model created the need to develop models in which the authentication and enrollment process is outsourced to a third party and hence the name: three-party model. In a three-party model, an RA assists the CA in verification of the information presented by the client. We did mention that an RA may also assist the user with various parts of certification request. If the client holds a cryptographic module, it may create a key pair and a certificate request. The client may present the certificate request to the RA, which after authenticating the client, forwards the request to the CA. If the client does not hold a cryptographic model, both the key pair and certificate request may be generated by the RA and forwarded to the CA after the RA has authenticated the client. The alternative way to get around the three-party model is when the client presents the certificate request to CA, which forwards it to the RA for outsourcing of the client authentication and waits for the RA to confirm the authentication before it processes the certificate request and issues the certificate. In either case, the point is that when RA is present, the RA is in charge of authentication.
- We should note that authentication does not have to be done electronically or as part of PKI management protocol. For instance, a user may authenticate itself in person to the RA by presenting a physical ID or submit the authentication credentials along with other material electronically to the RA. In either case, care must be taken that the CA does not issue any certificate for an unauthenticated client or any certificate including information that is not verified. The final certificate may be either delivered to the client by the CA directly or through the RA.

9.1.5 Comparison of Existing PKI Management Protocols

As previously mentioned, the functionality required from the PKI management protocol greatly depends on the administration policy and can vary from supporting only certificate requests to supporting electronic versions of all the transactions mentioned in the previous section along with additional functionality, such as key recovery.

In this subsection, we provide a comparison of PKI management protocols by focusing on the support for basic transactions such as certificate requests/responses and certificate revocation requests. We do not include the transactions supporting complicated PKI functions such as key recovery due to the expense involved in their implementation.

Since the certificate can be issued to both users and entities, we simply refer to the user or entity as the subject name, since the subject is the identity that appears on the face of the certificate. Another important consideration for protocol comparison, besides the support for basic transactions, is to decide whether the subject, i.e. the prospective certificate holder is to deal with the CA directly or through an RA. As one might expect, the use of two-party transaction model, in which the prospective certificate holder deals with the CA directly, makes for simpler protocols. On the other hand, more sophisticated and complex management protocols may support a variety of three-party transaction models, where a third party, generally an RA, assists the certification and certificate revocation process. We will go through some examples of three-party transaction models.

There are 5 PKI management protocols that are more prominent than the others:

- PKCS #10 with SSL: PKCS stands for Public key cryptography standards and SSL stands for secure socket layer.
- PKCS #10 with PKCS #7.
- CMP: Certificate management protocol.
- CMC: Certificate management using CMS (cryptographic message syntax or PKCS #7).
- SCEP: Simple certificate enrollment protocol.

The first two have surprisingly not had much support for PKI transactions specified for PKI earlier, while the others tend to provide much more functionality. In the following we provide more details on each protocol.

9.1.5.1 PKCS #10

Public key cryptography standards (PKCS) specifications were originally produced and copyrighted by RSA in conjunction with a small group of early adopters [RSAWEB]. Appendix A provides a list of PKCS documents, where [PKCSOV] provides an overview of PKCS standards.

PKCS #10 or certificate request syntax standard was later standardized by IETF as an RFC [PKSC102314]. The RFC describes a syntax for the certificate request messages created by entities that can create their own public key pair. The format for PKCS #10 is shown in Table 9.2.

As we can see, the contents in PKCS #10 certificate request is protected by a digital signature. The signature is calculated by the requester's private key and provides protection against tampering with the contents of the certificate request. The signature proves that the requester owns the private key corresponding to the public key included in the certificate. Without it the attackers can request certificates for public key pairs that they do not own.

We said that since a certificate includes both the identity and the public key of its owner, the PKI management entity (typically the RA) must perform verification of both identity and proof of private key possession. The problem is that the signature in PKCS #10 does not provide any identity verification, since anybody who can create a public key pair can request

Table 9.2 Certificate request format according to PKCS #10

Field name	Sub-field	Description
Certificate request information	Version	0 if according to RFC 2314 for PKCS #10
	Distinguished name (DN)	The distinguished name of the subject on the to-be-issued certificate
	Public key	Public key is included in the certificate, while the private key is held by the requesting entity
	[Attributes]	Information that requester wishes to be included in the certificate [optional]
Signature algorithm identifier [OID]		Object identifier syntax is defined by ASN.1 notation
Digital signature		Signature is generated with the private key corresponding to the public key included in the certificate

PKCS #10 certificate request message format is shown above.

a certificate on behalf of a legitimate user with identity included in the distinguished name (DN) field of the PKCS #10 request. One reason for this weakness may be that PKCS #10 was not originally designed for an electronic form of certification transaction. It used to be so that the certificate requester would take a paper copy of the request above to the RA in person. Even the IETF RFC 2314 keeps this as an option: "How the entity sends the certification request to a certification authority is outside the scope of this document. Both paper and electronic forms are possible."

Systems that deploy PKCS #10 certificate request must use one of a variety of methods for verification of the identity of the requester. For human users, this may be accomplished by physical contact with RA personnel or by using a pre-configured user ID-password authentication. For devices, a common approach is to augment PKCS #10 with SSL. Besides providing a means of creating and submitting certificate requests over the web, the SSL also tamper proofs the certificate request messages themselves and may be used for client authentication. This will be explained shortly.

Another issue with PKCS #10 is that, by itself PKCS #10 cannot be considered a complete PKI management protocol, since it only defines the syntax for a certificate request message, while providing no support for other transactions, such as responses to certificate requests, or messaging for revocation process. Most of PKCS #10 implementations use PKCS #7 to return the certificates. PKCS #7, as we will see later, is the common method for providing security for many protocols. Despite the wide practice of combining PKCS #10 and PKCS #7, there is no standard specification on how the combination is to be implemented.

In the following, we describe each of the two PKCS #10 augmentations (SSL and PKCS #7) briefly.

9.1.5.2 SSL Protection for PKCS #10

As mentioned above, PKCS #10 was originally designed for paper as the medium for conveying certificate requests, meaning that the requester would take the request personally to a CA or an RA and would authenticate herself in person before receiving the certificate. With the advent of web interfaces, it was better to have PKI users go through parts or the whole process of key pair generation, certificate request creation and submission using the web pages. Hence the need for authentication support over the web was created.

Secure socket layer (SSL) provides the means of server to client authentication as well as the establishment of a secure channel and shared key between the client and server. Once the server authenticates to the client using the server's own certificate and a secure channel is established, the SSL-based secure channel can be used to keep client's PKCS #10-based certificate request confidential and immune from tampering. There are several problems with the PKCS #10 and SSL combination:

- The SSL channel is established based on authentication of the server to the client. SSL does not provide native support for client to server authentication. Some implementations use SSL secure channel to pass client authentication credentials such as password and so on to the server. But this again means the certificate-requesting client must have been configured with proper authentication credentials to perform this authentication. It has also been stated that since SSL only provides stream protection and not individual packet integrity protection, it cannot be safely used as a basis for authentication to CA. Once SSL authentication is successfully performed, the CA has no knowledge of what identity has been used for this authentication and cannot safely say what identity can be put on the certificate.
- Adding SSL protection for the PKCS #10 does not solve the problem that there is still no specification for responses to certificate requests or for the certificate revocation process.

The way many implementations deal with these problems is that the certificate that is issued on behalf of the client is not delivered until the client authenticates itself to the PKI through some other non-web-based (not through SSL) means. Examples are either use of an out-of-bound method or use of three-party models that deploy RAs. An example of out-of-bound method authentication is explained earlier: sending an authenticator value to the client through email and expecting a specific email response from the client. Three-party models can be implemented in different ways:

- One example is instead of having the client establish the SSL connection to the CA, have the client present authentication material to RA, which in turn establishes an SSL connection to the CA to protect the certificate requests that travel between RA and CA. The CA sends the certificate to RA, which delivers it to the client.
- The other example is to have the client send the certificate request to the CA. But the CA does not process the request right away. Instead the CA forwards the certificate request to RA to handle the authentication process. Once the RA is satisfied with the client's authentication credentials, the RA sends a "go-ahead" response on certificate generation from the CA.

9.1.5.3 PKCS #7 Protection for PKCS #10

Cryptographic message syntax (CMS) was first standardized by RSA as PKCS #7 and then by IETF as RFC 2315 [PKCS2315]. PKCS #7 is a commonly used standard used for security protection of many protocols such as S/MIME and therefore its messaging is designed to carry data from other protocols, such as PKCS #10 certificate requests. The data from other protocols is carried either natively or with added cryptographic enhancements, such as signatures, encryptions, and so on. The message format for PKCS #7 is very simple: It consists of two fields:

- *Content*: This is the payload, or the data from other protocols, using PKCS #7 as an encapsulation mechanism.
- *Content type*: This field describes the type of the payload. PKCS #7 defines 6 content types: data, signedData, encryptedData and some other types that we don't mention here, since our interest is in defining the usage of PKCS #7 for PKIs.

RFC 2315 defines two classes of content types: base and enhanced. Content types in the base class contain "just data," with no cryptographic enhancements, while content types in the enhanced class contain a content of some type plus cryptographic enhancements such as signatures. Enhanced content types are typically encapsulated in the PKCS #7 format. Therefore, the content being enhanced (from other protocols) is called inner content, while the content that includes the cryptographic enhancements is called outer content. An example will clarify this shortly.

Many PKI deployments use PKCS #7 signedData format for signing PKCS #10 certificate requests (inner content in this case). The format for the completed format, i.e. a PKCS #10 certificate request encapsulated as signedData inside a PKCS #7 packet is shown in Table 9.3. Either the end client or the RA can do the signing, using their own private keys. However, the end client must already hold a current certificate with corresponding signature key to be able to sign. This means the end client cannot be the signer for initial certificate requests. PKCS #7 also allows multiple signers to sign the signedData content in parallel or in series. However, Table 9.3 does not show a case where multiple signers sign each other's signed contents (in a serial fashion). Here is how PKCS #7 used for protection of PKCS #10 requests:

- A PKCS #10 certificate request is generated, signed (by a client of an RA) and is inserted in the content portion of the PKCS #7 signedData as shown in Table 9.3. To add more clarity, the PKCS #10 request is shaded with gray color. Note that more information needs to be added to build the PKCS #7 content, since signedData includes many other fields as described below.
- The so-called signer, i.e. the entity that signs the data and vouches for it, signs the PKCS #10 request with a digest algorithm. Note that this signature is different from the signature included in the PKCS #10 request itself. The signature in PKCS #10 must be performed with the private key of the key pair for which the certificate is being requested for, while the signature included by the signer is done by the signer's own private key. Even when the certificate request was originally generated and signed by the client itself, the signer here can be an RA vouching for the client (after authentication has taken place, of course). Other optional data such as certificates and CRLs may also be added to data before running the digest algorithm, so that they can be protected as well. In Table 9.3, we show this by adding brackets around these fields. The signer also fills other fields that are required for signedData content format. These fields are version, a list of algorithms used for creating the digest. If optional certificates and CRLs were added to the data, a list of these certificates and CRLs is also added to the content. The signature (digest) itself is added to a field called signerInfo as explained below.
- To allow for the messages to be signed by multiple authorities or so-called signers, a specific field called signerInfo is added to the content. For simplicity, Table 9.3 shows a

Table 9.3 PKCS #10 encapsulated and signed inside PKCS #7 format

Field name	Sub-field	Sub-sub-field	Description
Type = signed Data			
Signed Data Content type	Version		= 1 for PKCS #7 according to RFC 2315
	Digest algorithm		A list of all the digest algorithm identifiers to be used by any of the signers. Each signer typically signs the content signed by previous signer along with other information it might wish to add and authenticate itself for the content
	ContentInfo = PKCS #10 certificate request	Version	As specified for PKCS #10 certificate requests
		DN	As specified for PKCS #10 certificate requests, the name for the prospective certificate holder
		Public key	As specified for PKCS #10 certificate requests: public key to be included in the certificate that is being requested
		[attributes]	As specified for PKCS #10 certificate requests
		Signature algorithm	As specified for PKCS #10 certificate requests
		Digital signature	As specified for PKCS #10 certificate requests: signature over the certificate request
	[certificates]		Optional certificates. This option can be used to disseminate certificates in a signed form. This can contain extended PKCS #6 and X.509 certificates and can contain chains from one or more root level CAs to all or some of the signers
	[CRLs]		Optional CRLs. This option can be used to disseminate CRLs in a signed form. The recommendation is that this CRL should relate to any of the certificates listed above as hot-listed
	SignerInfo	Version	
		Issuer and serial number	Identifies an issuer and an issuer-specific serial number for signer's certificate including signer's public key that is needed for verification of the signer's signature
		Digest Algorithm	This is signer-specific algorithm used by the current signer. Note any type of content can be signed by any number of signers in parallel

(*continued overleaf*)

Table 9.3 (*continued*)

Field name	Sub-field	Sub-sub-field	Description
		[authenticated attributes]	Optional attributes added by the signer. This is to allow each signer to include desired authenticated attributes
		Digest Encryption algorithm	Identifies the encryption algorithm under which the digest and other information are encrypted with the signer's private key
		Encrypted Digest (digital signature)	A signature produced over the signedData content and signerInfo optional authenticated attributes, if any. The message digest is produced by running the content digest and the attributes into the digest algorithm (see RFC 2315 section 9)
		[Unauthenticated attributes]	

case where only one signer has signed the message. The concept of having multiple signerInfo fields also allows each signer to add any extra signer-specific information that may be required for the processing of the message. This extra information can either be signed and authenticated by the signer (authenticated attributes) or added unsigned (unauthenticated attributes). SignerInfo field includes among other things, an identifier for the message digest that the current signer has used to compute its signature, along with a serial number for the signer's certificate and information on the issuer who has issued the certificate for the signer. This helps the recipient of the PKCS #7 message to locate the public key of the signer and thereby verify the signer's signature. Note that the digest algorithms used by different signers are different. The signerInfo includes the signature itself (calculated over content data and attributes that are to be authenticated), followed by unauthenticated attributes.

- The recipient, which is typically the CA, verifies each of the signatures by decrypting the encrypted digests, using that signer's public key obtained from the certificates for which the serial number and issuer are added to the message.
- Once the certification is complete, the CA can use PKCS #7 again to authenticate the messages that include the certificates back to the client or to the RA in signedData format.

The advantage of combining PKCS #10 with PKCS #7 is that more than one signer can sign the PKCS #10 certificate request message. This provides the flexibility of being able to have an RA signing a certificate request on behalf of a client. This is specifically important, when the client is required to authenticate to the RA (in paper or electronically) before the RA approves any certificate request on behalf of the client. Also a CA can always use PKCS #7 to authenticate its responses to the client, especially those responses that include certificates for the client (certificate response). Having implemented signatures for both requests and responses allows the client, the CA and even the RA to keep a record of the certification transactions. The problem with the PKCS #7 based approach is that although it provides an easy method for proof of possession for signatures keys, support of proof of possession for key management keys is more difficult. Also no specifications for certificate revocation requests are provided. Error and confirmation processes are not specified either.

9.1.5.4 IETF Certificate Management Protocol (CMP)

Responding to the shortcomings of all the existing protocols, such as PKCS #10, the IETF PKIX group decided to develop a do-it-all PKI management protocol. The outcome was a protocol defined by a combination of two RFCs: a certificate management protocol (CMP) in RFC 2510 [CMP2510] and a certificate request message format (CRMF) in RFC 2511 [CRMF2511]. This duo has an impressively comprehensive functionality set for 24 messages to do everything from basic PKI transactions, such as certificate request and response exchanges to the very sophisticated ones for key recovery. It also supports inclusion of RAs within three-party models as well as use of 4 different transport protocols. Furthermore, CMP messages are designed to multiplex several requests into one message. This will, for instance, help in situations where an organization is starting to roll out certificates by allowing RAs to make batches of certificate requests instead of sending them one by one to the CA.

The following summarizes the messages specified within CMP and at the same time gives an indication of the PKI management functions supported by CMP:

- *Certificate request and response messages*: CMP defines messages for both certificate requests and responses for delivering certificates to end clients. These messages also provide support for specifying the type of certificate, e.g. if the request is for an initial certificate or for a basic certificate from a client that already holds one certificate. Cross-certification requests and responses are also defined, so that other CAs can request and receive certificates from a CA.
- *Revocation request and response messages*: Used to request a revocation and to confirm that the revocation has happened or been denied.
- *Certificate and CRL distribution messages*: To announce issuance of a certificate or CRL.
- *Key recovery request and response messages*: To request and deliver private keys for key management.
- *Proof of possession challenge and response messages*: To prove possession of private keys used for key agreement or transport.
- *Confirmation and Error messages*: To provide confirmation or indication of errors.

Table 9.4 shows an example of how the formatting from CMP and CRMF are combined to build a certificate request. As one can see, CMP messages are comprised of 4 parts: header, body, and optional (shown with brackets) protection and certificates. While the CMP provides the header, the protection and certificate parts, the CRMF provides the actual certificate request format that goes inside the CMP message "body" section.

As one can see, in order to support many different functions and scenarios, CMP and CRMF messaging include many optional fields. This means, CMP is too accommodating for its own good: different implementations may not include the optional elements in full or to the same degree and consequently not interoperate with each other. CMP designers have made an attempt to overcome this problem by specifying a number of mandatory transactions that all implementations need to conform to. Another problem is that CMP transactions are

Table 9.4 Certificate request message according to CMP combined with CRMF

Field name	Sub-field	Description		
Header (according to CMP)	Version No			
	Sender	Name of sender. Can be in distinguished Name (DN) format		
	Recipient	Name of recipient. Can be in distinguished Name (DN) format		
	[message time]	Time of production of this message		
	[algorithm]	Algorithm used to calculate protection bits		
	[Sender key ID]	Sender key identifier		
	[Recipient key ID]	Recipient key identifier		
	[Transaction ID]	To match the responds and confirmations with corresponding requests		
	[Sender Nonce]	To provide replay protection, inserted by the creator of this message		
	[Recipient Nonce]	A nonce that the intended recipient of this message has previously used in a related message		
	[Free text]	To provide context-specific instructions, intended for human users		
	[General info]	To provide context-specific information, not primarily intended for human users		
Body (according to CMP)	Certificate Request according to CRMF	Certificate Request ID	Certificate request identifier, an integer value used for matching requests and replies	
		Certificate Template (defined in CRMF)	[Version]	
			[serial Number]	
			[Signing Alg]	
			[Issuer]	Name of issuer
			[Validity]	
			[Subject]	Name to appear on the certificate
			[Public Key]	Public key of the certificate holder
			[Issuer Unique ID]	
			[Subject Unique ID]	
			[extensions]	
		Controls		Attributes affecting issuance
	Proof of possession			
	Registration info			
[Protection]				
[Extra Certificates]				

deemed to introduce too many round-trips. Furthermore, the opponents of CMP claim that despite its complexity, CMP does not leverage the large installed base of PKCS #10 and PKCS #7. The controversy has led to diversions within IETF PKIX. The fact that PKCS standards entered the public domain from being exclusively owned by RSA made the divide within the community even deeper. Finally, the group started working on a new protocol that leveraged the existing PKCS standards more efficiently and CMC (explained below) was the result.

9.1.5.5 Certificate Management Using CMS (CMC)

Certificate management using CMS (Cryptographic Message syntax) or CMC protocol was the result of the newer specification effort by the IETF PKIX realized in the form of RFC 2797. CMC uses PKCS #10 for formatting basic certificate requests (for users who already hold a certificate), but still depends on CRMF for more generic certificate requests. CMS itself is standardized in RFC 2630 and provides message encryption and integrity protection for CMC messaging.

9.1.5.6 Simple Certificate Enrollment Protocol (SCEP)

Simple Certificate Enrollment Protocol is the successor of the certificate enrollment protocol (CEP). Both CEP and SCEP are the results of the joint development by Cisco and VeriSign to handle the communications between a PKI and network components such as routers, switches and other entities (such as VPN components). SCEP is designed to support the issuance of certificates to network devices. Although this is a proprietary approach, chances are it will be widely deployed. The protocol supports certificate requests, certificate and CRL queries and distribution of CA and RA public keys.

Both manual and pre-shared secret-based authentications are supported. Manual authentication means the CA operator must manually verify the requester's identity information through out-of-bound procedures, while pre-shared secret-based authentication is typically based on a password. Certificate revocation is also done manually: the device administrator calls the CA operator on the phone and after providing a challenge password previously assigned specifically for revocations, the CA revokes the certificate. SCEP supports LDAP and HTTP for access to certificates and CRLs. It has been claimed that SCEP supports CRL distribution points.

9.1.6 PKI Operation Protocols

PKI not only provides for issuing and revoking certificates (CRLs) for certificate users, but also delivers certificates and CRLs to certificate-using entities. The former set is generally handled by PKI management protocols (as explained earlier), while the latter is handled by PKI operation protocols. Provisions for delivery of certificates and CRLs through a variety of distribution procedures, such as LDAPv2, HTTP, FTP and X.509 are specified. The specifications include message formats and procedures for each environment. We do not go into any details here and refer the reader to RFCs 3494 and 2587 for operation protocols with LDAPv2 and 2585 for operation with FTP and HTTP.

9.1.6.1 PKI Certificate Discovery and Validation Protocols

As mentioned earlier, Online Certificate Status Protocol – OCSP (RFC 2560) was developed as a simple request/reply protocol that allows clients to ask an "OCSP responder" about the validity/revocation status of a certificate. OCSP responder returns signed real-time information about the inquired certificates. However, OCSP offers limited functionality and for that reason three protocols have emerged to resolve these limitations: OCSPv2, certificate validation protocol (CVP) and simple certificate validation protocol (SCVP).

9.2 PKI for Mobility Support

Now that we have covered the basic building blocks of PKIs and PKI management protocol, we can go over specific considerations when it comes to designing PKIs for networks that need to support mobile clients.

9.2.1 Identity Management for Mobile Clients: No IP Addresses!

By now, we know that the main purpose of a certificate is to tie a client identity to its public key. However, even though most clients would at most have a single public key pair, it is not so certain that they will always use the same form of identity in every interaction with the outside world. For instance, depending on the protocol stack layer the device is communicating with, it may use an IP address or a MAC address as an identifier. Cellular phones may use a phone number or similar forms of identity to connect to their networks. It is important that the identity listed on the face of the certificate is one that the client uses for identification and authentication signaling. Otherwise, presenting the certificate would be of no use. In Chapter 2 we explained why device and user authentication need to be separated. That means that the device and user need to be distinguishable from each other. The user must have an identity of a form that network devices can understand and verify. Furthermore, the user may have to use different devices to gain access to different networks. All this points to the fact that we need to distinguish between device and user certificates.

Until recently, in many networks the IP address could be synonymous with the device identity when it came to network operations, since the computing model has been one of computers being stationary at a secure lab or an office and using an IP address to communicate to the rest of the world. This is no longer true. The IP address for even stationary devices can be dynamically assigned through a DHCP server. Furthermore, users with laptops or other personal portable devices may need to stay attached to the network while maintaining their mobility. Mobile IP provides support for multiple IP addresses for a single device.

Hence, due to its dynamic nature, the IP address is no longer suitable as the identity in the subject name (see X.509 format earlier in this chapter) of device certificates. Note that, even if the IP address for the device were to be static, it is not assigned when the administrator is working on initiating the device with the network and possibly downloading security credentials (such as certificates) into the device. A certificate bound by the IP address must be issued and downloaded to the device after assignment of that IP address.

9.2.1.1 Certificate Subjects for Mobile Devices

In the following we provide a couple of alternatives for identities that can be used as certificate subject names and discuss their pros and cons.

Network access identifiers (NAI)

These may be a better candidate for identifying devices or users than IP addresses. To identify a device, one can define the NAI as MAC-Address@domain-name, where the MAC address can be chosen to be unique for the device and the domain name or realm name will be the realm to which the device will belong to, e.g. "Chicago police department". NAI is much more permanent than the IP address, since it is typically assigned based on affiliation to a domain rather than the current point of connectivity. Also NAI can be assigned at initialization by the network operator. However, NAI has issues with roaming or cases where a single client may be associated to multiple realms. This means the client may still need to obtain multiple NAIs. This means either the client must obtain one certificate per realm or obtain a multiple-identity certificate and know all its NAIs at the time it obtains the certificate. The problem that arises with the latter approach is that each realm may have its own certificate authority (CA). Another problem with using NAI as certificate subject name is that normally the NAI is assigned by the network operator. As a result, the CA operator cannot issue an NAI-based certificate until the NAI has been issued to the device by the network operator and properly verified by the CA operator.

MAC identifiers

Nowadays many devices are marked with IEEE MAC addresses. Normally the manufacturer receives a batch of consecutive IEEE MAC address and assigns one MAC address to each of the devices in the factory. This makes the MAC identifier a very attractive candidate for certificate subject name, since not only the manufactured-issued MAC identifier is unique for each device, it is also static regardless of the network domain or the operator it is going to connect to. Motorola goes as far as issuing and downloading the certificates into the cable modems during production. Despite all their advantages, using MAC identifiers as certificate subject names also has its disadvantages.

For legacy devices or devices that have not been assigned a unique MAC identifier, a system-specific identifier needs to be assigned so that it is unique in the system that it is going to be used. The certificate will be based on this ID. The term system has a broader meaning than the administrative domain, otherwise multiple certificates may be required.

The use of the MAC identifier for the certificate subject name brings up certain issues that must be resolved:

- The MAC identifier is a link layer identifier that is not recognized at higher protocol stack layers. A device can easily present a certificate with its MAC ID as identity to an entity that is a single hop away and hence can verify the sender's MAC address on the link. This opens the door to MAC identifier spoofing, unless proper layer-2 security mechanisms are in place. For nodes that are multiple hops away and can recognize the sender only by the IP address, there is no way to check the MAC ID that the sender presents in its certificate, unless the sender includes a specific extension including its MAC ID and signs it with its private key. After authenticating this client, the other end may still need to create the MAC

ID-IP address binding to be able to control the traffic from this client. Both of these issues add to the complexity of signaling and security management.
- If the certificate subject name is the device MAC identifier, only MAC ID can be used to look up the certificate for the device. This is an added inconvenience for domain name servers (DNS) that are in charge of providing mappings between fully qualified domain names (FQDN) and IP addresses. When other entities on the Internet need to communicate with another entity, a query can be made to the DNS using the FQDN to find the entity's IP address. However, when MAC IDs are used as identities on the certificates, such translation is difficult.
- A device may have more than one MAC ID. For instance, the laptop may have a MAC ID on the wired Ethernet that is different from the MAC ID of the WLAN NIC card inserted in the laptop. This can mean the device must have a different certificate from each of the MAC IDs associated with it.

9.2.1.2 Certificate Subjects for Human Users

Traditionally, users authenticate using secrets that are easy to remember by humans. Examples of such secrets are passwords, or pin numbers. As we will see in the next chapter, in mobile wireless environments, legacy authentication methods must be enhanced to meet the tougher security threats within those environments. For instance, exchanges based on the password alone would not be enough for a secure authentication. In some methods such as CHAP (Chapter 2), a module on the device uses the password entered by the user to calculate the content for authentication messaging (such as response to a challenge). However, as mentioned earlier, care must be taken to separate the credentials for the user from the credentials for the device. If the password remains on the device after being used by the device module, it can be hijacked by others if the device falls into the hands of unintended users. SIM-based authentication (Chapter 2) used in cellular systems is intended as a user authentication scheme. However, the subscribers rarely take the SIM cards out of the phones, so the SIM card may be considered as a device credential. Some more careful subscribers use a PIN to lock the SIM card. In such cases, the separation of user and device credentials is only as good as the entropy that a 4-digit PIN provides. Our job in this section is to convince the reader to use certificates for both users and devices.

User certificate is not an entirely new concept; many organizations have been using certificates for their employees or students for a long time and have well-defined policies for registration, issuance of certificates and publication of CRLs. However, issuing certificates for mobile users that roam between multiple administrative domains (organizations and/or operators) and deal with a variety of access networks using multiple devices can be a tricky task. We will come back to the issues involved. For now we will focus only on identity management for user certificates. The choice of certificate subject name for user certificate can vary depending on the scenario. It might make sense to have an NAI format of userID@domain_name as the subject name in the certificate. UserID can be anything from person's name or credit card number to a number assigned by the IT department of the organization to which the user is affiliated. In any case, the user must comply with both the security policy and the technology assigned by the network operator. Unfortunately, as mentioned earlier, NAI is domain-dependent, which means a user associated with multiple domains needs multiple NAIs and possibly multiple certificates, since each domain typically has its own trust anchor CA.

Being hard-core security advocates, we do not recommend storing user certificates on any device, unless the device is only carried by a single user. Even then, if the device is lost, the certificate must be revoked unless the private key is stored in a tamper-proof manner (tamper-evident is not enough) that will not reveal or even destroy the private key upon tampering. In cases where a user needs to use multiple devices, assuming that the user can use the same certificate with all devices, the user should carry a hardware token (smart card or USB dongle) including her certificate and private key. It should be noted that handling the private key in such manner may be less secure if the cryptographic module is still on the device, since the private key on the USB dongle must be downloaded into the device for use by the module.

9.2.2 Certification and Distribution Issues

Issuance and distribution of certificates can depend on many factors. The most obvious factor is whether the certificate is for a device or for a human user. We will treat the difference between user and device certificates in the following subsections. Here we provide issues that are common to both.

9.2.2.1 Validity Checking and CRL Distribution

One common issue is proper assessment of the validity of the certificate. Checking the validity based on the lifetime of the certificate is rather obvious, since the validity period is stated on the certificate. However, making sure that the certificate has not been revoked is less trivial, since it would require checking the certificate serial number against the CRL. Since the CRL tends to become large, the design goal is to not let the CRL grow too rapidly. Note, we are not saying that we are compromising security by not revoking the certificates when needed. On the other hand, we can have clever certification policies and profiles to attain that goal. For instance, if a device has a higher likelihood of being lost or stolen, it should have a certificate with a shorter lifetime. For instance, a personal organizer has a higher chance of being lost than a AAA server. A human user is more likely to change its affiliation with a service provider than the core router within the network. Natural expiry of the certificate increases the overall security of the network if the loss is not noticed. A short-lived certificate either never ends up on the CRL or if it was there, it will be removed after expiry. Another example is issuing very short-lived certificates for users that are going to be temporary within the domain. This way after the user leaves and the certificate is revoked by the administrator, the certificate will not stay on the CRL for long. Also certificate revocation is typically a rather involved process that may sometimes even be the result of a meeting by a disciplinary committee.

When multiple certificate profiles are created, it is possible to issue profile-specific CRLs as well. This will bring the size of each CRL down and help the certificate user in status checking, since this entity only needs to check the CRL for that profile. The final trick in downsizing the CRL is to have different CRL update intervals depending on the certificate profile. For instance, the certificate for the VPN server is less likely to be revoked than the certificate for laptops. Hence the CRL for the VPN certificate profile can be updated less often, which means the CRL needs to be pulled or pushed less often.

As mentioned earlier, mobile entities using bandwidth-limited links should not have to download the entire CRL for status checking. The network should provide request/response methods for status checking.

9.2.2.2 Roaming and Certification

As mentioned earlier, roaming presents new challenges for certifications. The first and foremost challenge is dealing with scenarios where multiple administrative domains are involved. Each administrative domain (an organization or an operator) in an autonomous trust domain is typically served by a separate CA and surrounding PKI elements. To accommodate roaming, these separate CAs must either conform to a hierarchy and a root CA, or must provide cross-certifications. Without such arrangements, mobile devices and users cannot use certificates from one CA inside a domain served by an independent CA. Whether the users acquire one certificate per domain, or use certificate chains with their original certificates or not, may depend on business and billing agreements between the user and those domains and between the domains themselves.

9.2.2.3 Device Certificates

As mentioned earlier, the easiest and most secure way is downloading certificates into devices during manufacturing. However, as we mentioned, the device ID may not always be known at the time of manufacturing, unless the manufacturer is customizing the device for a specific customer order. The next level of security would be that the manufacture still downloads a certificate based on a manufacture-assigned ID into the device. This certificate and the private key can later be used to swap for another certificate and identity assigned by the operator based on the same private key. This way, the private key is still not exposed. The downside is that the manufacturer certificate must be configured so that it can be used only for special purposes.

When neither of the solutions mentioned above are available, for instance, when the device is manufactured without a certificate, the devices are typically brought into secured initiation centers, where a public key pair and a certificate are issued and loaded into the device. Great care must be taken so that the private key is not exposed to the outside world during transit. If the device is capable of generating its own key pair, it signs a certificate request and submits the request to the CA to sign the certificate. The latter method is preferable, since it avoid exposure of the private key.

Device certificates typically have a longer lifetime than user certificates. Specifically, when the certificates are loaded into the device by the manufacturer, this lifetime is as long as the lifetime for the device itself. However, to make validity checking a manageable task, it is useful to create different certificate profiles depending on the lifetime of device certificate. As mentioned above, certificates for core servers and routers should have a longer lifetime than certificates for mobile personal devices.

9.2.2.4 User Certificates

Issuing and distributing user certificates can be trickier than doing that for device certificates. This is partly because it is not as easy to download a certificate into a user inside a secure facility (at least outside of Sci-Fi movies), partly due to the difficulty with on-line identification of users (remember the famous joke: On the Internet nobody knows you are a cat), and partly because typically there are tighter security policies when it comes to handling user certification.

User certificates are usually issued with the intervention of a human system operator, possibly the RA operator. Physical rather than electronic authentication (based on presentation of a proper ID, such as driver's license) with the RA operator may need to take place prior to issuance of a certificate request to the CA, even though the user may be allowed to perform the original certificate request online. Users may then receive an activation key from the system administrator to gain access to their certificates from the device they are downloaded onto.

Depending on the operator's security requirements, a user certificate may need to expire after 3 months or a year while a device certificate may last much longer.

9.3 Using Certificates in IKE

We did say at the beginning of this chapter that when we finish designing our PKI and certification procedures, we have still not completed the design and security and authentication services for our network and its clients. Use of certificates within security and authentication mechanisms must be designed properly.

One potentially huge market for PKI and certificates is IPsec-VPN vendors. In order to establish IPsec tunnels, most implementations rely on IKE for the establishment of a secure tunnel that allows IPsec negotiation and key generation to happen in a secure manner. We covered IKE in great detail in Chapter 4 and as we mentioned there, despite its powerful feature set, IKE does not exempt the designer from the task of having to deal with the initial authentication that is required between the two peers that need to establish the IPsec tunnel. We also mentioned the authentication alternatives that IKE provides, including the one based on pre-shared keys and the one based on public keys. The greatest benefit of a PKI is that it eliminates the need for pre-shared pair-wise keys between communicating peers. Hence it would be great if the two peers could use their certificates to each other in order to perform public-key based authentication without having to worry about establishing pre-shared secrets prior to IKE. This would be a great feature for VPN vendors. Showing foresight in this, the IPsec working group specified the content for many ISAKMP payloads for carrying certificates and certificate-related information (as described below) and even enumerated a large number of certificate types to be used, without defining the syntax and semantics of how such certificate-related information should be understood and treated.

On the other hand, the PKIX working group in IETF provided a large set of certificate mechanisms for a broad range of applications for Internet protocols, such as key exchange and certificate management without providing any guidance for IPsec use.

Having two oversized frameworks (ISAKMP and PKI) seemed to lead to under-specified interaction procedures causing a lack of interoperability, since each of the implementers could choose an arbitrary number of options from each of the PKI profiles for IKE use and ignore the fact that another implementer could have used a different set. Therefore, it seemed necessary to develop a standard for using PKI technology in the context of IPsec by profiling the PKIX framework for use with ISAKMP and IPsec, on one hand, and clarifying the use of different ISAKMP payloads when it comes to using certificates, on the other. To mention some of the specific problems:

- IP addresses are not valid as certificate subject names (identities). There needs to be a way to provide non-IP address identities within the certificates and yet provide the same level of security that coupling the source IP address with the identity of the sender provides.
- Certificates generally have long validity periods, e.g. in the order of years. However, the VPN environments may need to make sure the certificates generated and presented for the purpose of remote access are short-lived. This means that mismatching lifetimes for IKE, the IPsec SAs and the certificates could pose possible security threats. Also dealing with certificate revocations is unspecified within IKE.
- The ability to use certificate chains is important in some scenarios.

Despite all this need for co-existence, it seemed that the VPN consortium and the PKI forum seem to have ignored each other for a long time. Several attempts have been made by the VPN consortium to reduce the complexity of the use of certificates for IKE. The consortium even tried to take the PKI profiling work through the IPsec working group in IETF. Finally, a new IETF working group, called pki4ipsec WG [PKI4IPSECWEB], was formed to deal with these issues. In the following we provide the major highlights of this work. It should be noted that this work is still in its infancy and some of the material presented in this section is subject to change due to ongoing progress.

We described ISAKMP in Chapter 4 and provided definitions for many of its payloads. However, ISAKMP also defines a number of payloads for dealing with certificates when it comes to key exchanges. We will go through those in more detail here to illustrate the ISAKMP and PKI interaction problems.

- We defined the ISAKMP ID payload in Chapter 4. However, the Identification (ID) payload is important when using certificates, since it is used to carry the identities claimed by each peer during the key exchange. As mentioned earlier, there is a need to carry non-IP identities within certificates. ISAKMP allows the use of many non-IP address identities (such as FQDN-based identity and user-FQDN-identity according to RFC 822). Whatever identity is used within the ID payload must be trustable, since the peer uses this identity to perform peer authentication and possibly to find the certificate in the directory as well as to look up the authorization and security policies. For those reasons, the identity presented in this payload must be tied to the sender's possession of some keying material. We will describe what this means in using certificates in the context of IKE.
- ISAKMP defines a Certificate (CERT) payload to allow a peer to transmit certificate/s or certificate-related material during an exchange, whenever directory services are not available (Table 9.5). The actual data included in the certificate payload is defined by a field called "certificate encoding", which is basically a type indicator. ISAKMP defines 10 types of data to be carried by the CERT payload. But not all of these types are related to PKIX and X.509 certificates. ISAKMP also defines a Certificate request (CERTREQ) payload to allow a peer to request a set of certificates or CRLs from its peer (Table 9.6). Hence part of the profiling specification is to define which of these data types are relevant for IKE-PKI interaction. Also, ISAKMP does not describe how the certificates are requested or

Table 9.5 Certificate Payload Format according to ISAKMP RFC 2408

0 1 2 3 4 5 6 7	0 1 2 3 4 5 6 7	0 1 2 3 4 5 6 7 0 1 2 3 4 5 6 7
Next Payload	reserved	Payload length
Cert. type	Certificate Data	
Certificate Data		

Table 9.6 Certificate Request Payload Format according to ISAKMP RFC 2408

0 1 2 3 4 5 6 7	0 1 2 3 4 5 6 7	0 1 2 3 4 5 6 7 0 1 2 3 4 5 6 7
Next Payload	reserved	Payload length
Cert. type	Certificate Authority	
Certificate Authority		

provided or how irrelevant certificates or multiple certificates or certificate payloads without CA information are handled.

- ISAKMP defines a Hash (HASH) payload is defined to carry a pointer to the certificate. The pointer can then be used to locate the actual certificate.

9.3.1 Exchange of Certificates within IKE

9.3.1.1 Certificate Data Type Profiling for ISAKMP

As mentioned, ISAKMP allows a peer to request a certificate or certificate-related material from another peer by including a Certificate Request Payload in the key exchange signaling. The second peer responds by including the requested material inside the Certificate Payload. However, part of the profiling work is to specify exactly what certificate-related data types could be used in these payloads. Only the following certificate-related data types (certificate encoding) are now allowed to be used for the ISAKMP Certificate Request Payload and the ISAKMP Certificate Payload. In most cases, one observes that Jon Postel's advice of robust protocol design: "Be liberal in what you accept and conservative in what you send" is followed very carefully in the use of certificate types in ISAKMP profiling:

- *X.509 signature certificates*: These are certificates that are used for signing purposes. If an entity receives a CERTREQ that includes this data type but does not support signing certificates, it should ignore the request. The profiling recommends against sending unsolicited certificate data (CERT payloads). In other words, an entity that has not sent any CERTREQ payloads during an exchange, should not expect a CERT payload. The

profiling work does not consider other types of certificates such as X.509 key exchange certificates. It is allowed to send multiple certificates as long as they are sent in multiple payloads.
- *Certificate revocation lists (CRL)*: Since ISAKMP limits the maximum size of 64K for CERT payload, many CRLs cannot be transferred within ISAKMP. The profiling work recommends against generating CERTREQ payloads, where the certificate type is "certificate revocation list". If a peer receives a CERTREQ that includes a CRL as the certificate type, the peer should ignore the request.
- *Authority revocation lists (ARL)*: ARLs are basically the same as CRLs, except that they only list CA-certificates that have been revoked. The treatment of this data type is similar to treatment of CRLs.
- *PKCS #7 wrapped X.509 certificate*: This type only indicates that the certificate-related material is wrapped inside a PKCS #7. The profiling work recommends against including this type inside a CERTREQ payload. Implementations receiving a CERTREQ including this type may accept the request and treat it as if it were an X.509 signature certificate.

9.3.1.2 In-Band Versus Out-of-Band Exchanges

One of the main goals of the profiling work is to allow the IKE peers to perform an in-band exchange of certificates for authentication purposes. However, the profiling work aims to provide the flexibility of allowing exchange of certificates (using the ISAKMP payload) that are irrelevant to the ongoing IKE authentication. Another feature would be to make the protocol use keying material, obtained from out-of-band methods rather than those that can be obtained from the exchanged certificates. Also despite its main goal, the ISAKMP-PKI profiling work aims to allow the IKE peers to obtain certificates and related materials through out-of-band methods, when efficiency requires it. In such cases, the IKE implementation should be able to ignore CERTREQ payloads, if policy allows.

When in-band certificate exchange is desired, the peer must state this desire by sending at least one CERTREQ. As mentioned earlier, a peer that does not send any CERTREQ should not expect to receive any certificates. Also a peer that does not receive any CERTREQ payload should not send any CERT payloads. Again, the "be liberal in what you accept" philosophy recommends providing the ability to deal with CERTREQ payloads that include unsupported certificate types and gives guidelines on what to do in those cases. The implementations should tolerate reception of duplicate identical CERT payloads or receive irrelevant certificates, but should not send duplicate CERT payloads.

Completely irrelevant certificates should be tolerated, since there may be a legitimate reason for them. An example is when the sender wishes to hide its true identity or other important information that is typically sent unencrypted by sending multiple certificates. Another reason for receiving seemingly irrelevant certificates is when several certificate chains follow the trust chain to the same trust anchor CA to the end entity.

9.3.1.3 Certificate Authority and Certificate Chains

The peer requesting a certificate must specify the CA they trust as the trust anchor for the certificates they expect from the sending peer. The requesting peer must populate the Certificate Authority field with the subject name of the trust anchor. If the policy of the requesting peer can allow for multiple trust anchors, one CERTREQ per trust anchor must be generated. The specification forbids sending certificate requests with an empty certificate authority field. The entity receiving a certificate request with "X.509 certificate-signature" as its certificate type must make sure that the certificate it is sending is acceptable to the requesting peer, i.e. it is not issued by a trust anchor different from the one specified in the CERTREQ. If not, the sending entity must include each certificate in the chain, from the sending entity's certificate to the certificate from the issuer whose name matches the Certificate Authority field within the CERTREQ.

9.3.2 Identity Management for ISAKMP: No IP Address, Please!

Regardless of what type of identity is used in the ISAKMP identification payload, this identity is used as a look-up key for both policy and certificate directory look-ups. Inclusion of this payload is mandatory in IKE phase 1 messaging, to the extent that if the identity included in this payload does not match an identity on the certificate presented by the peer, the receiving peer must abort the ongoing key exchange. Therefore, great care must be taken in providing trustworthiness for this identity. The ISAKMP specification puts several limitations on the use of the IP address inside its ISAKMP ID payload: first, this IP address must match the source IP address of the peer. Second, this IP address must match exactly one of the identities listed on the certificate presented by the peer. As mentioned earlier, from the PKI management standpoint, this poses a strong requirement for certificate generation: the certificate must be generated after the device has acquired an IP address. For mobile devices with possibly dynamic IP addresses, changing the IP address that was used as certificate subject name would require obtaining a new certificate.

Due to obvious difficulties in accommodating the requirements associated with use of an IP address as ISAKMP identification payload, the profiling work recommended against using the IP address for ISAKMP identification payload. The specification, however, mandates that IKE implementations must allow the configuration of local policies requiring the peer source address to appear on the certificate. Several other forms of identities are profiled in the specification, of which the most important ones are FQDN IDs and user-FQDN IDs (RFC 822) to support host-based and user-based access control lists configured in the VPNs supporting hosts and users without fixed IP addresses.

Using the identity for policy look-up is a bit more flexible than the case for identity verification and authentication. In cases, when the access control list at the VPN is based on the IP address, the IP address of packets from the peer may be used as policy look-ups as long as a mapping between that IP address and one of the verifiable identities can be made, since the IP address is not very trustworthy. This means even if the IP address is recognized by the VPN gateway and the policy to be applied to the peer's traffic can be determined, the policy is not implemented on the IPsec traffic until the IP address and identity are validated. When FQDN-based identities are used in conjunction with DNS to arrive at IP addresses, care must be taken so that the security of DNS database and its procedures can be guaranteed through DNS security mechanisms [DNSSEC2535].

To accommodate the need for a secure identity, X.509 certificates are allowed to carry multiple forms of identity for the owner. The specification also allows the implementers to pick an identity from those listed on the certificate that is different from the identity included in the ISAKMP ID payload for their policy look-up. This is to accommodate cases where a peer does not have knowledge of what identity the other peer uses for its policy look-ups and thereby has an inappropriate identity inside the ISAKMP payload.

9.4 Further Resources

In 1999 the PKI forum formed a group called OASIS to foster standard-based interoperable PKI solutions for e-business. The group is sponsored by several technology and banking companies. For more information, the reader is referred to the PKI forum and OASIS websites:
http://www.pkiforum.org/ and http://www.oasis-open.org/home/index.php.

We did mention many of the new IETF PKI specifications in brief, sometimes by only providing their RFC numbering. The reader is referred to the IETF PKIX working group website for more information and the actual specifications:
http://ietf.org/html.charters/pkix-charter.html

For more information as well as the current status on profiling of PKI for IPsec and IKE, the reader is referred to the pki4ipsec working group website [PKI4IPSECWEB].

For more information on government recommendations on using public keys for authentication, the reader is referred to [FIPS196].

9.5 References

1. [X5093280], R. Housley et al, "Internet X.509 Public Key Infrastructure Certificate and Certificate Revocation List (CRL) Profile", RFC 3280, IETF, April 2002.
2. [OCSP2560], M. Myers et al., "X.509 Internet Public Key Infrastructure Online Certificate Status Protocol – OCSP", RFC 2560, IETF, June 1999.
3. [FIPS196], US Department of Commerce, "Entity Authentication Using Public Key Cryptography", FIPS 196, Federal Information Processing Standards publication, February 1997, http://www.mirrors.wiretapped.net/security/info/reference/nist/fips/fips-196.pdf.
4. [RSAWEB], RSA Security®, http://www.rsasecurity.com/rsalabs/pkcs/
5. [PKCS102314], B. Kaliski, "PKCS #10: Certification Request Syntax, Version 1.5", RFC 2314, IETF, 1998.
6. [PKSC2315], B. Kaliski, "PKCS #7: Cryptographic Message Syntax, Version 1.5", RFC 2315, IETF, 1998.
7. [PKCSOV], B. Kaliski, "An Overview of PKCS Standards", An RSA technical laboratories note, RSA, November 1993.
8. [CMP2510], C. Adam, and S. Farrell, "Internet X.509 Public Key Infrastructure, Certificate Management Protocols", IETF, RFC 2510, March 1999.
9. [CRMF2511], M. Myers et al. "Internet X.509 Certificate Request Message Format", IETF, RFC 2511, March 1999.
10. [ISAKMP2408], D. Maughan et al., "Internet Security Association and Key Management Protocol (ISAKMP)", RFC 2408, IETF, November 1998.
11. [PKI4IPSECWEB], IETF PKI4IPsec working group website URL, http://www.ietf.org/html.charters/pki4ipsec-charter.html.
12. [DNSSEC2535], D. Eastlake, "Domain Name System Security Extensions", RFC 2535, March 1989.
13. [PKIGMAN], P. Guttman, "PKI: It's Not Dead, Just Resting", IEEE computer magazine, August 2002.

9.6 Appendix A PKCS Documents

A list of PKCS documents is provided below. The documents are provided from the links in the RSA web page at [RSAWEB]. Also an overview on these standards can be found in [PKCSOV]:

- PKCS #1: RSA Cryptography Standard
- PKCS #2: incorporated in PKCS #1
- PKCS #3: Diffie–Hellman Key Agreement Standard
- PKCS #4: incorporated in PKCS #1
- PKCS #5: Password-Based Cryptography Standard
- PKCS #6: Extended-Certificate Syntax Standard
- PKCS #7: Cryptographic Message Syntax Standard
- PKCS #8: Private-Key Information Syntax Standard
- PKCS #9: Selected Attribute Types
- PKCS #10: Certification Request Syntax Standard
- PKCS #11: Cryptographic Token Interface Standard
- PKCS #12: Personal Information Exchange Syntax Standard
- PKCS #13: Elliptic Curve Cryptography Standard
- PKCS #15: Cryptographic Token Information Format Standard

10

Latest Authentication Mechanisms, EAP Flavors

10.1 Introduction

As previously mentioned, wireless links introduce a new set of problems for network security design. It is much easier for outsiders to observe the communications over the air than over a physical wire. This problem is even more serious when, as is often the case, the boundaries of the wireless network coverage do not coincide with the physical boundaries of the enterprise. An eavesdropper can simply sit in her car in the building's parking lot and listen to the exchanged data over the air by simply tuning her receiver to the communications performed wirelessly inside the building.

The feasibility of eavesdropping makes user authentication mechanisms that pass user's credentials in the clear, such as password authentication protocol (PAP), vulnerable to eavesdropping. The passive attacker can simply capture the user identity and password while they are being passed during authentication exchanges and later impersonate the user by simply replaying the captured identity and password pair for the network. Even challenge response methods are vulnerable to eavesdropping. The attacker, after recording both the challenge from the network and the response from the user in the other direction, can launch an off-line dictionary attack. The attacker does this by testing a large set of keys to arrive at the response from the challenge and that way guesses the password.

Another issue that makes authentication in wireless environments more complicated is the user's lack of trust in the network providers arising from the rapid proliferation of hot spots providers and the lack of tamper-resistance in these networks. As shown in Figure 10.1, in many cases, such as at an airport, a coffee shop or a hotel, the access point is in a public area outside the geographic boundaries of the service provider's network. The user typically uses the wireless link owned by an access network provider (such as hot spot provider) operating

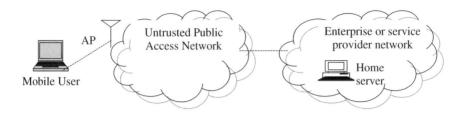

Figure 10.1 Remote connection of mobile nodes to their home network through untrusted access networks

in the public area to get to the service provider's network (be it a commercial network or the network for the company the user work for). The user does not have any prior business or administrative relationship with the access network provider. She simply needs to use the access network to get to a network that she trusts, such as her company network. Neither the user nor the access network provider network has the other party's credentials available at the initial interaction. Hence the user must be protected against illegitimate intermediate operators. Any hacker can set up a rouge access point and spoof as the access point provided by the less technical savvy owner of the coffee shop and thus collect user identity, credit card numbers, install Trojan horses into user's laptops, and so on.

At the same time that the intermediaries cannot be trusted, due to the large proliferation and the inadequate measures for protection of authentication credentials, spoofing of the legitimate backend servers that belong to the user's service provider network or company network are becoming easier too. This means the user must also be able to trust the entity that claims it is the home server (either a AAA server or an application server) for the user.

Hence, it is no longer wise to design an access control method that only requires the user to authenticate to the network. Not only should mutual authentication be required, but also authentication methods should be designed in a manner that protects the client's credentials from eavesdropping by the intermediaries. This includes additional requirements on identity protection (anonymity) and location privacy during authentication.

As a great variety of access technologies are becoming available to provide wireless connectivity, the need to design more flexible security and authentication mechanisms is becoming more pronounced. As mentioned in Chapter 2, extensible authentication protocol, EAP, is gaining more and more popularity due to the flexibility it provides in deploying newer and more diverse authentication mechanisms and algorithms without requiring major upgrades to the network access devices. In Chapter 3 we provided the highlights of EAP key management and distribution framework and explained how the various phases of this framework are being designed to cater to a variety of access network and client types in future.

We also mentioned that use of EAP to carry authentication message exchange for a variety of mechanisms is being more and more wide-spread. This to the point that the IETF EAP standard RFC 2284 [EAP2284], which was a short specification presenting EAP as an extension to PPP (see Chapter 2), has now been deprecated and replaced with a respectable 60-some page RFC 3748 [EAP3748]. As we mentioned in Chapter 3, EAP has been used extensively by the IEEE community in the design of 802.11i security procedures [IEEE80211i]. For all those reasons we devote this chapter entirely to discussing a variety of EAP authentication specifications, such as EAP-TLS, EAP-TTLS, EAP-SIM. We also discuss some industry *de facto* standards such as Cisco LEAP, due to their wide deployment.

10.1.1 EAP Transport Mechanisms

As mentioned in previous chapters, EAP was originally designed to support network access and authentication mechanisms in environments where IP messaging was not available. For that reason the EAP was designed to specifically run over data link layers, such as PPP and Ethernet type links. Transport of EAP over PPP links was explained in Chapter 2. In this section we complete that discussion by explaining the transfer of EAP over IEEE LAN links.

To support three-party authentication models (Chapter 2), in which the network edge device (NAS) out-sources the authentication process to a backend authentication server (AAA server), support for EAP signaling has been added to AAA protocols such as RADIUS and Diameter. Even though this was explained in Chapters 6 and 7, we provide a generic description of how EAP is carried over AAA protocol to allow an independent reading of this chapter.

10.1.2 EAP over LAN (EAPOL)

As mentioned earlier, EAP has been designed to support access control and authentication signaling over lower layer communications protocols. For dial-up networks and cellular networks, PPP is needed for data link layer (layer 2) framing of data packets, since the framing service was not provided by the physical channel in those systems. That meant EAP messages needed to be encapsulated inside PPP frames. We described the procedure for this encapsulation in Chapter 2.

EAP over LAN (EAPOL) was designed for IEEE 802.1X and the following specifications to provide a way to lay EAP over LAN protocols. Since IEEE 802 LANs provide framing services, the EAP over LAN (EAPOL) protocol concept was very simple: encapsulate EAP messages inside EAPOL frames (as shown in Figure 10.2).

Even though this could be accomplished by a single frame type, namely the EAPOL packet frame (described below), the 802.1X committee decided to add some more EAPOL message types to carry out a few tasks other than just encapsulating EAP messages that are generated elsewhere (i.e. in the supplicant or the backend server).

Overall, five types of EAPOL frame types (or as interchangeably called here: EAPOL messages) were defined, of which *only the "EAPOL packet" frame carries EAP messages*:

- *EAPOL-Start*: The general assumption with EAP is that the authenticator somehow senses the presence of the user. In wired LANs, this is easily done, since the user physically connects its LAN cable to a port on the hub or switch. After sensing the connection, the network asks for the user's identity through an EAP-request identity message. On the other hand, in wireless environments, sensing user connections are not that simple. Also, when the supplicant first connects to the LAN, it does not know the address of the authenticator, or even whether an authenticator is present. To provide a wireless equivalent for the wired network-sensing mechanism, the 802.1X designers provided a new EAPOL frame: EAPOL start frame. Upon request for access, the supplicant sends an EAPOL start frame to a group-multicast MAC address reserved for 802.1X authenticators, to let the authenticator know it requires access. When an authenticator receives the EAPOL-start frame, it can start the EAP messaging, which is typically started by asking the user for user identity through an EAP-Request identity encapsulated inside an EAPOL packet frame (described below).

- *EAPOL-Key*: Originally this message was intended to support combining authentication and key distribution processes by allowing the authenticator to send keys (for encryption and other purposes) to the supplicant. However, since the 802.1X specification was not clear on how such keys can be protected during the transfer, this message is hardly ever used as it was originally designed. Newer mechanisms such as 802.11i [IEEE80211i] and RSN [EDARB80211] use a modified version of EAPOL-key message. The EAP key management framework that is being developed in IETF [EAPKEYID], once complete, will probably provide a more formal standard on how keys are managed within the EAP framework. It should also be noted that often EAPOL-key frames are exchanged for messaging that relates to key generation but do not actually carry any keys.
- *EAPOL-Packet*: This message serves the main purpose of EAPOL, namely to provide a container to carry EAP messages. As shown in Figure 10.2, the EAPOL packet frame consists of an Ethernet MAC header, a packet type that identifies the type of EAPOL message, and the packet body itself, which is the EAP message to be carried. All types of EAP messages (EAP request, EAP response, EAP success and EAP failure) are carried by this EAPOL frame type.
- *EAPOL-Logoff*: This message is used by a supplicant to indicate its wish to disconnect.
- *EAPOL-Encapsulated-ASF-Alert*: This message has not been used in many places and is only added here for completeness.

10.1.3 EAP over AAA Protocols

EAP is becoming a widespread framework that handles authentication and key management that deploys the backend AAA servers and the AAA infrastructure using the three-party authentication models. In these models the NAS acts as a pass-through, meaning that the NAS speaks layer-2 protocols with the client (supplicant), while communicating with the AAA server through a AAA protocol. Hence as shown in Figure 10.3, the EAP signaling must be

Ethernet header	version	EAPOL packet type	Packet body length	Packet body

Figure 10.2 EAPOL frame format

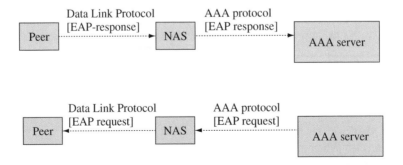

Figure 10.3 Transfer of EAP packets between the peer and AAA server

encapsulated within layer-2 protocols on the client side and within the AAA protocol on the AAA server side. Both RADIUS and Diameter provide specific support for transport of EAP, as we explained in Chapters 6 and 7. The reader is referred to those chapters or to RFC 3579 [RADEAP3579] for RADIUS and the Internet draft specification [DIAEAPDR] for Diameter for more details.

From a transport point of view, when EAP is encapsulated inside a AAA protocol packet, it will enjoy the same transport guarantees that the deployed AAA protocol receives from its underlying transport protocol (UDP for RADIUS and TCP or SCTP for Diameter).

Since EAP was intended to run over links where IP services are absent, it was designed to provide some native reliability features, such as support for retransmissions and duplicate packet elimination. However, EAP was specifically designed to be a protocol with a simple message set to carry exchanges of short messages for authentication purposes. The fact that the EAP message set only includes 4 messages (EAP request, response, success and failure messages) attests to this. For those reasons, EAP is not suitable to transport large amounts of data traffic and hence lacks sophisticated transport capabilities, such as packet ordering. Therefore, when run over link layer, the link layer protocol must provide EAP with packet reordering services.

Another transport issue with EAP is that EAP is a lock-step protocol, when it comes to request/response pairs. In other words, when a request is sent from the authenticator to the peer, another request cannot be sent until a response is received from that peer. Note that this is different from retransmissions needed in case of transport loss, for which EAP has provided provisions. On the other hand, the EAP success and failure messages are not acknowledged and hence provisions must be made for lost success and failure messages, if so required.

The final transport issue that needs consideration is fragmentation. Fragmentation typically needs to be performed when the maximum transmission unit (MTU) size is smaller than the protocol packet size. When an EAP authentication mechanism uses large packet sizes (such as packets carrying digital certificates), fragmentation may be required. Take an example where the MTU size is around 1400 bytes, while a certificate chain consisting of 3 certificates each around 1 KByte is to be sent. This means the original packets would need to be divided into multiple fragments. EAP is not designed to support fragmentation and re-assembly natively, which means if EAP peers negotiate an authentication mechanism that requires big packets but does not provide its own fragmentation and re-assembly services, they will run into difficulties. Some of the more advance EAP-based mechanisms such as EAP-TLS (explained later) do provide fragmentation services.

10.2 Protocol Overview

The EAP message set and format were explained in Chapter 2. As seen there, EAP message set consists of only 4 messages, of which only EAP request (from server/NAS towards supplicant) and EAP response (from supplicant to server/NAS) carry data, while EAP success and EAP failure messages only carry the result of the authentication procedure.

The simplicity of the EAP message set is quite reasonable, if we bear in mind that EAP is only meant to carry the exchange messaging for a specific authentication method that is performed between the peer and the EAP server (typically the AAA server) and at the same time allow a dumb NAS to act as a pass-through on the path between the peer and the server. The NAS only need to understand the EAP success and failure message, while allowing EAP requests and responses to pass through without understanding their content or meaning. A type field within the request and response messages is used to indicate the type of data as well as the type of authentication mechanism that is being used. Examples of lower type values used in the request/response packets are shown in Table 10.1, while the use of type field to indicate specific authentication mechanisms is shown later on in this section. The EAP framework specification calls these mechanisms the "EAP methods" and refers to messaging specific to those mechanisms as method-specific messaging.

A typical EAP exchange (sometimes referred to as EAP conversation) proceeds as follows:

- The authenticator (NAS) sends an EAP request to the peer. This request can be a request for authentication credential or simply a request for some sort of information. The type field in EAP request message indicates what sort of information is being asked for. A prime example is asking for the user's identity. EAP requests carrying Identity type are usually called EAP Identity request messages. When human user interactions are required, optional displayable messages for the user may be included in the messaging. As noted in Table 10.1, if information exchange at this point is performed in clear-text form, exchanging user identity during this early phase is not recommended, especially if

Table 10.1 Examples of type field values used for EAP request and response messages

Value	Type	Description
1	Identity	Used by the authenticator to query the Identity of the peer and by the peer to respond with her/its identity to the authenticator. Since identity request and response are sent in the clear, this may not be a desirable approach to obtain peer's identity when privacy is a concern. In such cases it is recommended that each EAP method specifies a safer mechanism to obtain identity. Alternatively, the peer may respond with decorated identities that are different from the peer's true identity.
2	Notification	Used to convey a displayable message from the authenticator to the peer. One example is when the user types her identity incorrectly and needs to be notified by the authenticator to try again. Another example is notification about a password that is about to expire. Unless prohibited by the authentication method, the peer responds to a Notification request with a Notification respond.
3	NAK (response only)	A legacy type that can only be used in EAP response messages. Examples are when a peer needs to indicate to the authenticator that it cannot accept the required authentication type. In such cases the peer response contains suggestions on one or more authentication types. NAK type is not used for error indications.
>=4	Authentication methods	Explained later on

Note: Type 4 and above are used to indicate specific authentication methods.

user privacy protection is a concern. It is recommended that the identity of the client is communicated to the authenticator through method-specific signaling that provides confidentiality for messages carrying user identity. We will discuss EAP-TTLS as a prime example of such signaling. Some EAP methods also provide per-packet authentication, integrity and replay protection.
- The peer responds to the EAP request, with an EAP response message of the same type. For instance, if the request was an Identity request, the response would be an Identity response. If the type field in the EAP request had suggested an authentication method, the peer either responds to the needs of the authentication method with a EAP response message of the same type, or responds with an EAP response message of type NAK (type 3), indicating that the suggested authentication method is not acceptable. In such cases, the peer's response contains suggestions on one or more authentication methods. This is accomplished by including the type values for those methods within the type-data field of the NAK response message. We will talk about the authentication types in more detail shortly.
- From this point on many EAP request/response pairs may be exchanged between the peer and the authenticator to carry method-specific conversations. As mentioned before, after sending each EAP request, the authenticator (NAS) waits until it receives a response from the peer. The NAS does not have to understand each authentication method and can simply act as the pass-through agent for the EAP server. This will also facilitate upgrades of authentication methods as well as the credential management, since peer credentials do not have to be maintained (or "cached") at the NAS or even revealed to the NAS.

As mentioned above, EAP types 4 and higher are used to indicate the authentication method used by the EAP messaging and for that reason are referred to as EAP method types. Except 20, almost all method types from 4 up till 45 are already allocated to various EAP authentication methods. The latest list of EAP method types can be found on the IANA website for EAP numbers [EAPIANA], from where we cite some of the more popular ones in Table 10.2.

Type values 4–6 represent authentication methods that are described by the main EAP RFC [EAP3748]. For that reason we provide a brief description of these methods here. In the rest of this chapter we will provide more details of some of the other popular methods, such as EAP-TLS, EAP-TTLS, LEAP and PEAP.

Type value 4 indicates that authentication messaging relates to an MD5-challenge as an authentication method, which is very similar to PPP-CHAP (explained in Chapter 2).

Table 10.2 Popular EAP method types

Type	Method	Type	Type
4	MD5-challenge	23	UMTS Authentication and key agreement
5	One-Time password	25	PEAP
6	Generic Token Card (GTC)	26	MS-EAP Authentication
9	RSA Public Key Authentication	32	Secur ID EAP
13	EAP-TLS	38	EAP-HTTP Digest
15	RSA security Secur ID EAP	43	EAP-FAST
17	EAP-Cisco Wireless (LEAP)	46–191	Available via review by Designated expert
21	EAP-TTLS	192–255	Reserved + Experimental

The request message carries a challenge, while the response either carries a reply to challenge (type 4) or a NAK (type 3) if the peer does not accept the MD5-challenge as authentication mechanism.

Type value 5 indicates that messaging relates to one-time passwords (OTPs). We explained OTPs in Chapter 2. Here, the request contains an OTP challenge, while the response includes a reply.

Type value 6 indicates that messaging relates to authentication method using a generic token card implementation that requires user input. The request contains a displayable message and the response contains the token card information necessary for authentication.

In the remainder of this chapter we will explain how some of the more sophisticated authentication methods work. However, before doing that it is useful to go through how the EAP framework is used to carry a generic authentication method both in order to understand the remainder of the chapter better and to help the reader with her own design of future EAP authentication methods.

10.3 EAP-XXX

The EAP framework is simply a facilitator for providing a forum that allows the peer and the EAP server to negotiate the authentication method and later perform the authentication exchange itself through a dumb proxy authenticator (NAS).

The EAP itself does not perform the actual authentication; rather, EAP is augmented with an authentication method that has its own requirements and procedures. EAP request and response messages are used to carry the information required for this authentication method between the peer and the server until the EAP server indicates to the NAS success or failure for the authentication process. In this section we will explain the mechanics of carrying the messages for a generic authentication method.

Let us say that during the initial EAP request and response exchange, the peer and the EAP server agree to choose the authentication method XXX, which may or may not be a mutual authentication process. When EAP is deployed to help the peer and the authentication server to perform authentication mechanism XXX, it is customary to say that an EAP-XXX authentication has been performed, where XXX is the actual authentication method, whose exchanges are carried over EAP messages. Examples of EAP-XXX are EAP-TLS, EAP-TTLS and EAP-SIM. Also Cisco has provided LEAP (lightweight EAP) and PEAP (protected EAP). We will come back to these processes later on. But for now let us see how generic EAP-XXX messages are treated in the EAP framework.

Figure 10.4 shows how method-specific messages for authentication method XXX are carried over EAP. As the message arrives at the lower layers and is delivered to the EAP processing entity, based on the code field in the EAP message, the entity knows what sort of EAP message (request, response, etc.) it has received. When the EAP packet is a request or response and includes a type field, based on the type field, the EAP entity knows what authentication method has to be performed and delivers the message to the process designated for that authentication method. In cases, where the identity has been exchanged through an earlier EAP-Identity request/response process, the identity may be passed to the authentication method processor.

The final EAP success or failure messages do not include any type field and hence can be processed by a generic EAP entity, such as a NAS. This is a powerful feature, since it allows

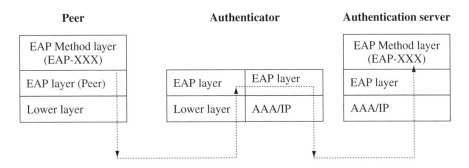

Figure 10.4 EAP layering model for carrying messages for authentication method XXX (EAP-XXX Authentication)
Note: Authenticator acts as pass-through most of the time.

the authenticator (NAS) to simply peek into the EAP code field and only look for success or failure to allow access to the peer, while passing any other EAP messaging that carry method-specific information without having to understand its content. All the authenticator needs to know is how to pass EAP messages with code 1 (request), 3 or 4 (success or failure) to the peer and EAP messages with code 2 (response) back to the server. This is the core value brought by EAP framework: When new authentication methods are introduced, no software upgrades are required at the network edge devices acting as authenticators, which are typically less sophisticated and large in numbers. Only the peer and the authentication server need to be upgraded to support the new authentication method.

The model shown in Figure 10.4 represents the three-party authentication model that is widely spread today. However, there may be rare cases, where the authenticator is capable of performing one or two authentication mechanisms locally and hence the three-party model needs to be simplified to a two-party authentication model. In that scenario, one can use the simpler model shown in Figure 10.5. The model shown also applies to cases where EAP signaling may be used in a peer-to-peer mode, such as some key management applications.

The authenticator, by looking at the type field within the EAP packet, should be able to know whether it should participate in a two-party authentication and act on the message or simply forward it to the peer or authentication server as it does in a three-party model.

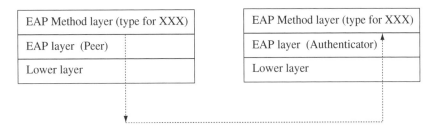

Figure 10.5 EAP layering model for EAP-XXX in a two-party model (when authenticator acts as EAP server or end point)

10.3.1 EAP-TLS (TLS over EAP)

EAP was initially developed as an extension to the point-to-point protocol (PPP) to provide more enhanced authentication procedures over PPP links. Even though EAP was later on developed as a framework to provide messaging support for a variety of authentication mechanisms, many of these mechanisms do not provide mutual authentication, encryption or key management services. Providing mutual authentication as well as key management mechanisms over PPP links were the original drivers behind the EAP-TLS specification work [EAPTLS2716].

Unlike IPSec, which generally depends on an external authentication and key management mechanism such as IKE, as we saw in Chapter 4, transport layer security (TLS) comprises of two layers: a handshake layer that performs authentication, cipher suit negotiation and key exchange and a record layer that provides end-to-end security protection at the transport layer. This makes TLS a strong candidate as an authentication method for the EAP framework, especially since TLS is capable of using digital certificates for identity verification. Ability to use certificates makes TLS handshakes suitable for access control in wireless environments, where a roaming client, that has no prior trust relationship with a network, can authenticate and subsequently exchange proper key materials for a secure communication channel with the network. As we will see shortly, the combination of EAP and TLS means that EAP-TLS inherits the security advantages of TLS and the flexibility of EAP, such as accommodating backend servers.

10.3.1.1 EAP-TLS Architecture and Message Format

As mentioned earlier, EAP-TLS was originally designed to support authentication over PPP links, which are obviously single hop point to point links, where the authenticator resides at the network end of the PPP link. We mentioned that EAP can be carried over PPP or LAN link (EAPOL) between the supplicant and the authenticator. On the other hand, TLS is also a peer-to-peer protocol, in which the TLS handshake and communications happen directly between the ends of the transport session. So if the TLS messaging can be encapsulated inside EAP packets that themselves are carried over a PPP or a LAN link, then the EAP framework can accommodate TLS as an authentication and key management method and thereby EAP-TLS is born.

Note that TLS was originally designed to be carried over a transport protocol and all TLS data, even those during handshake phase are carried as TLS records. Table 10.3 shows how TLS messaging (TLS records) can be encapsulated inside EAP packets. Note that the stack for EAP-TLS follows the model shown in Figure 10.4, where the EAP-TLS messaging is carried as what is called EAP-XXX layer in that figure. The reader is referred to Chapter 2 for description of EAP message format.

As mentioned previously, EAP requests and EAP responses include a type field that indicates the type of information being carried between the supplicant and authenticator or authentication server and type values equal and higher than 4 are used to indicate specific authentication methods. EAP data type of 13 indicates that TLS data is being carried for EAP-TLS authentication. This way, EAP-TLS provides a mechanism to carry TLS messages over all the media that can carry EAP messages. Server-to-client TLS messages are simply embedded in EAP-request messages, while client-to-server TLS messages are embedded in EAP-response messages.

Table 10.3 EAP-TLS Request/Response Packet Format

Field	Subfields		Description
EAP Header	Code		Type of EAP packet (1 for EAP-request, 2 for EAP-response)
	ID		One octet long to match requests and responses
	Length		2 octets long, showing the length of the entire message
EAP Data	EAP data Type = EAP-TLS		Sets to 13 for EAP-TLS packet
	EAP-TLS data	Flags	L for Length, M for More fragments, S for EAP-TLS Start Message and R for Reserved
		TLS Message Length	4 octets long, contains the total length of the TLS Message
		TLS data	Contains encapsulated TLS packets in TLS record format

One may wonder at this point: most architectures today are three-party models that deploy backend authentication servers or AAA servers. How do we fit a client–server protocol such as TLS to the three-party model? Even the answer to this question lies in encapsulating the TLS messages inside EAP messages:

The NAS (authenticator) does not have to understand any of the TLS handshake or record layer details. All it needs to know is whether TLS messaging is complete or not. For this, EAP-TLS provides a simple wrapping mechanism around TLS messaging, by adding EAP-start, EAP-identity to the beginning of TLS handshake and EAP-success/failure messages to the end of TLS handshake. The NAS only needs to understand these wrapper messages as stated before.

Now that the reader is asking smart questions, another issue is: TLS is an end-to-end protocol that must run over a reliable transport protocol. How can TLS go through 3 parties and multiple hops, many of which do not implement TCP or SCTP? Again the answer lies in encapsulating TLS messaging in EAP messages:

We described how EAP can be carried over a link layer protocol between the supplicant and authenticator and then over a AAA protocol between the authenticator and the backend server. As long as TLS records are embedded inside EAP messages, we are OK over any media that can carry EAP. As seen in Table 10.3, some additional flags are added between the EAP header and the TLS message to address transport problems such as fragmentation and start problems.

The three-party model for EAP-TLS authentication is shown in Figure 10.6. Here the TLS is still performed in a peer-to-peer manner between the client (supplicant) and the server acting

Figure 10.6 EAP-TLS in a three-party model

as the TLS server (typically also a AAA server), while EAP is performed in a three-party fashion over the NAS acting as a pass-through.

However, one issue is that often a security association between the NAS and AAA server must exist to support the transport of sensitive material, especially when EAP-TLS is used for the distribution of keys to the NAS.

10.3.1.2 Protocol Overview

Since EAP-TLS specification was initially prepared for use over PPP links [EAPTLS2716], our protocol overview here is also based on a two-party model, in which the client deals with an authenticator that also acts as an EAP server. In previous sections, we described how the model can be extended to a three-party model, where the client authenticates to a TLS server (co-located with a AAA and EAP server) through a NAS acting as a pass-through.

The messaging for EAP-TLS procedure is shown in Figure 10.7 and performed as described in the following. As customary in EAP, messages from the server to client are carried in EAP-request messages, while all the messages from the client are carried in EAP-response messages. All request and response messages include a type field with a value equal to 13, indicating EAP-TLS as authentication method in use. As we will see shortly, EAP-TLS places great emphasis on the use of session ID to improve the efficiency of the protocol. By using the session ID properly, the server and the client can avoid repeating the authentication and key exchange handshakes when the established keys are still active. This is explained in more detail below:

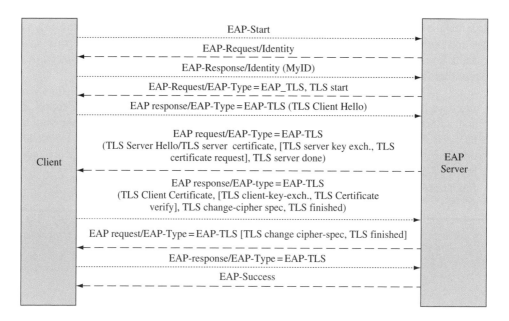

Figure 10.7 EAP-TLS messaging procedure

- When EAP is being used over a PPP link, the process starts with the client sending a PPP LCP acknowledgement, showing client's agreement to use EAP as the authentication process. For other links, the starting mechanism depends on the messaging protocol available for that link. For instance, for 802 links, the process may start with the client sending an empty EAP message (such as EAP start) to indicate the start of the process. After this, the actual EAP-TLS exchanges start.
- The first actual exchange provides the server with information on user identity. This exchange starts with the server sending an EAP-request/identity packet to the client, which in turn responds with an EAP-response/identity packet.
- Once the client identity is provided to the EAP server, the TLS signaling starts with the server sending an EAP-TLS/start packet. The EAP-TLS/start packet is simply an EAP-request packet (with type 13 for EAP-TLS) that contains no EAP-data except that the "S" bit inside the flags field is set to indicate a start packet.
- The client responds with an EAP-response packet that encapsulates a TLS Client Hello message. The TLS message is formed according to TLS record layer format. As described in Chapter 4, the TLS Client Hello message is the start of the TLS handshake exchange and contains a set of cipher suits supported by the client as well as a possibly non-zero session ID.
- The server responds with a EAP request packet that encapsulates a TLS Server Hello message, containing relevant TLS records such as TLS Certificate, server_key_exchange, certificate_request, server_hello_done and/or Finished handshake messages, and/or a TLS change_cipher_spec message. The certificate request record is included if the server requires the client to authenticate itself with a public key certificate. If a session ID suggested by the client exists, it means that a mutual authentication and key exchange has taken place previously and the parameters for a previously established session are still valid. The server chooses to resume the session and hence must choose the cipher suit that was previously negotiated. On the other hand, if the session ID suggested by client is zero or unrecognized by the server, the server issues a new session ID to indicate start of a new session.
- The client responds with an EAP response message that includes a number of TLS records in its data field. The TLS records include TLS change_cipher_spec message and finished handshake message, and possibly certificate, certificate_verify and/or client_key_exchange handshake messages, depending on the scenario: If the server has indicated the resumption of a prior session, it means shared session keys have already been established and therefore new shared keys need not be established. In this case, the client only sends change_cipher_spec, which indicates that cipher suits and keys are agreed upon and the TLS state machine can move from handshake exchange to record layer. No client_key_exchange records are needed here. If, on the other hand, the server indicates establishment of a new session, it means that new shared keys must be established and for this to happen, the client needs to include a client_key_exchange record in its EAP response and the exchange must continue until the key exchange is complete and a secure record layer is established. This is again indicated by exchange of change_cipher_spec messages from both ends and a Finished message by the server (normal TLS handshake messaging).
- Once TLS handshake is complete, the user sends another EAP response message with empty EAP-TLS to indicate this and the server responds by EAP-success, ending the authentication.

10.3.1.3 Drawbacks with EAP-TLS

EAP-TLS may not be suited to every client–server interaction scenario and may have some security consequences as described below:

- *Need for client certificate*: EAP-TLS is designed for scenarios where server and client need to mutually authenticate each other and is specifically suited for cases where both server and client use certificates for this mutual authentication. However, for scenarios where symmetric authentication methods are not required, e.g. when the client is not required to authenticate itself to the server or is not capable of using certificates to authenticate itself to the server, EAP-TLS has less applicability. Requiring certificates from the user can be especially burdensome, since it adds to the load of certificate authorities (CA) and the complexity of certification and CRL creation processes.
- *User versus device certificates*: EAP-TLS relies on client certificates for client authentications. With the increased level of user mobility and the increased demand on network connectivity through a variety of devices, the need for the distinction between device authentication and user authentication is becoming more pronounced. Other examples are scenarios where many users share the same device, such as a device that is used by field personnel, or scenarios where one user may use multiple devices to access the network. Certificates, stored in the device only, authenticate the device in which the certificate is stored, not the user holding the device, unless there are specific mechanisms to lock and unlock the user certificates from the device. There is no provision in EAP-TLS to distinguish between user and device certificates.
- *User identity protection*: EAP-TLS does not protect the privacy of the user. The user identity is revealed to the outside world both through the EAP-identity message and through the certificate face value.
- *Protocol efficiency*: Certificate-based EAP requires lengthy protocol exchanges, which are not desirable for mobile users that need to be re-authenticated after a handover. TLS session resumption (look) is a method to provide a seamless handover, but it does not help initial authentication delays.
- *NAS support*: EAP-TLS was designed for two-party authentication models, where the client authenticates to a PPP authenticator rather than a backend server. Extending the model to accommodate a backend server can have security implications, since the key negotiated between the EAP server and the peer must be transmitted to the authenticator safely (as explained in EAP key management framework, Chapter 3).

For the reasons mentioned above, a new method, called EAP tunneled TLS, EAP-TTLS is being developed as described in the following section.

10.3.2 EAP-TTLS

No doubt when any certificates are deployed in the network, some sort of PKI must be implemented and maintained. However, as we saw in Chapter 9, managing a system with only server certificates may be a lot less complex than a system that deploys both server and client certificates, especially when both users and devices have to be certified separately. The number of clients is far bigger than the number of servers and this may require more elaborate facilities to process and maintain data related to key/certificate generation and revocation. This is the first reason why network designers shy away from authentication mechanisms that require client certificates.

Furthermore, when the client can either be a device or a user, different procedures must be implemented for issuance and management of user versus device certificates. As explained in Chapter 9, users must be authenticated in person and registered prior to obtaining certificates. The policies and profiles for user certificates and device certificates are usually different. All this will drive the costs for a PKI system up and may also require a higher degree of user sophistication as well. This is the second reason why deploying an authentication mechanism that allows the clients to support only less sophisticated legacy methods such as PAP, CHAP, or the MD5 challenge for authentication to the server, is desirable.

Finally, when an authentication method that requires client certificates such as EAP-TLS is deployed, unless the operator has the PKI know-how, services from a third-party PKI provider must be acquired and paid for. For instance, Motorola has a long-standing tradition of providing PKI and security services for its cable modem products. Cable modem boxes are shipped with device certificates with a cost of a fraction of a dollar. Motorola also has the capability to provide the PKI management functions (explained in Chapter 9) for the operation of these certificates at low cost. However, when an operator needs to refer to third party PKI providers, it may have to pay several hundreds of dollars per active certificate. The potentially large number of clients in the system can drive the cost of network security up very rapidly.

On the other hand, there are cases when client authentication is not important. For instance, when the client needs to obtain some public data from an information server, the client may not be required to show its/her authenticity, while the server needs to prove the legitimacy of its identity and the information it is providing. In such cases only the server needs to authenticate itself to the server. Another prominent example of this asymmetric scenario is the e-commerce application that we explained early on in Chapter 4. The web server needs to prove that it belongs to a legitimate business, so that the user can trust it with its credit card information. The need for asymmetric authentication mechanisms was another main driver behind the design of an asymmetric authentication mechanism that, in contrast to EAP-TLS, does not require the client to authenticate to the server with exactly the same mechanism as the server authenticates to the client.

The problem that arises is that the legacy authentication mechanisms, especially when deployed over wireless links, create their own set of problems, ranging from well-known security flaws such as password sniffing and dictionary attacks to more recent concerns such as user privacy breaches. Not only the passwords can be sniffed or replayed, but also the user identities, location and activities can be exposed to a passive listener.

A new clever extension to EAP-TLS is being designed in IETF to address such concerns [EAPTTLSDR]. In EAP-TTLS, just as in TLS, the server still uses a certificate to authenticate itself to the client. However, in order to support legacy mechanisms for client authentication, EAP-TTLS takes advantage of the TLS in the following way:

The TLS handshake moves forward with only the server authenticating to the client. The client does not authenticate to the server immediately, but waits with its authentication until a secure TLS tunnel is established. The TLS record layer is established based on server authentication only. Once the record layer is established, the client then using the secure tunnel can follow any of the traditional authentication methods such as PAP, CHAP, MS-CHAP through legacy RADIUS infrastructures, without worrying about its identity or password being sniffed in the process or replayed by rouge users later on. The client also uses EAP for authentication if desired.

This added flexibility and security measure is highly desirable in wireless environments. The secure tunnel established by the TLS handshake can even be used for cipher suite negotiation and key exchange for secure client-wireless AP in a way that initially is hidden from the access point. This is explained in the next subsection and is especially useful if the client initially does not trust the AP and access network provider. Note that, EAP-TTLS supports the traditional TLS mutual authentication using client certificate as well, if so desired.

10.3.2.1 EAP-TTLS Functional Elements

Aside from the reasons mentioned earlier, another main design criterion for EAP-TTLS is to support the establishment of secure connections even in roaming environments, i.e. when the client needs to connect to a domain that is administratively different from the one the user is affiliated with (scenario shown in Figure 10.1). This would require of the client to authenticate through and establish pair-wise keys with random wireless access points belonging to a domain that has no trust relationship with the client. Hence, a sane security practice dictates that the client's authentication credentials be protected from the initially untrusted access points and other intermediaries between the client and the authentication server until such trust relationships are established at a later point.

To support such needs, EAP-TTLS specification defines the concept of a TTLS server. The TTLS server is a AAA server with which the client establishes the secure TLS channel and engages in EAP-TTLS exchange. The TTLS server may be able to authenticate the client. However, in many cases, the TTLS server is not the client's home AAA server and thus needs to refer to that server for client authentication. The generic architecture for EAP-TTLS signaling is shown in Figure 10.8.

It should be noted that the model shows only logical entities; the TTLS AAA server and the home AAA server may be physically co-located in the same way that the TTLS server may be physically located at a NAS. On the other hand, there may be one or more AAA proxy servers between the NAS and the TTLS server and between the TTLS/AAAF server and the home AAA server.

As with EAP-TLS, EAP-TTLS signaling between elements in the architecture can run over any protocol that is capable of encapsulating EAP messages, such as PPP or EAPOL on the client side of the NAS and RADIUS or Diameter on the server side of the NAS.

The following assumptions apply for the interactions between the functional elements within the EAP-TTLS architecture:

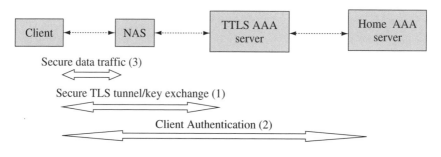

Figure 10.8 EAP-TTLS architecture

- *Client-access point relationship*: Generally, the client and the AP have no security relationship prior to the start of the EAP-TTLS negotiations. In fact, EAP-TTLS is designed to protect the client password or the challenge/response handshake between the client and AAA server from eavesdropping by untrusted APs (model in Figure 10.1). The secure channel between the two is established as a result of completed EAP-TTLS key management procedures.
- *Client-NAS communications*: Running EAP over PPP or EAPOL is the same as assuming that there is only a single hop between the client and the NAS. This means the NAS must be implemented at the first hop, which could be within a foreign network domain. Even though this seems a rather trivial point, it can pose a restriction on the application of EAP-TTLS. For instance, if a user connects to a hotel WLAN AP to get to its company's VPN gateway, the VPN gateway cannot act as a NAS, since it is at least 2 hops away from the client. This means we cannot use EAP-TTLS as an authentication mechanism between the VPN gateway and the client, unless we can design a new encapsulating protocol that can carry the EAP between the client and the VPN gateway. The other issue is that EAP-TTLS is designed to be initiated by a link layer protocol.
- *Key exchange and NAS–server relationships*: One result of the EAP-TTLS is that the client and the NAS arrive at a secure channel that supports data encryption and authentication between the client and the NAS (as shown in Figure 10.8). However, the keying material required for this channel is actually first derived between the client and the TTLS server rather than between the client and the NAS. The client will arrive at the keying material as a result of the EAP-TTLS process, while the NAS is unaware of this process. The TTLS server must then transfer the keying material to the NAS over a AAA protocol. This means the level of the security assurance provided by the AAA protocol for this transport is crucial for the integrity of the whole process. RADIUS specifications only provide shared secret procedures to provide such assurance. This not only requires the existence of a pre-established security association between NAS and TTLS server, it may not need the level of required security for many networks. Diameter may provide better support in this respect.
- *TTLS server–HAAA server interactions*: Note that EAP-TTLS only protects communications between client and TTLS server, which means when user credentials need to travel from TTLS server to home AAA server (typically over a AAA protocol), again that transit must be secured. For RADIUS this is typically provided in the form of shared secrets between adjacent AAA entities. This may be an inadequate level of protection. In Diameter, IPSec security associations can be used.
- *AAA proxies*: Whenever AAA proxies are implemented, a security association between adjacent servers may be required, unless the AAA protocol allows end-to-end security protection.
- *AAAH server–client relations*: The home AAA server must be aware of the credentials that the client is going to present for authentication. For instance, if the client is going to use passwords, the home AAA server must be able to authenticate the client based on the password. *The AAAH does not need to support EAP-TTLS, unless it also acts as TTLS server*; it only needs to support the legacy client authentication mechanisms that are deployed.
- *TTLS server–client relations*: The TTLS server is, on the other hand, not required to have a pre-established security association with the client. The secure channel between the TTLS server and client is established based on the server authenticating to the client using the server public key certificate and proof of private key possession. In rare cases, the client

may also authenticate to the TTLS server using client certificates, but that is not necessary. This secure channel is the result of the completion of the first phase of EAP-TTLS negotiations (the TLS handshake).

- *Cipher suit negotiations*: To make sure both the NAS and the client agree on a common cipher suite for the secure channel they are about to establish, both the client and the NAS can send their cipher suite proposals inside specific EAP-TTLS AVPs and AAA protocol fields, respectively, to the TTLS server. The server then needs to find a match, typically giving priority to the client's preferences and communicates the results to both the NAS and the client.

10.3.2.2 Messaging Overview

EAP-TTLS specification is not standardized at this point. The latest revision of the specification includes two versions: EAP-TTLSv0 and EAP-TTLSv1 [EAPTTLSDR]. We describe only the earlier version here, since EAP-TTLSv0 enjoys wider support from the vendors. We suffice by mentioning that EAP-TTLSv1 allows authentication during the TLS handshake rather than after the handshake as in EAP-TTLSv0.

The messaging for EAP-TTLS follows the generic model for EAP-XXX authentication method in Figure 10.4. In other words, in the same manner as EAP-TLS, EAP-TTLS messaging is carried at the EAP-XXX layer over EAP layer. However, the difference becomes clear once we remember that the goal of EAP-TTLS was to accommodate the secure exchange of legacy and less sophisticated client authentication. Also remember that EAP does not provide any confidentiality services for the EAP-XXX layer above it. Therefore a secure channel is needed through which the client can perform simpler authentication mechanisms such as PAP and CHAP. Running TLS over EAP (with EAP-TTLS type) under the legacy protocols provides this secure channel for these protocols. This justifies the to-the-uninitiated-eye awkward EAP-TTLS stack, shown in Table 10.4.

This way, EAP-TTLS packets encapsulate TLS records, which themselves carry encrypted authentication or key exchange messages. Encapsulation of TLS records within EAP-TTLS packets as well as encapsulation of EAP-TTLS packets within EAP packets is very similar to that shown for EAP-TLS in Table 10.3 with a few exceptions: The most obvious exception is that within EAP request and response messages, type 21 is used for EAP-TTLS encapsulation. There are a few other exceptions relating to the way EAP-TTLS handles fragmentation and messaging start, but we refer the reader to the IETF draft for the details.

Table 10.4 Deploying legacy client authentication mechanisms (such as PAP, CHAP) over secure TLS channel in EAP-TTLS framework

PAP, CHAP, MD5-CHAP, EAP-MD5
TLS
EAP-TLS
EAP
PPP, EAPOL, RADIUS, Diameter

One point about EAP-TTLS messaging is worth mentioning and that is the way EAP-TTLS tunnels the information between the client and the TTLS server through the TLS record layer, once the TLS secure tunnel is established: the designers of EAP-TTLS have chosen to use the Diameter attribute-value-pair (AVP) format to carry the data that is to be encrypted through TLS record layer. Note that this does not mean implementation of Diameter as a AAA protocol is required for EAP-TTLS deployments. It is simply a matter of borrowing a clever design from Diameter for the transfer of information through TLS. The AVP format includes an AVP code to identify the attribute, some flags, information on length of the entire AVP, and the data related to the attribute (Table 10.5).

In summary, to protect the sensitive information with EAP-TTLS framework, such information must be encoded according to the AVP format and passed to TLS record layer.

10.3.2.3 Protocol Overview

Even though it may be obvious at this point, we choose to mention it again: since EAP-TTLS deploys the TLS features to protect client authentication, EAP-TTLSv0 exchange inherits the two phases of TLS. Hence EAP-TTLS negotiations comprises of two phases: TLS handshake phase and TLS tunnel phase.

During the *TLS handshake phase*, the TTLS server authenticates to the client using its certificate. The cipher suites for the upcoming TLS secure channel (record layer) are negotiated and activated during this phase. This phase typically starts with the access point prompting the client for its identity, using an EAP-request/Identity message. However, to protect the client's true identity, the client does not respond with its actual identity. The client typically responds with the information that helps the NAS to route the traffic to the realm that can carry the EAP-TTLS negotiations (such as @mytls.com). The client reveals its true identity only to the secure TLS channel in the next phase. Following the EAP-Identity messages, the TTLS server sends a specific EAP request/start packet to indicate the start of EAP-TTLS. When the client sees this message, it starts the TLS handshake. Upon completion of this phase, the server and client share keying material (such as the TLS master secret) for the protection of subsequent communications. Note that the level of protection provided by TLS record layer depends on the strength of the negotiated TLS cipher suites. Also note that the keying material generated during this phase is only for the TLS record layer, i.e. the TLS session between the client and the TTLS server (as shown in Figure 10.8). The keying material for the protection of client–AP traffic is generated later on.

The second phase of EAP-TTLS negotiations (called the *TLS tunnel phase*) uses the TLS record layer established in the first phase to protect the messaging in the second phase.

Table 10.5 AVP format used in EAP-TTLS borrowed from Diameter specification

0 1 2 3 4 5 6 7 0 1 2 3 4 5 6 7 0 1 2 3 4 5 6 7 0 1 2 3 4 5 6 7	
AVP code	
Flags	AVP length
Vendor ID (optional)	
Attribute data	

The second phase accomplishes the main functions intended for EAP-TTLS, such as client authentication, negotiation of cipher suites for client–AP communications and key material distribution. Again, note that the secure tunnel ends at the TTLS server (not at the AAAH server); the TTLS server de-tunnels the encrypted AVPs and determines whether they are related to client authentication or to other purposes.

We will go through an example on how the client authentication according CHAP is performed later on. For now, we note that, when the data is related to client authentication, if the TTLS server does not have access to client credentials, the TTLS server is not able to perform the client authentication. In such cases, it forwards the de-tunneled data to the AAA server using a secure AAA protocol transaction. This architecture removes the burden of supporting and understanding EAP-TTLS from the legacy AAA servers, since only the TTLS server deals with EAP-TTLS signaling. At the same time the TTLS server does not have to deal with the details of client authentication. The TTLS server simply waits for the AAAH server to approve or reject client authentication and issues EAP-Success or Failure based on the results from the AAAH server. Note, however, that if client authentication requires multiple exchanges with the client, the TTLS server needs to forward the exchange data to the intended recipient. For instance, for CHAP authentication, the TTLS server forwards the challenge issued by the AAAH server towards the client.

Any keying material that is required for the protection of subsequent data connection between the client and the AP is generated in the second phase of EAP-TTLS. However, this keying material is still generated from the TLS master secret, client random value and server random value that were generated between the client and the TTLS server during the first phase of EAP-TTLS negotiations (see TLS handshake in Chapter 4). The keying material is first generated at the TTLS server but is transmitted to NAS only when the TTLS server receives an indication of a successful client authentication from the AAAH server. The TTLS sends this key material to the NAS over a AAA protocol (rather than over EAP-TTLS) encrypted using the security association that exists between the NAS and the TTLS server. Once the transfer is complete, the client and the NAS share a security association and key material can active this association to secure their communications.

10.3.2.4 Session Resumption: EAP-TTLS Support for Mobility

EAP-TTLS specification also provides support for a quicker negotiation if the client and TTLS server have already established a session earlier. Assuming that the session is not stale and the server has not completely erased all the records on the session, the client can invite the server to resume a previous session by including the identifier for the session (session ID) in a client hello message, which is the first message in TLS handshake.

As mentioned earlier, the TLS session is established between the client and the TTLS server and hence the identity of the current NAS (or AP) does not affect any of the session characteristics. After handing over to a new AP, the client simply resumes the earlier session with the server (the TTLS server is the same) without the need for a new tunneled authentication to the TTLS server. Note that the TTLS server does not retain all the information regarding the key distribution for the client and therefore, the client must send some of the related information to the server. Also the TTLS server must still convey the session authorization information, such as the maximum allowed bandwidth, the maximum session time, and so on, to the new AP. The reader is referred to [EAPTTLSDR] for more details.

10.3.2.5 Example: CHAP over EAP-TTLS

Now that we have described an overview of the EAP-TTLS protocol, we finish the discussion on this protocol by going through the details of EAP-TTLS negotiations supporting a client that uses CHAP for authentication. Figure 10.9 shows the messaging sequence.

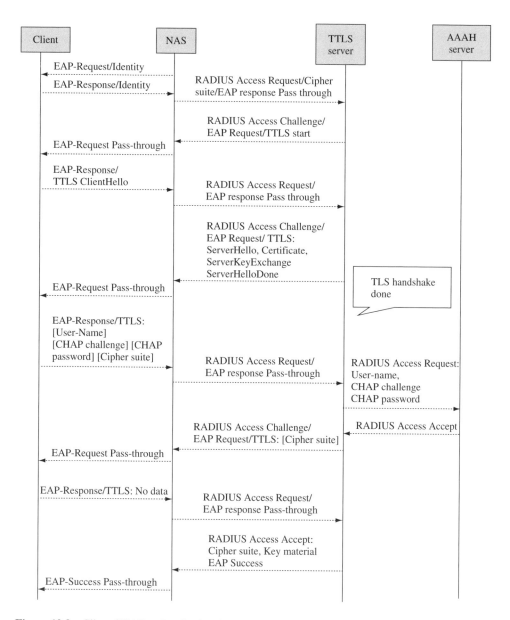

Figure 10.9 Client CHAP authentication through EAP-TTLS messaging
Note: Brackets indicate data that is encrypted using TLS record layer and is invisible to the NAS.

The TLS handshake phase is performed as described earlier. The end of this phase is marked in Figure 10.9. To spice things up, here we have shown an example of how EAP-TTLS can be used for cipher suite negotiation and key distribution for the secure communications between the client and the NAS (see the last assumption in the section describing EAP-TTLS functional elements). As shown, both client and NAS can indicate their preferences to the TTLS server and let that server act as the mediator. This explains what is shown in the picture as the "cipher suite" suggestion added by the NAS to the RADIUS access request following the EAP-Identity request/response pair. The client, on the other hand, waits for the establishment of the TLS record layer before sending its "cipher suite" preferences to the TLS server (as explained below). More details on the cipher suite negotiations for client–NAS communications are found in Figure 10.9 and the step-wise description of EAP-TTLS/CHAP below.

Once the TLS record layer is established, client authentication starts. Note that the challenge is not issued by the NAS. In fact the challenge and response exchange may need to be protected from eavesdropping and dictionary attacks from the untrusted NAS. Both the client and server generate 17 octets of challenge material that includes CHAP challenge (16 bytes or 128 bits) and CHAP identifier (one byte). Now the reader may ask what sort of challenge is this, if it is generated by the client itself? The answer is that the challenge is generated from a so-called "ttls challenge" based on the TLS master key established during the handshake phase. This means the client and the TTLS server can generate the same challenge material without any extra handshakes (making the "H" within "CHAP" sort of meaningless!). Also, note that the AAAH server is unaware of the CHAP material initially.

Hence the exchange following completion of TLS handshake is performed as follows:

- In response to the EAP-Request pass-through following the TLS handshake, the client generates the CHAP material: the CHAP challenge and the CHAP password (the response to the challenge) and includes each in an AVP form consistent with that used by the AAA protocols along with its user-name and sends all of this in an EAP-response (with TTLS type) to the TTLS server. The client may alternatively send its preferences on cipher suite to be used for protection of the communications between the client and the NAS. All the AVPs are encrypted using TLS record protocol.
- The NAS sees this as an EAP request and encapsulates the message inside a RADIUS access request (without being able to see the encrypted AVPs) sent to the TTLS server.
- The TTLS server decrypts the CHAP authentication data and sends them to the AAAH server using a RADIUS access request message and the security method provided for RADIUS messaging between the TTLS server and the AAAH server.
- Upon successful verification of CHAP authentication data, the AAAH server issues a RADIUS accept message and sends it to the TTLS server.
- Once the TTLS server gets the indication of a successful client authentication from the AAAH server, if the cipher suite negotiation and key distribution for client–NAS communications are supposed to be performed through other means, the TTLS server can at this point send an EAP-success to the NAS, which simply passes it through to the client. The description of tunneled CHAP using EAP-TTLS in [EAPTTLSDR] does not include these services and stops here. In the following, we describe how EAP-TTLS and

the EAP key management framework can be combined to provide secure client–NAS communications.
- It is convenient to use EAP-TTLS facilities and the EAP key management framework to help the client and NAS with cipher suite negotiation and key distribution. To do this, instead of generating an EAP-Success to indicate the end of negotiations, the TTLS generates another EAP-request (with TTLS type 21) and includes the result of cipher suite negotiation (if successful) for the client. The TTLS server uses the TLS tunnel to send this information in encrypted format through the NAS.
- The NAS acts as a pass-through and sends the request (including the encrypted cipher suite data) to the client, which in turn responds with an EAP-response containing no data. This indicates to the server that the client has accepted the results.
- Upon receiving the response from the client (through the NAS), the TTLS server sends an EAP-Success, and includes the cipher suite information along with the key material for the client–NAS link to the NAS along with the AAA protocol message that carries the EAP-Success message to the NAS. This communication is protected through the security mechanisms provided for the AAA protocol carrying messaging between NAS and TTLS server.
- The NAS understands the EAP-Success. It extracts its own data and forwards the EAP-Success to the client.

It should be noted that EAP-TTLS is still work in progress in IETF. Furthermore, it has gained wide popularity for many wireless scenarios. This means there may be many different and non-interoperable varieties of this messaging implemented out there. The messaging shown above is only an example to illustrate the concept.

10.3.3 EAP-SIM

GSM cellular systems deploy a smart card called a subscriber identity module (SIM) card to carry out user authentication. We explained SIM-based authentication in Chapter 2 briefly. However, to save the reader the round trip, we provide a short overview here again. Fundamentally, the GSM authentication method is a challenge response method, where the subscriber unit is presented in a 128-bit random challenge (called RAND). The subscriber unit, using the designated cryptographic module in the inserted SIM card, responds with a 32-bit response value called SRES. The strength of the SIM-based authentication method lies in both using a secret key and a secret algorithm, when calculating the SRES from RAND. The cryptographic algorithm is typically an operator-specific confidential algorithm, deployed in the SIM card on the client end and in the authentication center (AuC) on the network side. The secret key, Ki, is stored on the SIM card and in the AuC and is otherwise unknown to both the user and the rest of the operator network. Furthermore, the GSM client SIM card and the AuC independently calculate a session encryption key, Kc, for encryption of the mobile phone conversations over the wireless link. The set of (RAND, SRES, Kc) is normally referred to as GSM triplets in the industry.

One major difference between SIM and other authentication algorithms is that normally the algorithm is known and hence the rest of the burden for providing adequate protection is put on the strength of the key and the execution of the procedure.

EAP-SIM is a new authentication method that uses the EAP framework to use a SIM-based authentication algorithm between the client and a AAA server. However, EAP-SIM has been extended for use by WLAN clients. The advantages of EAP-SIM over conventional SIM are twofold: first, the EAP exchange can take place with a AAA server (Figure 10.10), as opposed to an AuC, which is specialized by cellular networks to only do authentication. Second, EAP-SIM provides a mutual authentication, allowing for network to client authentication. EAP-SIM is described in an Internet draft [EAPSIMDR].

One issue is that since, the GSM triplets (required for client authentication) can only be calculated by the AuC, the client authentication in EAP-SIM needs to involve the AuC, which means the model extends beyond the regular three-party authentication model. Another main difference between EAP-SIM and other authentication methods involving a AAA server is that the need for the pre-established secret between the client and the AAA is replaced by the need for coordination between client SIM card and AuC.

The client authentication is carried out as follows: once the AAA server determines that the user intends to authenticate using EAP-SIM method, the server must fetch the GSM triplets from AuC. The AAA extracts the RAND value from the triplet received and constructs an EAP-SIM challenge message that contains the RAND value as a challenge to be issued to the client. The client calculates the response (SRES) to the challenge and the session encryption key (Kc) using the challenge (RAND) from the AAA server. To avoid dictionary, the client does not return the SRES, since an attacker could observe both the RAND and SRES and start an offline crypto-analysis. Instead, the client and server rely on message authentication mechanisms to ensure the client's possession of the SIM card and the identity and key (Ki) associated with it. The client calculates a 160-bit message authentication code (MAC) over its response message EAP-SIM challenge response (or parts of it) and sends the value of the MAC to the server. This value is referred to as MAC_SRES. The AAA server, on the other side, calculates its own MAC_SRES over the message received from the client and compares it with MAC_SRES sent by the client. If they match, both client authentication and message authentication have been completed in one step.

For completeness, it should be mentioned, that even the EAP-SIM challenge message from the server towards the client is authenticated using a MAC function. The result of the MAC function called MAC_RAND is added to the message for integrity protection. The client will respond to the EAP-SIM challenge, only if the MAC_RAND calculation on the client side matches the value sent from the server.

The protocol from the client to the NAS depends on the link layer technology that is used, e.g. EAPOL over 802.11 WLAN links or PPP. The protocol from client to the NAS depends on the link layer technology that is used, for instance, in the case of 802.11 WLAN links, the EAPOL is used, while for dialup links, PPP is used.

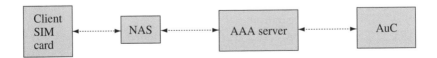

Figure 10.10 Architecture for EAP-SIM authentication

10.4 Use of EAP in 802 Networks

We mentioned that for wireless links EAP is carried over link layer protocols such as PPP or EAPOL. IEEE 802 LANs provide their own inbuilt framing support for traffic the way PPP does for dial-up and cellular data links. The link characteristics are known in advance and network protocols and features are carried inside LAN 802 headers, so the LCP or NCP type functions of PPP are not needed either. For those reasons, authentication functionality was one of the most important features missing from 802 protocols when it came to controlling access over 802 links. This was the motivation behind the creation of 802.1X: to protect the network from unauthorized access by users trying to access Ethernet connections that were physically accessible to the public. Hence authentication and access control at the network point of presence became the main goal of 802.1X.

It should be noted that the recent popularity of 802.1X-based methods for providing security within 802.11 WLANs does not mean that 802.1X is only designed for 802.11. On the contrary, 802.1X is designed for all 802 type links. In fact, 802.1X design started before 802.11 specifications started with wired connections in mind, even though the latest version of 802.1X [8021X2004] was published yesterday (time of writing: 9:20 pm December 15, 2004) to facilitate the use of 802.1X with 802.11.

10.4.1 802.1X Port-Based Authentication

The 802.1x model is a logical model consisting of a switched Ethernet hub with multiple logical switches, where there is one logical switch per user. The reason for insisting on the term "logical" is that, for wireless networks, such as 802.11 networks, there is no physical switch as opposed to wired 802 networks. 802.11 is like an access point that is dealing with multiple users trying to connect. As a user tries gain access to the network, it is as though it is trying to logically connect to the hub. A software port called the port access entity (PAE) is created for the user, where each port includes an authenticator and a switch (shown in Figure 10.11). The switch is open or disconnected to illustrate the controlled ports, i.e. ports that are used for variety of applications and services. This switch remains open (port closed) until the authentication is complete and will close (port open) when the user is authorized to use these services. The uncontrolled port is always open to allow for transit of authentication exchanges. For this reason, 802.1X is referred to as a port-based authentication.

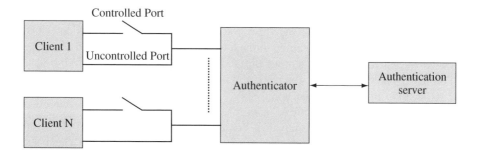

Figure 10.11 Authentication according to 802.1X model

10.4.1.1 EAPOL in 802.1X and Interaction with RADIUS

The logical point to implement 802.1X is the access point. However, the AP generally only includes the authenticator and not the authentication server (all the user information). So the AP only marginally understands the EAP messages carried by 802.1X and it is difficult for the authenticator to understand all the authentication methods negotiated during the EAP message exchange. Therefore, the authenticator simply forwards the EAP messages to the authentication server (RADIUS server) as described in generic EAP-XXX model. Note that 802 in general includes any type of layer-2 network, but IEEE does not get involved with higher layer protocols. As a result, even though some RADIUS interactions are defined, the amount of interaction with RADIUS server is nowhere near those for EAP-TLS and EAP-TTLS. As explained earlier, the EAP messages between supplicant and authentication server are encapsulated in EAPOL messages from supplicant to authenticator (using 802.1X), and in RADIUS when going to the authentication server.

10.4.1.2 Security Flaws of 802.1X, WPA/RSN and 802.1aa

The issue with 802.1X (since it is designed for wired networks) is that it is mostly concerned with access control, i.e. initial authentication which only provides a one-time protection, whereas in wireless 802.11 networks, the user's channel can easily be hijacked and therefore we need ongoing security protection, such as encryption as well as message integrity protection (and time synchronization) and most of this needs to be set up at the time of initial authentication. Several task groups and specifications such as IEEE 802.1aa, IEEE 802.11i, IEEE 802.1AE, and IEEE 802.1af have attempted solutions to these flaws. We refer the reader to the IEEE standardization process and the large amount of literature on the subject and do not get into its details due to the lack of stability in the specifications.

10.4.2 Lightweight Extensible Authentication Protocol (LEAP)

To facilitate wireless network access for mobile users, Cisco developed a lightweight EAP authentication variant that allows a user to use its Cisco wireless LAN (WLAN) network interface card (NIC) to perform a simple Windows-type user name and password login to gain access to the network as if it was connected through wired Ethernet. The result is what is called Lightweight EAP, LEAP, that can support both Windows Active Directory databases and RADIUS infrastructures.

LEAP takes some ideas from the Microsoft version of CHAP, commonly known as MS-CHAP in that it uses a challenge/response mechanism for authentication. However, LEAP provides a mutual authentication, through which both the authentication server and the client challenge each other. From a protocol standpoint, on the client side, LEAP is designed based on IEEE 802.1X standards and integrates well with the 802.11 MAC procedures such as the association process. On the wired side, LEAP attempted to be consistent with the earlier specification of transport of EAP over RADIUS [RADEXT2869].

LEAP implements the Wired Equivalent Privacy (WEP) security mechanism, which is a symmetric key-based encryption mechanism to protect traffic between the client and the WLAN AP. We will not go through WEP in this text due to the massive criticism that WEP has received. Although LEAP uses WEP for wireless link encryption, it provides its own key management services that are meant to address WEP's known weaknesses against attacks. For instance, LEAP provides the ability to generate temporal short-term session keys.

In the following, we provide a summary of the LEAP authentication process. In should be noted that LEAP is a proprietary protocol, whose details are not published. Hence any publication that is not from Cisco contains the result of guesswork and reverse engineering:

- WLAN client (NIC) associates with the WLAN access point (AP)., which at this point blocks all other requests from the user.
- The RADIUS server and the client perform a two-way challenge/response authentication: the RADIUS server authenticates the user first and then the client authenticates the server. In the process, the user is requested to perform a log on and hence types its user name and password as if she was connected to a wired Windows network. The client (network interface card) performs the necessary conversions on the password (see below) and engages in an 802.1X exchange with the RADIUS server.
- The RADIUS server derives pair-wise (intended between client and WLAN AP) WEP session keys locally.
- The RADIUS server delivers the WEP session keys to the WLAN AP using RADIUS signaling. RADIUS shared secrets between the server and the WLAN AP are used for secure transfer of these keys. The RADIUS server sends the keys along with an EAP success notification to the WLAN AP.
- The WLAN AP sends an EAPOL-success message to the end client, indicating the success of the authentication process.
- The client derives the same pair-wise WEP session key as the RADIUS server did previously. This way, the session keys are never exposed to the wireless link.
- The WLAN AP sends an EAPOL-key message (no keys are included) to indicate to the client that the keys can be activated.
- The client and the WLAN AP activate the WEP session and use the established WEP keys to protect the ongoing traffic.

LEAP avoids sending the passwords in the clear by deploying an MD4 (message digest 4) hash function to convert the password to secret values that can be sent over the wireless link. The use of one-way hash functions reduces the risk of password exposure. Furthermore, performing the MD4 hash twice on the password creates a so-called Windows NT key that is the proper format for use with existing Windows-based authentication servers (Figure 10.12).

LEAP is claimed to provide a per user per session key as opposed to the earlier WEP solutions, but discloses the user identity during the authentication handshake. Also since LEAP is a password-based solution, it has been criticized for its weakness against dictionary attacks.

Unfortunately, since LEAP is a three-party authentication model, it does not explicitly authenticate the access point to the client and hence is vulnerable to man-in-the-middle

Figure 10.12 Use of user-passwords to arrive at NT key in LEAP

attacks, in which a rouge access point may be able to impersonate a real access point in its dealings with the mobile client.

10.4.3 PEAP

Protected EAP (PEAP) is a more recent protocol that has been developed for the same premises as LEAP, namely for WLAN environments where the client does not have access to certificates. However, unlike LEAP, PEAP provides a protected EAP negotiation (as the name suggest) to hide the user identity and her authentication exchange. PEAP is very similar to EAP-TTLS in the sense that it requires server certificate-based authentication to establish a secure TLS channel during its first phase, to let the client EAP authentication happen during its second phase. However, the main difference between EAP-TTLS and PEAP is that while the way EAP-TTLS carries any authentication material through AVPs over TLS records allows it to use any legacy authentication protocol (PAP, CHAP, MSCHAP) as well as EAP-based methods, PEAP only allows authentication methods that are defined for use with EAP. This may or may not pose a restriction on PEAP in the long run and but in the short term while the legacy authentication mechanisms are not supported by EAP, it is a drawback for PEAP. Also it seems that PEAP has not enjoyed as much popularity in use as EAP-TTLS has.

We do not go too much into the details of PEAP and refer the reader to the IETF draft that was written by Cisco and Microsoft (the two main contributors behind PEAP) [PEAPDR] and is still work in progress.

As in EAP-TTLS, PEAP avoids sending the user identity unprotected. During phase 1 of the exchange the EAP-Identity packets can simply carry information on the user's company or backend server (e.g. anonymous@mycompany.com). This is helpful in cases where an AP in a public lounge serves the users from many different companies. The real identity of the user is exchanged during the protected exchanges of phase 2.

Cisco PEAP uses one-time password (OTP) or logon passwords for client authentication. OTP databases from RSA security or other vendors can be used. Microsoft PEAP uses MS-CHAP 2 for client authentication and hence is limited to Windows NT domains.

10.5 Further Resources

A wealth of information on authentication methods including EAP-based authentication methods can be found at the IETF PPP extension working group website:
http://ietf.org/html.charters/pppext-charter.html

A newly formed IETF working group is working specifically on EAP standardization issues such as EAP key management framework and use of EAP for network discovery. This group has also specified the new design for EAP. The website is found at
http://ietf.org/html.charters/eap-charter.html

Furthermore, a new specification on the requirement for usage of EAP for wireless LANS is now commonly used a criterion for gauging correctness of any deployment-specific key management architecture that is based on the EAP key management framework [EAP80211REQ].

More information about newest authentication methods can be found at IEEE 802 standards web page at
http://ietf.org/html.charters/eap-charter.html.

More information on Wi-Fi alliance can be found at
http://www.wi-fi.org/OpenSection/index.asp.

Finally, Edney and Arbaugh have written an excellent book [EDARB80211] on the latest 802.11 security procedures such as WPA and RSN as well as older security procedures. In the process they explain many other concepts such as TLS, RADIUS and so on.

10.6 References

1. [EAP2284], L. Blunk, J. Vollbrecht, "PPP Extensible Authentication Protocol (EAP)", IETF standards track, RFC 2284, March 1998.
2. [EAP3748], B. Aboba et al., "Extensible Authentication Protocol (EAP)", IETF standards track, RFC 3748, June 2004.
3. [IEEE80211i], Institute of Electrical and Electronics Engineers, "Supplement to Standard for Telecommunications and Information Exchange Between Systems – LAN/MAN Specific Requirements – Part 11: Wireless LAN Medium Access Control (MAC) and Physical Layer (PHY) Specifications: Specification for Enhanced Security", IEEE 802.11i, July 2004.
4. [EDARB80211], J. Edney, W. Arbaugh, *Real 802.11 Security, Wi-Fi Protected Access and 802.11i*, Addison-Wesley, March 2004.
5. [EAP80211REQ], D. Stanley et al., "EAP Method Requirements for Wireless LANs", IETF, Internet draft accepted for Informational RFC, draft-walker-ieee802-req-04.txt, August 2004.
6. [DIAEAPDR], P. Eronen et al., "Diameter Extensible Authentication Protocol (EAP) Application", IETF work in progress, Internet draft, draft-ietf-aaa-eap-07.txt, June 2004.
7. [EAPKEYID], B. Aboba et al., "Extensible Authentication Protocol (EAP) Key Management Framework", IETF work in progress, Internet draft, draft-ietf-eap-keying-03.txt, July 2004.
8. [EAPIANA], type numbers for EAP specified by IANA, http://www.iana.org/assignments/eap-numbers.
9. [RADEXT2869], C. Rigney et al., "RADIUS Extensions", IETF, RFC 2869, June 2000.
10. [RADEAP3579], B. Aboba, P. Calhoun, "RADIUS (Remote Authentication Dial In User Service) Support for Extensible Authentication Protocol (EAP)", RFC 3579, September 2003.
11. [EAPTLS2716], B. Aboba, D. Simon, "PPP EAP TLS Authentication Protocol", IETF, RFC 2716, October 1999.
12. [EAPTTLSDR], P. Funk, S. Blake-Wilson, "EAP Tunneled TLS Authentication Protocol (EAP-TTLS)", IETF Internet draft, draft-ietf-pppext-ttls-05.txt, April 2004.
13. [PEAPDR], A. Palekar et al., "Protected EAP Protocol (Version 2)", IETF Internet draft, draft-josefsson-pppext-eap-tls-eap-07.txt, October 2004.
14. [8021X2004], "Standard for Local and Metropolitan Area Networks – Port-Based Network Access Control", IEEE, Dec. 14, 2004 (supersedes 802.1X-2001).
15. [EAPSIMDR], H. Havarinen, et al., "Extensible Authentication Protocol Method for GSM Subscriber Identity Modules (EAP-SIM)", IETF individual draft, draft-haverinen-pppext-eap-sim-15.txt, November 2004.

11

AAA and Identity Management for Mobile Access: The World of Operator Co-Existence

11.1 Operator Co-existence and Agreements

Once, there was a time when carrying a mobile phone of half the size of a car battery was considered a revolutionary way of communicating. As cellular phones shrank in size and information started flowing in a variety of forms, such as short messaging, files, pictures, staying connected through wireless channels has started posing new demands. People are starting to ask, why not connect their computers to the Internet or their work network wirelessly when they are out of their home or office environment? Partly the limited bandwidth capacity, partly the costly set-up of a cellular network (both in the form of spectrum and equipment), started a new trend: new access technologies that are cheaper and offer more bandwidth than cellular are being developed. Unfortunately many of these networks, due to their engineering characteristics, may have limited scope and feasibility. For instance, 802.11 networks still lack the support for high-speed spatial mobility that cellular networks provide, while the relevantly high bandwidth services that these networks provide make them excellent candidates for local area networks with limited coverage and support for mobility. A hotel could install such a local network to offer high-bandwidth connectivity for its guests in the lobby, while the guests still expect to use their cellular phones for conversations or low-bandwidth data transfers while driving outdoors. When those same guests arrive home, they may log in to the Internet using their home cable operator.

11.1.1 Implications for the User

As we can see, the limited scope of each access technology forces the user to gain connectivity through a variety of network technologies. From a user's perspective, this can mean several things:

- The user may simply want to use a communication service provided by the access network. For instance, a user may want to use her cellular phone to place a call, accessing a cellular network, or use her 802.11 card to connect her laptop to her company network. On the other hand, the goal of the user may be to connect to a service provider on the Internet. In this case, the user is using the access network to simply gain connectivity to the service provider's network.
- The user may have to use different communications devices to gain connectivity through each type of access network. Each of these access networks typically has specific requirements on identification and security. This means the user needs to obtain a set of network-specific identity and credentials to be able to connect to that particular access network. For instance, to connect to a third generation (3G) cellular network, besides the cellular phone, the user needs to be provisioned with a phone number, an IMSI (International Mobile Subscriber Identity) and a SIM card, while to connect to a WLAN using LEAP authentication, the user needs a WLAN interface card along with a user name and password that the WLAN administrator recognizes. First of all, this means the user needs to obtain a set of identity and credentials for each of the access networks she is connecting to. Second, the user must trust the device that she is using to connect to that network not to reveal her identity and credentials to the outside world. Third, a user dealing with many access networks needs to keep track of all the identities and credentials she needs for each access network and use them to start each communication session.
- When the point of using the access network is to gain access to a service offered by other networks (such as a company enterprise network or the Internet), the user needs yet another set of network level identity and credentials for each of those secondary networks. Take the example of a user that is using a hotel hotspot WLAN to remotely connect to her company enterprise network. The user needs to connect to the company VPN gateway through the WLAN access point provided by the hotel. In this case, besides having to follow the hotel network security procedures, the user needs to comply with the security and identification mechanism set by the VPN policies. For instance, the user may have to remember a user name and carry a smart card containing her security credentials. On the other hand, when the point of using the access network is to gain access to some sort of online service on the Internet, the user then needs to comply with the security and identification mechanisms required by the online service provider. For instance, nowadays the user may like to use online banking services, book trips, or deal with her medical insurance provider through the Internet. In such cases, the user needs to present elements of her real-life identity, such as her real name, social security number, or identification numbers, such as commerce ID or insurance number that are provided by that organization.
- From a networking perspective, these identities are application-layer specific and not network specific. The organization that is offering the online service is aware that the user is using some sort of access network to gain access to its site. Therefore, the organization

should protect the user's information from outsiders, including the access network provider, as far as possible. The implication of all this is that the organization needs to set up a secure connection directly with the user (through the access network), so that the communications that are carrying the user's real-life identity, such as her social security number, can be protected. To do this, many organizations require a secure login to their websites. Only after the user has typed an organization-specific user name and password, is her connection secure and then the user can safely enter her sensitive information into her browser. From the user's perspective, the problem is the user needs to remember yet another set of user name and password for each online service she is dealing with. A user dealing with many online services needs to keep track of the identity and credentials for each site.
- Finally, from the application level point of view, the user may need to enter the same real-life information (at least for an honest user), for every online service. For instance, the user may need to enter her shipping address or credit card number every time she purchases something online. If there was a method through which the user could present this information in a secure manner without typing it again every time, it would provide a great deal of convenience for the user and a great amount of satisfaction for the shopping community obtaining revenue from this shopping spree. One prime solution for this identity management problem is provided by Pay Pal. Each eBay customer can sign up with Pay Pal and enter her name, address, credit card number and possibly other preferences only once. From this point on, any monetary transaction between the shopper and sellers can be handled over Pay Pal, without the user having to enter her credit card information again.

As we can see, from the user perspective, dealing with multiple sets of identity and credential information for access networks, service networks, application and online service providers can be a big hassle. Creating identity management solutions will provide a great deal of convenience for the user.

11.1.2 Implications for the Operators

Now that we have considered the user perspective, let us look at the issues from the network and service provider perspective. In a materialistic and non-dotcom world, where businesses need to operate based on revenue, any trust relationship is typically related to a business relationship. When operators are involved, it is hard to imagine a security and trust relationship between the administrative domains without a billing agreement in place. Hence identity and credential management should go hand in hand with billing and accounting management. To provide the added level of convenience for the user, many of the business agreements and accounting procedures between the networks should be achieved in a manner that is transparent to the user. We will talk about this in more detail shortly.

Witnessing the proven success of some of the access network technologies, such as WLAN and intimidated by of the seriousness of claims made by other upcoming technologies, such as IEEE 802.16, many traditional network providers (e.g. cellular, cable network operators) are now realizing that there will be a point where their legacy technologies need to co-exist with these newer technologies. Now the problem is that, if these newer access technologies are offered by newcomers or other providers, it translates to lost pieces of the revenue pie. Therefore, for major access service providers there will be two alternatives:

- The first alternative is to expand their existing networks with complementary networks offering these new access technologies. This involves buying, setting up and maintaining extra infrastructure equipment and networks and is rather costly. One example is the existing cellular operators that are supplementing their relatively low-bandwidth but wide area networks with small local area networks that are capable of delivering high bandwidth at hot spots. The easier version of the solution is also to buy up the new operators offering these new technologies, assuming the operators are small. However, the capital of each operator can reach only so far and eventually with the growing number of new technologies and newcomers, the traditional operators are forced to resort to the second alternative.
- The second alternative is to hold on to the cash reserve and instead strive to achieve a business agreement with an operator that is providing a complementary set of technologies. This solution is especially attractive when operators are of comparable size or operate in different geographic areas.

We believe that co-existence according to the second alternative is highly likely to be the prevailing model. This way neither operator nor access technology risks customer attrition and the user can enjoy multiple access technologies depending on her needs. This is how the next generation of networks (NGN) provides the notion of converged networks, where many different technologies are converged to offer well-rounded solutions to a roaming user. When dealings between a set of co-existing operators as business affiliates exist, the user can also enjoy the convenience of receiving a single bill, while the network operators enjoys a protected revenue stream.

Before going into the technical challenges involved with the operator co-existence, we would like to first cover another case for operator co-existence as envisioned by the Alliance for Telecommunications Industry Solutions (ATIS). This case is not due to the availability of different access technologies, but due to the way the user receives a variety of services. ATIS predicts a set of scenarios, where a service provider (such as a content provider or an application provider) is different from the access network provider [ATISNGN]. A common case may be a news web site or a gaming web site that provides services to the users only through the Internet and the user needs to connect through an Internet provider or a cellular operator to access the web site for the service provider. ATIS does recognize our previous case as well, i.e. the case where each service provider is offering only a subset of all access technologies and the user may need to use an access technology that is only offered by an access network provider different from its own service provider. Regardless of the case for operator co-existence, the operators need to have the proper business and trust relationships in place. From a technology standpoint, establishing and deploying these business and trust relationships translate into terms that are familiar by now: solving authentication, authorization and accounting problems. Operator co-existence translates into multi-domain AAA problems. Since there are many ways to set up the relationships in a multi-domain scenario, co-existence will also take different forms and the AAA solutions need to be tailored specifically to each case:

- *Bilateral relationships*: In this case, the two operators have a direct peer-to-peer agreement.
- *Brokered relationships*: In this case, two operators deal with each other through a brokerage service.
- *Alliances*: In this case, each operator has a multi-way relationship with several members within an alliance as needed.

In the following we describe each of these scenarios in more detail from a AAA perspective.

11.1.3 Bilateral Billing and Trust Agreements and AAA Issues

As mentioned earlier, this model can be deployed for two scenarios:

1. Two co-existing operators providing complementary connectivity service to the user across their access networks.
2. A service provider using an access network provider to reach the user.

In the first case, the agreement is a type of roaming agreement, meaning that the user can use both networks but have a business relationship with only one network (the home operator). The user typically also has an initial and permanent security relationship with only the home operator, but establishes a temporary trust relationship with the new operator. This new trust relationship is transitive, i.e. it is derived from the long-term relationship between the two providers and the relationship between the user and her home operator, when needed. By trust relationship, we typically mean that mutual authentication has taken place (the identities of both sides are confirmed) and a proper set of security mechanisms and related keys have been established to protect the communication channels. The user does not have any business relationship with the visited network. This is shown in Figure 11.1.

In the second case, where a service provider needs to co-exist with an access provider to reach the user through the access network, the user typically has a separate relationship with both the access network provider and the service provider. We will not consider that case here since the billing model is less sophisticated in that case (the poor user simply

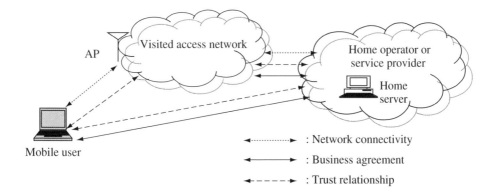

Figure 11.1 Business and trust relationships for a roaming user

pays two separate bills every month), while the trust model is similar to what we described for co-existing access providers. For that reason, let us focus on the AAA issues that arise for the first case, assuming a bilateral agreement exists between two access network operators:

- As we said earlier, the idea is to allow the user to only deal with one single monthly bill from her "home" operator, regardless of the operator she is using. This means the user has a business relationship only with her "home" provider. As a result of the business relationship, a trust relationship is established between the user and the home operator, meaning that the user receives a set of credentials only with the home operator.
- To support a single bill feature for the user, there needs to be a behind-the-scenes business agreement between the operators, so that the visited operator providing service to the roaming user can collect and present verifiable usage records to the home operator.

For this scenario to happen, several things need to be done:

1. The serving (visited) and the home provider must agree on a charging scheme, so that they each know how much a specific service provided with a specific quality at a specific time costs. The visited network collects the usage information based on this scheme and sends it to the home network in the form of accounting packets. Based on a billing agreement and the aggregate usage information, the home operator collects revenue from the user and pays the visited operator. To provide scalability, the visited network may opt to send bulk accounting information on a large number of users visiting from the home operator during a specific period of time. AAA protocols in conjunction with roaming protocols need to achieve this functionality. But as well as accounting there are other things that need to happen.
2. In order to warrant a payment from the home operator, the visited operator must obey the authorization policies set by the home operator. Let us see what this means: honoring the business agreement between the user and the home operator, the home operator sets up an authorization profile for the user. This profile shows what level of service the user is allowed to receive. Whenever the user requests a service, the user authorization profile stored at the home network must be consulted before an authorization directive can be issued. When the user is making the request from a visiting operator's domain, the visiting operator must consult the home operator to receive this authorization directive. The visited operator needs to make sure that the services it provides to the user are not beyond the user–home operator agreement. If such violations occur without consulting the home operator first: good for the user and bad for the visited operator.
3. Authorizing the user for service in the visited operator domain may be achieved in the following way: the visited operator can forward the initial service request to the home operator. This request may even include a simultaneous authentication request, but we will deal with that later. The visited operator only acts as a proxy and sends the request in the form of a AAA request message to the home operator. However, to do this, the visited operator needs to know who the user's home operator is. This is part of the identity management problem that we will describe in the next subsection. After receiving the request, the home operator makes the authorization (and typically authentication) decision and forwards the results back along with specific service provisioning information to the visited operator.

4. Obtaining authorization from the home operator and granting the user service involves policy and service level agreements along with the billing agreements between the operators. Based on these agreements, the visited operator may have to perform some translations of quality of service or other parameters before forwarding the provisioning information to its edge device (NAS) providing the service to the user.
5. Finally, we come to most obvious thing that needs to happen prior to authorizing and granting service to the user and that is authenticating the user. We will explain the details of the authentication and identification later on. For now, we will say that the user does not have a trust relationship with the visited network. This means the user needs to authenticate with her home network along the path provided by the visited network and in such a way that her credentials and maybe even her identity are not revealed to the visited network. At the same time the visited operator must be configured to allow this traffic over its network (possibly based on the agreement with the home network).
6. Once the home network authenticates the user, the home network verifies the legitimacy of the user for the visited network. Now at this point the user still does not trust the visited network. The authentication process and the home operator must facilitate establishment of this trust. The home network can based on the trust relationship that it has with the visited network and recent verification of user's credential help the user and visited network to establish a trust relationship. This is explained in more detail below.

11.1.3.1 Identity Management and Security Issues

To forward the user's authorization and authentication requests to her home operator AAA, the visited network needs to know who the home operator for the user is. This means the identity the client presents to the visited operator must reveal the identity of her home operator. An example of such identity is the network access identifier (NAI) that is in form of Joe@myisp.com. The realm part of NAI (myisp.com) reveals the home operator domain and allows the visited network to forward the request over the proper routing channels to the home operator. Typically the home operator is able to map the client NAI to a true client identity and thereby make the AAA decision on the request. However, since the request may travel over multiple AAA forwarding proxies, the integrity of the messaging needs to be protected from tampering by man-in-the-middle attacks. Furthermore, this messaging may need to be secured against eavesdropping by all the intermediaries as well. When user privacy is an issue, not only the content of the request messaging needs to be protected from eavesdroppers, but anything that reveals the user's location or identity needs to be hidden as well. Traditional security mechanisms such as IPsec are either incapable of providing this protection (IP address is in the open) or create issues with forwarding. We mentioned that the operators can use the NAI instead of the client's true identity. When privacy is a concern, use of NAI is not completely secure, since the mapping between NAI and user identity can soon be discovered by attackers trying to find the user's true identity and location. Methods are being developed in IETF to allow the networks to use pseudo and/or temporary identities or anonymous identities (anonymous@myisp.com) for accounting signaling. In Chapter 10 we described EAP-TTLS as a way of providing encrypted channels to protect the messaging that includes user's true identity. We will come back to this point shortly. The point we are trying to make here is that all the above points to the need for proper identity management.

The second issue that is independent of the routing problem described above is that each operator may assign the user an identity that follows the administrative policies or the access technology specific to that operator network. When dealing with that provider, the user is required to identify herself in such a way that she is recognized by that provider. This means, along with the identity given by her home operator, the user is assigned yet another identity from each provider network she visits. With proper arrangements the user does not need to keep track of the identities and credentials for both operators and needs to be aware of only one single identification and authentication scheme: the one required by the home network. In order to comply with each operator policy and technology requirements, identity management procedures, including translation mechanisms that are transparent to the user are required. When pseudo identities are used, these identities are also added to the mix.

Identity management is only one part of the security problem. As mentioned earlier, since the user does not have a trust relationship with the visited network, the initial authentication must happen with the home network and the messaging that carries the user's identity and credentials may have to be protected from the visited network. Supporting legacy user authentication mechanisms such as PAP and CHAP would not be adequate for roaming environments, since these methods do not have the ability to hide the user identity and credentials from the visited network and other communication media in a proper manner. The lack of an initial trust agreement with the visited network makes the authentication to the home network more complicated. In Chapter 10 we described IETF methods such as EAP-TTLS that establish secure pipes (TLS record layer) between the user and the home network to carry the authentication exchange without exposing sensitive material to the visited provider.

Another identity-related issue arises when the initiating accounting mechanisms from the visited network use one form of pseudo identities, while the authentication signaling uses another form of pseudo identity. Typically at some point during the interaction with the home domain AAA server, the home server and possibly other servers will re-write the anonymous pseudo identity to an identity that can be tied to the client. This is important since the AAA server needs to forward authorization directives about the user's traffic and privilege to intermediate proxies (including those inside the access operator domain). However, if the accounting and authentication signaling does not travel over the same path, serious translation issues will arise.

Drawbacks for the Bilateral Model

Bilateral agreements are a lot like pair-wise security keys: they only support two-party relationships. As the number of involved parties grows, scalability issues arise very quickly. The home operator needs to reach billing agreements with many different roaming service providers. The home operator also needs to establish trust relationships and security associations with each of these operators. Furthermore, the home operator typically needs to perform identity management functions, such as translations and key management functions. This does not seem to be a sustainable model in the long run.

11.1.4 Brokered Billing and Trust Agreements

One of the ways to solve the scalability problems of the bilateral model is to introduce brokerage services. The brokered model is shown in Figure 11.2.

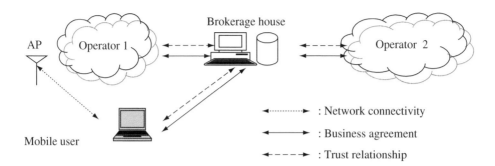

Figure 11.2 Brokered model to support operator co-existence

In an ideal, but probably theoretical, brokered model, the user only has a business agreement and trust relationship with the brokerage house. As the user roams from one access network operator to the next, or uses an access network to reach a service provider, each provider forwards the service and authentication request from the user to the brokerage house. The brokerage house acts in a way similar to the way the home operator acts in a bilateral model. So most of the problems described in the previous section still need to be solved for this case as well. The brokerage house performs the identity management, translation, authentication, authorization functions and hosts an accounting and billing server to produce the user bills and perform revenue sharing between the operators. If the brokerage house also hosted a PKI service with Certificate Authorities that all operators and users trusted, then many of the trust issues could be solved in an integrated and secure manner. However, all that seems rather idealistic for now.

Issues with the Brokered Model
- There are not many full brokerage houses of that kind today.
- Even if such a brokerage house did exist, it needs to be trusted by all the involved operators from a security standpoint. This is because the brokerage house acts as a single source of trust and needs to be trusted to keep all security-related material for the user and to understand the security policies of each operator to be able to perform proper credential management and translations for each operator. The brokerage house must be trusted not to impersonate any of the users to get service from any operator. Even when the brokerage is fully trusted, an attack on the brokerage house will jeopardize the security of all the operators and the user.
- The brokerage house also needs to be trusted from a business standpoint. It needs to be able to understand the billing, authorization and service level policies for each operator and not authorize the user for services beyond what she is allowed by her service contract. The brokerage house must also be able to understand the accounting logs produced by each operator and apply the proper billing scheme (according to that operator) to produce the final bill.
- The brokerage house can be a single point of failure, especially if it serves as an authorization clearing house. If the brokerage house servers fail, all the involved operators will fail to provide services to the users. When the brokerage house acts only as a billing settler,

a failure may lead to potentially large business and litigation disputes. The brokerage service and its communications with other partners must be designed with proper fail-over and survivability mechanisms.
- Not a lot of standardization has taken place to facilitate the functions described above.

At this point in time, billing and authorization are sensitive issues, since they have a direct impact on operator revenue. This means the user is still associated with a home operator, which defines the user's service and authorization profile and handles the user's bills. It is, however, more acceptable to outsource some of the security functions such as certificate issuance to external PKI vendors acting as brokers. We still believe that the brokered model has a good chance of appearing in future converged networks.

11.1.5 Billing and Trust Management through an Alliance

Due to the sensitivity of some of issues raised for the brokered model, a more viable model for co-existence may be one based on an alliance. The alliance is built based on a community of trusted parties. As long as each new member fulfills certain obligations and performs a set of initial functions, the new member can be trusted. The trust model within an alliance can be formed in several different ways:

- The new member may build a trust relationship with every member of the alliance through mutual authentications and key exchanges. This is very cumbersome in a large alliance and suffers from the same scalability issues as the bilateral model does.
- The new member may build a trust relationship with a few members initially or on as-needed basis (e.g. based on users' services requests). This is a more scalable model. However, the problem is that some trust relationships have to be built in real time. For scenarios such as those involving high-mobility users, signaling delay is a major concern and service performance can suffer.
- The new member may build a trust relationship with a few members and use the trust relationship between those members and their partners in alliance to broaden its reach. In this case, the trust between the new member and the secondary set of partners is only transitive, i.e. based on the trust between the member's immediate partners and their partners. Transitive trust has its own obvious problems, as we will discuss later on.

From the service point of view, the service models can be different too:

- One case would be that some members provide only network access services, while others provide application services and a third group deals with brokerage services.
- In the most general case, various members of the alliance may perform a variety of functions. For instance, one member may only provide an access network with a particular technology along with some identity translation functions, while another member provides a number of application-related services (such as gaming) and deals with the authorization issues.

The upside of building alliances with distributed responsibilities for services is that the problem of the single point of trust and the single point of failure is avoided.

In the interest of not repeating ourselves, we will go through a short description of Liberty Alliance as an example of the alliance model next and during that discussion provide more details on pros and cons of the alliance-based model.

We believe that in the long run, the billing and trust model based on a bilateral model will not prevail due to scalability and administration issues. We also believe that the brokerage model will only be viable if it is provided by an unbiased and well-trusted source. Furthermore, it will have business and trust implications, unless it is supported by fail-over mechanisms. The viability of the model based on the alliance will depend on how the trust and business relationships between the individual members are realized. Ultimately, an alliance deploying multiple specialized brokerage services may be the organic survivor of all the models.

11.2 A Practical Example: Liberty Alliance

The motivation for Liberty Alliance is simple: to help make interactions between a user and a variety of service providers easier. When dealing with different organizations and businesses, the user would not have to present her identities or state her preferences for the service again and again. After the user has provided her identity and credentials and configured her preferences, the providers' systems will handle identity and preference management issues between themselves and offers the user a single sign-on experience. When privacy is concerned, the user's information is protected either through the use of aliases or partial identities, or by simply only sharing the minimum necessary information. For instance, take the example of when a service provider only needs to know the user's postal zip code to be able to provide weather information. The user's full and true identity should not be revealed for this application.

The definition of identity provided by Liberty Alliance is broader than usual and includes attributes and preferences as well. The attributes and preferences may go into specifying a large number of parameters such as type of devices the users have or the processes that are used in conjunction with providing service to the user. Even though Liberty Alliance has been enhancing the user experience as the main goal of the specification work, it clearly states that enabling businesses to maintain and manage their customer relationships without third-party participation is also one of its goals [IDWSFOVER]. As we will see below, by creating trust relationships with each other, the service providers can share user information and that way allow a single sign-on experience for the user. One of the main differences compared to traditional single sign-on frameworks is that Liberty Alliance facilitates a decentralized approach to authentication and authorization for multiple providers.

Liberty Alliance defines a concept called federated network identity management architecture [LIBIDARCH]. The architecture consists of several modules:

- Liberty Identity-Federation Framework (ID-FF): This module specifies identity federation and management and can be used on its own or in conjunction with existing identity management systems to solve the problems related to network identity (authentication being one example). The functions that provide these solutions are called identity services in the Liberty Alliance literature.
- Liberty Identity Web Services Framework (ID-WSF): The ID-WSF module defines a framework for creating, discovering and consuming identity services. We will provide more details on identity services later on, when describing this module.

- Liberty Identity Services Interfaces Specifications (ID-SIS): These are a collection of services that use the functions of the ID-WSF. Avoiding the jargon a bit, we can say that ID-SIS are services that are more tangible to the user, such as contact book, calendar and location-based services, while ID-WSF are network functions enabling these services. Thinking in design terms, the ID-SIS services sit on top of ID-WSF services. The first ID-SIS available will be Personal Profile Identity Service [LIBIDARCH].

In the interest of keeping this discussion short, we will not go through the top level services experienced by the user due to their large variety and the extensible nature of ID-SIS module. Since the focus here is security and AAA type services offered by the network, we will focus more on the concepts and mechanisms defined by the ID-FF and ID-WSF frameworks that enable implementation of those user experienced services.

11.2.1 Building the Trust Network: Identity Federation

Liberty Alliance offers two powerful concepts, namely, Federated Network Identity and circle of trust to tackle the identity and trust management issues:

- Federated Network Identity links a variety of user identities together and call those identities "linked identities". Federation of identity (also called account linking) means a user that has multiple accounts at different sites (from service providers) can federate (link) these accounts for future authentications and sign-ins. The first requirement is that these service providers must be so-called Liberty-enable sites, i.e. they must be able to support the Liberty Alliance identity architecture and protocols. The other important requirement is that these providers must trust each other and the information they are sharing. In other words, they must be members of the same circle of trust.
- A circle of trust is formed by a group of service providers that trust each other and can share linked identities (user information). The trust is partly based on well-defined business agreements between the service providers and partly based on the exchange of identity and security information between the providers before they start sharing user information. Besides the providers, the circle of trust includes the user as well, since the member providers are required to notify the user and ask for her consent prior to sharing user-related information with each other (linking user identity information). Figure 11.3 shows an example, where a user is a member of two circles of trust to handle her work and personal web-related matters.

11.2.1.1 Identity Services

Before going into the roles of various entities in a circle of trust, it is useful to describe the notion of identity service a bit more technically. Identity Service is an abstract notion and can include any web service that does any of the following: retrieve information about an identity, update information about an identity, or perform some action on behalf of some identity.

In the following example we try to illustrate the Liberty Alliance notions of identity service, identity provider and single sign-on, all at once, since single sign-on can be seen as a main application for identity services.

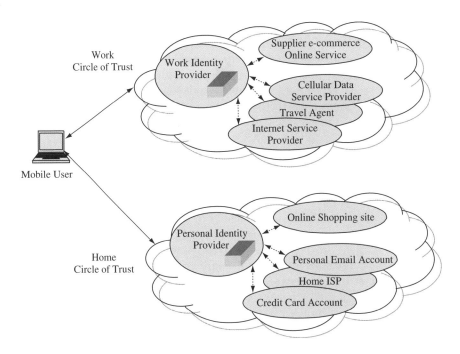

Figure 11.3 A user dealing with two circles of trust for work and personal web services

Alice wants to navigate through different sites on the web. So she first logs on to her ISP using her name and password. Alice may also opt to give her ISP her credit card information, so that every time she needs to buy something from XYZ.com website, she does not have to enter her credit card information again. Based on the agreement between the ISP and XYZ.com, the ISP prompts Alice, asking her whether the ISP can share Alice's credit card number with XYZ.com during the purchase. The ISP can be seen as an identity service provider for Alice, since it holds Alice's credit card information. Assume Alice buys a lot of baby-related items, such as clothing and toys from Cheapbabyoutfit.com. She may like to register her address at Cheapbabyoutfit.com and let this website act as a provider of her shipping address information, whenever any other commerce site needs her address information. In this case since the Cheapbabyoutfit.com acts as a provider of a piece of Alice's identity (her address), Cheapbabyoutfit.com is also considered as an identity provider for Alice, except that it provides a part of Alice's identity that is different from the one provided by her ISP. Cheapbabyoutfit.com must register Alice's address by performing a dialog with Alice once and then every time another service provider within a circle of trust that includes Cheapbabyoutfit.com and Alice needs Alice's address information, Cheapbabyoutfit.com only prompts Alice's web browser asking: "Do I have permission to link you to identity information with XYZ.com?" By giving consent to Cheapbabyoutfit.com, Alice allows this site to share her address information with XYZ.com. Providing Alice's credit card information or shipping address is considered a basic identity service, offered by Alice's ISP and Cheapbabyoutfit.com, respectively.

The Liberty Identity Web Services Framework (ID-WSF) specifies many more complicated identity-related services. One important example is the so-called Discovery Service. Discovery

Service defines how various Service Providers can dynamically discover a user's (principal) registered identity services and in the same sense who provides it. For instance, in the previous example, when Alice browses different web sites, part of her identity (credit card information) may be retrieved from ISP and other parts (shipping address) from Cheapbabyoutfit.com when needed, as well. This means Cheapbabyoutfit.com needs to register itself as the provider for Alice's shipping address. This way, other service providers after consulting with the Discovery Service (based on specified Discovery protocols) can find out where they can retrieve a registered identity service for Alice. Typically an access control policy is defined to share the information related to each identity service [IDWSFOVER] and that helps a provider to find out which Identity Provider stores and releases what particular part of the user's identity and what a service provider needs to present to have access to that information.

As mentioned earlier, the definition of identity in Liberty Alliance also includes service attributes defined for or by the user when receiving a service. For this reason, identity services include attribute sharing services as well. Permission-based attribute sharing, i.e. obtaining the user's consent before sharing the attributes is very important. ID-WSF defines protocols that support dialogs between attribute providers and service providers that are trying to provide a customized service to a user based on the user's indicated attributes and preferences. These dialogs provide a facility for obtaining the user's permission to be obtained and therefore are part of the Liberty Alliance specifications.

11.2.1.2 Circle of Trust

We provided a brief description of a circle of trust earlier. As seen in Figure 11.3, a circle of trust includes all of the following: user, Identity Provider and Service Provider.

1. *User (called Principal)*: The web service architecture [IDWSFOVER] states that the user only needs to have a common Web browser. However, it is also stated that today's common web browsers do not provide support for Liberty Protocols and may need further enhancements.
2. *Service Providers (SP)*: These are the provides offering web-based services to the users.
3. *Identity Provider (IDP)* is either a specialized provider of identity services, or a service provider that performs some identity services for the user along with the services that it provides as a service provider.

11.2.1.3 Building the Circle of Trust

Obviously, building a circle of trust requires the establishment of security and trust relationships between individual players (user, service providers and identity providers). As mentioned earlier, the service providers and identity providers notify each other when a user's identity is being federated. As we will see later on, another feature provided is that, when linking a user's account, each organization can also communicate the type of authentication that should be used for the user sign-on.

Identity federation involves the exchange of user identity information between two providers. To guarantee the security of such exchanges [IDWSFOVER], a number of requirements are placed on these exchanges:

- The exchange of information between the service providers and identity providers must be secured with confidentiality and integrity protection. First of all, this means that each of the two providers must have authenticated its own identity to the other party. The framework supports real-time (during ongoing sessions) establishment of trust relationships between providers through real-time exchange of X.509 certificates and other provider data. This means protocols to carry certificates and do mutual authentication are needed. The alliance is pledging to use IETF work such as XML signatures, SSL/TLS and security work in OASIS (a PKI group) and World Wide Web Consortium.
- For exchanges between the user agent and a provider (service or identity provider), establishing security and trust means the provider must present its own verifiable identity (authenticates) before the user agent presents her credentials or other personally identifiable information to the provider.
- The ability to support the exchange of minimum set of authentication information is required. The minimum set could include information on authentication status, instance, method and eventual use of temporary or persistent pseudonyms.
- The framework also supports the use of pseudonyms or anonymous services for identity protection, when required. A simple example of a service that is provided over the web today is one that anonymously provides information on weather, movies or restaurants for a specific area. A user should not have to reveal her full address or phone number. Zip code information should suffice. The pseudonyms should be unique within each federation but in a way that does not enable outsiders to derive the user's true identity from the pseudonym.

Providers are also obligated to notify each other when they have terminated a user's account. Each service or identity provider gives each user a list of her federated identities that the provider has stored.

11.2.2 Support for Authentication/Sign on/Sign off

One of the strongest features of Liberty Alliance is the provision of a user single sign-on. Typically an identity provider (IDP) acts as a single sign-on provider for the user. This requires establishment of a trust relationship that starts with a mutual authentication between the user and the identity provider. The user trusts the identity provider to keep her identity information safe and to share it only based on her consent. The circle of trust members also trust the IDP to keep an accurate state regarding the identity and assertion of authentication for the user and to let them know when such an assertion is no longer valid (the user has signed out or is no longer trusted).

When an identity provider has the information regarding the user, when the user tries to log on to another service provider web site, the identity provider is consulted. The first time the service provider needs to use the services of the identity provider, the identity provider needs to perform the identity federation as follows. After receiving the request for identity service from the service provider, the identity provider asks the user if the user wishes to federate her identity stored at the identity provider with the local identity required at the service provider.

The issue that arises is that each of the service providers refers to the user with a different identifier. In fact, they may even use different aliases (or handles) for the identities for a particular user, instead of using those identifiers. When each provider uses a different identifier for the user account, how could different providers identify the user for which they need to exchange information? One way to handle this issue would be to introduce a globally unique identifier for the user. This, however, proves to be rather impractical. Instead the identity federation service provided by the identity provider provides the translation. An example of how this translation is done by an identity provider is shown in Figure 11.4.

This way the identity provider can also act as a broker, if the two service providers need to exchange user information. However, the issue with this brokered way of user identification is that the two service providers cannot federate the user identities directly and must go through an identity broker for user information. If John Smith wishes the service providers to exchange his information directly, he must explicitly federate his identities with the two service providers. An alternative to using the brokered approach is to establish a direct bilateral relationship between the two providers. Instead of going through a designated broker, the two providers directly federate the user's identity information. Even though this may seem convenient, it may have security and practical implications. Consider Figure 11.5, where a chain of service providers have bilateral agreements with each other.

SP1 is using A1 as an alias for John Smith when interacting with SP2, while SP2 is using A2 for John Smith when interacting with SP1. At the same time both SP2 and SP3 use B1 and B2 to refer to John Smith through their bilateral agreement. Without SP2 in the middle, SP1 and SP3, who do not have a direct federation agreement about John Smith, cannot

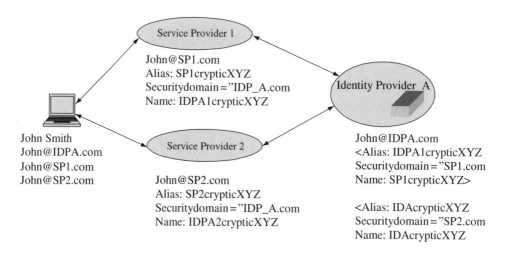

Figure 11.4 Translation between identities used at different service providers (close imitation of Figure 15 in [IDFFOVER])

Figure 11.5 Chain of service providers providing transitive relationships

Use of Multiple Levels of Authentications

exchange his information with each other: The trust relationship and identity federation would in that case be transitive. A link in the chain cannot be skipped over.

Single sign-on can happen at any Liberty enabled site within the circle of trust as long as the user has given consent so that her identity at that site with be linked with the identity federation provided through the identity services. It should however be noted that some events may require a more stringent identity verification than others. For instance the user in Figure 11.4, can only use her user name and password to get onto her work ISP network, but when she needs to order a shipment from a supplier from the supplier's e-commerce web site, she may need to authorize the release of her credit card number or other commerce information. Even though the supplier's e-commerce and the user's work ISP are part of a circle of trust, the supplier may have indicated to the ISP that when the user is entering orders, she must be prompted for a second level of authentication exchange. In this case a service provider has requested the user's identity provider to re-authenticate the user, possibly through a different authentication class.

Global Sign-out

A global sign-out procedure is also defined, so that when the user *logs out* of a Liberty-enabled site, she is automatically signed out of all the sites linked with that *session*.

11.2.2.1 Enabling Protocols

As we saw earlier, one important function of Liberty Alliance is to provide support for identity federation by facilitating the discovery and use of identity services within a trusted community of service providers and users. The Liberty Identity Web services framework (ID-WSF) defines the functions and protocol for discovery and invocation of identity services through the use of SOAP. SOAP stands for Simple Object Access Protocol [SOAPV1.2] that is standardized by World Wide Web Consortium (W3C). SOAP is a lightweight protocol for exchange of information in a peer-to-peer (decentralized) manner. The ID-WSF specification allows for the discovery of identity services, exchanging information on authorization for use of the identity service, and the identity service invocation itself.

Communications channels required between service providers and identity providers are created through the user agents deploying web-based redirect messaging [IDFFOVER]. The idea of redirecting is used in many places. In this case, a user agent sends a request to a service provider. The service provider then redirects the user agent to the identity provider by providing the location of the identity provider in its response to the user and at the same time it provides enough information on its own location, so that the identity provider can in turn redirect the user agent to the service provider when needed.

One method for providing web redirect services is the use of HTTP-redirects. Hypertext Transfer Protocol (HTTP) is one of the most famous protocols from W3C and is an application layer protocol used for exchange of complex text over the Internet. The Internet peers use HTTP request (GET) and HTTP response messages to exchange information. The information to be transferred is appended in the form of Uniform Resource Identifier (RFC 2396) elements to the HTTP request (GET) and HTTP response messages between the user agent and service or identity provider.

In Liberty Alliance, the service provider and identity provider find each other through HTTP redirect messages that they exchange with the user and include the location information for themselves and the other party within the URI fields added to their HTTP conversations with the user (as explained earlier). Note that the use of URI is not secure, since the data goes in clear text. SSL or TLS should be used to protect the clear text messages. Also, web redirect procedures use cookies on the user's browser as a way of keeping state. If cookies are not user specific, another user using the browser could use the stored cookie to obtain services she is not entitled to.

11.2.3 Advantages and Limitations of the Liberty Alliance

Many of the advantages of the Liberty Alliance framework should be obvious based on the description provided so far. Here is a compact list:

- Liberty Alliance frameworks provide the user with a single sign-on service to the user web logins, not only from the authentication standpoint, but also from the personalization aspect: the user needs to deal with the exchange relating to her profiles, preferences and policies on usage and sharing of her identity information only once.
- Liberty Alliance is generous with the choice of authentication mechanism and technology, since different identity providers may need to choose different technologies or be bound by different set of regulations (depending on the sensitivity and financial impact of the data to be exchanged). Liberty Alliance also allows multi-tiered authentication, which means when a higher degree of assertion for the user's identity is required, the user can be asked to re-authenticate at that higher layer. An example is a user that has once authenticated with the ISP using her user name and password but then needing to authenticate with her bank again to perform financial transactions.
- Liberty Alliance allows network administrators to manage user information more efficiently. It allows businesses to conduct transactions with authenticated customers and partners in a more efficient and to-user transparent manner.

Liberty Alliance has a well-defined scope. However, from an overall network and service architecture perspective, this scope is rather limited. Despite its value in providing the powerful concepts and solutions for identity management, the Liberty Alliance solution is not a comprehensive solution to provide mobility and security and must be combined with solutions provided elsewhere. Here are some examples:

- The main goal of Liberty Alliance seems to be facilitating application layer authentication for web logins. Mobility is supported in a virtual sense: the user navigates between web sites for different service providers. Physical mobility in the sense of changing geographical location or method of connectivity (changing the type of access network or operator) is not discussed. From the network perspective, security procedures such as authentication and key management are required at all layers of the stack, from layer-2 authentication and key management procedures to call control level authentication. This would limit the scope of Liberty Alliance when it comes to providing single sign-on to networks that require credential translations for security procedures at a variety of stack layers. Many such procedures are standardized by organizations such as IEEE, 3GPP, IETF, and so on.

- User authentication is only part of the overall authentication problem. Network management protocols such as Mobile IP or SIP signaling require message authentication. Message authentication for network management signaling does not seem to be part of Liberty Alliance.
- Problems related to network and AAA operation, such as authorization and enforcement signaling from the decision point to the enforcement point (network edge device providing the service) are not covered. The same goes for the problems related to establishing remote network connection through VPN connections.
- The fact that the user's information is not stored centrally at one location and is scattered over an ensemble of providers can cause some practical discovery problems. Even though Liberty Alliance provides solutions for these discovery problems at the application layer (using web-redirects), this approach may not be appropriate or relevant for lower layer security problems.
- Again, the framework requires support of a range of authentication methods for mechanisms to identify and classify these authentication methods. Examples of the authentication methods, mentioned in the specification, are username/password or certificate-based SSL or Kerberos. However, the framework considers a number of authentication-related problems beyond the scope of its current version (1.2) of specifications:

    - Both the user and the identity provider need to ascertain what authentication method the other party is using and potentially negotiate towards a method that is commonly supported by both parties. Liberty Alliance currently considers the method detection and negotiation exchange beyond the scope of its single sign-on design. This would limit the interaction of users and arbitrary identity providers and thereby limit the scope of Liberty Alliance solutions when it comes to supporting handovers where the users are physically moving (rather than navigating with their browsers). The methods should possibly be augmented by IETF EAP-type negotiations.
    - In the same manner, the Liberty specifications do not provide guidance on how a service provider can detect the authentication methods and profiles supported by an identity provider for a user and assumes that such actions are performed out-of-band. Even when the methods used by the other party are known, the interactions of identity providers and service providers typically require an exchange of certain metadata (set of information) such as X.509 certificates for the providers and other service point information. It is assumed that such data is exchanged prior to identity information exchange. This may be an indication that Liberty Alliance protocols are not suitable for scenarios, such as handovers, where expediency is a factor.
    - The service providers, that are part of a circle of trust and use the services of an identity provider, typically receive the user credentials from that identity provider. However, as mentioned earlier, there may be cases (such as the bank example), where a service provider requires a re-authentication with a higher degree of assertion. It is not clear at that point whether the identity provider or the service provider itself performs this action. Liberty Alliance specification leaves the choice to the implementers.

11.3 IETF Procedures

Even though most of the protocols for managing network-related problems come from IETF, the strength of IETF seems to be in the design and standardization of protocols that deal with specific network-related issues. The IETF rarely deals with system-level issues, such as developing complete AAA and identity management architecture. So it would be surprising to see frameworks similar to Liberty Alliance deploying multiple protocols and procedures to handle multiple problems, to emerge from IETF. As we saw in many chapters of this book, one powerful framework that comes close is EAP. The EAP framework provides authentication and key management frameworks, while also hinting at edge device discovery and security association establishments (Chapter 3). As we saw in Chapter 10, EAP also allows the user to use pseudo identities in sensitive environments. We also showed how mechanisms such as EAP-TTLS, that allow the user to perform authentication with her home network over untrusted access networks are examples of authentication and identity management protocols that are suitable for multi-operator environment. It is, however, not clear how EAP can be used for authorization and accounting purposes. This means in its current state, EAP is not ready for use as a AAA and identity management framework. It seems, however, that a wave of proposals is coming into IETF that deal with identity management issues and the use of EAP for that purpose. It remains to see what the outcome will be.

On the other hand, groups that deal with the standardization of RADIUS extensions are working on identity management issues from the aspect of accounting. Take a scenario where the user identity is important for the roaming operation but the true identity cannot be revealed to the intermediary networks or proxies. On the other hand, the local operator providing services to a roaming user, and the intermediaries providing routing services need to be paid. These entities need to know who to contact and who to charge for the usages. Furthermore, the interim proxies may have policies on usage privileges such as the number of simultaneous sessions and may need the user's identity to be able to enforce these policies for the user's traffic. This means the accounting infrastructure must be able to refer to the user in a way that is unique and unambiguous. Within the RADIUS extension working group there have been proposals dealing with this problem. The proposals suggest that a specific billable identity (called the chargeable user identity, CUI) that can be used by the accounting infrastructure is used to refer to the user, when privacy is a concern and the user's real identity cannot be known. However, there are a number of problems:

- Only the home RADIUS server has the knowledge of the user's identity and is able to generate a unique chargeable identity for the user as well as a unique binding between the chargeable identity and real identity.
- Since the authentication and accounting infrastructures use different identities for the user and may even route their packets differently, mismatches between the way each infrastructure refers to the user and hence rewrites the identity field within various messages may occur and the complete solution needs to deal with this problem.
- The process should not undermine the anonymity of the user, which means the CUI assignment must be temporary. If the CUI alias is used for a long period, it will defeat its purpose, since it may still be used to identify the user.
- The CUI must be communicated absolutely in the clear, especially when methods such TTLS and PEAP that hide identities are used.
- When privacy is concerned, the user name may not include a billable identity (this was being debated). The user may use different class/attributes with the NAS.

In Chapter 9, we described the importance of PKI in providing and managing certificates and trust relationship. We also discussed the identity used on the certificate (subject name) and issues related to management of identity for device and user certificates. Identity-related issues in the use of certificates for IKE and IPsec establishment were also discussed in that chapter.

11.4 Further Resources

Research and engineering in the field of AAA and identity management for mobile access are very active and for that reason much of the work being done today is not considered stable enough for inclusion in this book. But as we mentioned earlier, the work of many standard bodies and for a need to be gathered together to provide the user with the ultimate seamless experience. We refer the reader to the work being done in IETF, Liberty Alliance [LIBERTYWEB], ITU and other standard bodies for the ongoing progress.

11.5 References

1. [ATISNGN], "Part 1: NGN Definition, Requirements and Architecture", Issue 1.0, ATIS Next Generation Network Framework, November 2004.
2. [IDFFOVER], T. Wason, "Liberty ID-FF Architecture Overview, V 1.2", Liberty Alliance Project.
3. [IDWSFOVER], J. Tourzan, Y. Koga, "Liberty ID-WSF Web Services Framework Overview, V 1.0", Liberty Alliance Project.
4. [LIBERTYWEB], Liberty Alliance Specifications web site URL, http://www.projectliberty.org/resources/specifications.php#box1.
5. [LIBIDARCH], "Introduction to the Liberty Alliance Identity Architecture", Rev. 1.0, Liberty Alliance Project, March 2003.
6. [SOAPV1.2], M. Gudgin et al., "SOAP Version 1.2, Part 1: Messaging Framework", http://www.w3.org/2000/xp/Group/2/06/LC/soap12-part1.xml, January 2005.

Index

3GPP2, *see* Third Generation Partnership Project 2
802.11I   236, 238, 260
802.1X   43, 55, 237, 238, 259, 260

AAA
  AAA application   6, 19
  AAA Architecture group   10, 19
  AAA key for Mobile IP signaling   167, 181
  AAA key in EAP key management framework   52, 56
  AAA server   6, 7, 20–1, 44, 52, 167, 181, 193
  AAASA   179
AA-answer (Diameter NAS)   165
AA-Mobile node Answer   190
AA-Mobile node Request   190, 194
AA-request   164, 165
Access control   21, 41
Access control list
  host-based access control list (PKI for IPsec)   231
  user-based access control list   231
Access network discovery   115
Access point
  light-weight access point   114
  new access point   115, 123
  old access point   115, 123
Access router
  candidate access router   115, 118, 119
  current access router   115
  target access router   115
Access technology   7, 56, 118, 266, 272
Accounting
  application   13
  data   15
  event driven models   15
  interim   14, 18
  management   13
  metrics   14
  proxy   14
  records   17
  reliability   17, 140, 162,163
  request   128
  server   14
  server fail-over   18
Administrative domain
  home administrative domain   10, 167
  visited administrative domain   10, 113
AES, *see* American Encryption Standard
Agent advertisement
  agent advertisement challenge extension   109, 184
Agent sequence   12
American Encryption Standard   50
Anonymous key exchange   39
Anti-replay protection   36, 65,108
Application
  application specific information   20, 22
  application specific module   19, 20
  diameter application   149, 151, 162
  identifier (diameter)   151, 158
  server   11, 65, 236
Application-specific module   19, 20
AR, *see* Access router
ARL (pki4ipsec)   230
ASM, *see* Application-specific module
Asymmetric key algorithms   50
ATIS   268
Attribute
  attribute hiding   132
  attribute value pair (diameter)   154, 155, 156, 165, 190, 253
  vendor specific   130, 144, 176, 197
Auditing   13
Auth-Grace-period (in Diameter NAS)   166
Authentication
  device authentication   2, 31
  message authentication   4, 33

Authentication (*Continued*)
   mutual authentication   5, 33, 40
   port-based authentication   259
   user authentication   2, 26, 27, 283
Authentication extension
   generalized Mobile IP authentication extension   184
   home-foreign authentication extension   108, 184, 200
   MN-AAA authentication extension   183, 184, 197
   mobile-foreign authentication extension   108, 184, 195
   mobile-home authentication extension   108, 180, 184, 195
Authentication header   73, 74
Authentication model
   three-party authentication model   6, 29, 42, 55, 237, 261
   two-party authentication model   6, 44, 57, 243, 248
Authentication server
   backend authentication server   43, 55, 127, 237
Authentication token   33, 36
Authenticator   6, 27, 32, 44, 55, 57, 74, 107, 131, 141, 240, 243, 259
Authority revocation list (pki4ipsec)   230
Authorization
   authorization application   8, 10
   response   162
Authorization-lifetime AVP   166
Authorization-only (in Diameter NAS)   166
AVP
   AVP code   156, 164, 253
   CHAP-algorithm AVP   165
   CHAP-Auth AVP   165, 172
   CHAP-challenge AVP   165
   CHAP-ident AVP   165
   CHAP-response AVP   172
   destination host AVP   156, 159
   destination realm AVP   157, 159
   MIP-feature-vector AVP   193
   MIP-MN-AAA-Auth   191, 193
   NAS authentication AVP   165
   NAS authorization AVP   165
   origin host AVP   154, 156, 172
   origin realm AVP   156
   password-retry AVP   165
   result-code AVP   155, 157
   user-password AVP   165

Backoff mechanism   141
Backward compatibility   129, 130, 144, 152, 170, 185
Berkley Internet Name Domain (DNS)   112, 170, 224
BET, *see* Bi-directional edge tunnel
Bi-directional edge tunnel   116
Bilateral agreements   143, 272, 280
Billing
   non usage sensitive billing   17
   server   13
   usage sensitive billing   17
BIND, *see* Berkley Internet Name Domain (DNS)
Binding update   114
Bootstrapping, *see* Mobile IP, bootstrapping

CA, *see* Certificate authority
Candidate access router   115, 118
Candidate access router discovery protocol   116, 118, 119
Capability discovery   118, 119
Capability exchange   159, 163
Capability negotiation   143, 151, 159, 169
Capability pre-filtering   120
CARD (Candidate access router discovery)   116, 118, 119
   CARD reply   119
   CARD request   119, 120
   preference sub-option   120
   requirement sub-option   120
Care of address
   collocated care of address   102, 189
   FA CoA   102, 189
CcoA, *see* Collocated care of address
CDMA2000   26, 196
CEP, *see* Certificate enrollment protocol
CERT   228
Certificate
   certificate payload   228
   certificate request   84, 206, 211, 217, 219, 220, 229, 247
   certificate revocation   206, 210, 211, 225
   certificate revocation list   210, 230
   certificate status checking   207
   certificate type   93, 227
   device certificate   226, 248
   subject name   205, 211, 223
   user certificate   224, 226
   X.509 certificate   22, 197, 205
Certificate authority
   trust anchor certificate authority   207, 230

Index

Certificate enrollment protocol 213, 221
Certificate management protocol 219
Certificate management using CMS 213, 221
Certificate request message format 214, 219
Certificate revocation list 210, 230
Certification 206, 212, 225, 226
CERTREQ 84, 228
Challenge 109, 184
   agent advertisement challenge extension 109, 184
   Mobile-Foreign challenge extension 109, 184, 193
Challenge handshake authentication protocol
CHAP, *see* Challenge handshake authentication protocol
Chargeable user identity 284
Cipher suite 56, 58, 93, 252, 256
Cipher suite independence 58
Circle of trust 276, 278
CMC, *see* Certificate management using CMS
CMP, *see* Certificate management protocol
CoA, *see* Care of address
Collocated care of address 102, 189
Command, *see* Diameter
Command code, *see* Diameter
Congestion control 18, 124, 157
Connection (Diameter) 95, 136, 150, 157, 162
Context data block 122, 125
Context transfer
   context transfer activate request 125
   context transfer data 125
   context transfer data reply 125
   proactive context transfer 124
   reactive context transfer 125
   time efficient context transfer 121, 123
Context transfer protocol 121
Contractual relationship 10
Converged networks 268, 274
Cookie 82, 282
COPS 147
Correspondent node 100, 106
Cost allocation 13
CRC, *see* Cyclic redundancy checks
Credit control
   credit control server 152
CRL, *see* Certificate revocation list
CRL server 207, 210
CRMF, *see* Certificate request message format
Crypto period
Cryptographic key

CTP, *see* Context transfer protocol
Cyclic redundancy checks 4, 26

Data encryption standards 68, 75, 85
Data integrity protection 4, 33, 74
Denial of service attack 107, 123
DES, *see* Data encryption standards
DH, *see* Diffie-Hellman
Diameter
   AA-answer 164, 165
   AA-Mobile node Answer 190
   AA-Mobile node Request 190
   AA-request 164
   accounting 149, 151, 162
   AMA, *see* AA-Mobile node Answer
   AMR, *see* AA-Mobile node Request
   application 149, 151, 156, 159, 162, 167, 188
   application identifier 151, 158
   AVP 153, 155, 157, 164, 190, 193, 253
   backward compatibility 130, 170
   base protocol 148, 150, 156, 164
   capability negotiation 143, 151, 159, 169
   client 149, 152
   command 171, 190
   command code 154, 189
   credit control 151, 155
   EAP 152, 167
   EAP answer 167, 168, 245, 256
   EAP application 163, 167
   EAP request 42, 167, 237, 244
   HMA, *see* Home agent MIP-Answer
   HMR, *see* Home agent MIP-Request
   Home agent MIP-Answer 190, 195
   Home agent MIP-Request 190, 194
   mandatory bit 156
   MIP-Feature-Vector AVP 193
   Mobile IP application 152, 165, 167, 188, 190
   Mobile IP AVPs 188
   NASREQ application 152, 167
   node 151, 152
   peer table 158
   re-authentication 164, 166
   re-authorization 164
   realm based routing table 158
   server 149, 151, 153, 162, 165, 189
   session 160
Diameter Agent
   proxy agent 153
   redirect agent 153
   relay agent 142, 153
   translation agent 153, 166, 171

Dictionary attack   29, 37, 235
Diffie-Hellman
   content encryption keys   59
   key encryption keys   48, 60
Digest   4, 27, 34, 37, 39, 41, 68, 216
Discovery service   277
DNS   170, 224
DOI, *see* Domain of Interpretation
Domain   14, 208, 226
Domain name (dns appendix)   172, 223, 224
Domain name servers   224
Domain of Interpretation   80
DoS, *see* Denial of service attack
Downgrade attacks   41

EAP
   EAP authentication method   44, 58, 241
   EAP authentication phase   26, 44, 56
   EAP discovery phase   56
   EAP key management   43, 54, 55, 57, 238
   EAP key transport phase   56
   EAP methods   240, 241
   EAP request   42, 138, 237
   EAP request identity   43, 237, 246
   EAP response   42, 138, 238, 246, 255
   EAP server   44, 55, 57, 168, 242, 246
   EAP-Message attribute   139
   EAP-SIM   236, 257
   EAP-TLS   43, 244, 248
   EAP-TTLS   40, 248, 250, 252, 254, 255, 262, 284
   EAP-XXX   242
EAPOL   237, 260
E-commerce   5, 40, 92, 249, 281
Encapsulating security payload   73, 74
End to end identifier   154
End-to-end security   156, 160, 168
Ephemeral key   48
Extensible authentication protocol, *see* EAP

FA, *see* Foreign agent
Failover   168
Fast Mobile IP   115
Feature profile type   122
Federal Information Processing Standards   33
Federated Network Identity   275, 276
FIPS, *see* Federal Information Processing Standards
Flooding attack   60
FMIP, *see* Fast Mobile IP
Foreign agent   100, 102, 106, 152, 177, 189

Forward tunneling   76, 101
FPT, *see* Feature profile type
FQDN, *see* FQDN
Fragmentation   239
FreeRADIUS   145
FreeS/WAN   97
Fully qualified domain name   172, 224

Generic authentication framework   41, 55
Group key   48
GSM
   GSM triplets   257

HA, *see* Home agent
HAAA, *see* Home AAA server
Handover
   fast handover   114, 117
   heterogeneous handover   114
   homogeneous handover   114
   layer 2 handover   115
   layer 3 handover   113, 116, 120
   low-latency handover   115, 120
   mobile-controlled handover   114
   network-controlled handover   114
   seamless handover   117, 248
   smooth handover   117
Hash
   hash algorithm   4, 29, 34, 37, 50, 107
   hash function   4, 34, 50, 74, 87
   keyed hash function   34, 87
   secure hash algorithm   36, 75
Header compression   26, 121, 123
HMAC   34, 75, 107
HoA, *see* Home address
Home AAA server   10, 12, 189, 250
Home address   99, 111
Home agent
   dynamic home agent assignment   111
   local home agent   111
   redirected HA   113
   requested HA   112
Home agent MIP-Answer   190, 195
Home agent MIP-Request   190, 194
Hop-by-hop identifier   154
HTML form   40
HTTP   38, 221, 281
HTTP basic authentication   38
HTTP digest authentication   39

IAB, *see* Internet architecture board
IANA   43, 122, 156

Index   291

Identity federation   276
Identity services   275, 276
IESG, *see* Internet Engineering Steering Group
IETF
   AAA working group   10, 19, 147, 150,
      162, 172
   IPsec working group   78, 228
   Mobile IP working group   118, 176
   PKIX working group   209, 227
IKE
   aggressive mode   83, 87, 89
   IKE authentication   80, 86, 88, 197, 230
   IKE phases   80, 81, 82
   IKE pre-shared keys   52, 53, 62, 66, 67,
      88, 89, 198, 227
   IKE SA   80, 82, 86
   main mode   81, 88
   new group mode   83
   phase 1   81, 83, 86, 88, 89, 195, 231
   phase 2   57, 80, 82, 87, 161
   quick mode   82
Information-theory code   4
Initialization vector   48
Internet architecture board   25
Internet Assigned Number Authority, *see* IANA
Internet Engineering Steering Group   130
Internet Engineering Task Force, *see* IETF
Internet Key Exchange, *see* IKE
Internet Protocol Security   73
Internet Research Task force   10
Internet security association and key management
    protocol, *see* ISAKMP
IP in IP encapsulation   100
IPsec
   AH, *see* Authentication header
   authentication header   74
   domain of interpretation   80
   encapsulating security payload   74
   ESP, *see* Encapsulating security payload
   inbound processing   79
   integrity checksum value   74
   outbound processing   78
   SA   10, 54, 57, 77, 78, 85, 177, 178
   security association   10, 54, 57, 77, 78, 85,
      177, 178
   transport mode   76
   tunnel mode   76
IRTF, *see* Internet Research Task Force
ISAKMP
   ISAKMP aggressive exchange   83
   ISAKMP exchange type   84

ISAKMP header   84
ISAKMP identity protect exchange   83
ISAKMP payload   80, 83, 84, 86, 228
ISAKMP phases   80, 81
ISAKMP SA   80, 84, 87
ISAKMP security association payload
   80, 81, 86

KDC, *see* Key distribution center
Kerberized internet negotiation of keys   66
Kerberos
   Kerberos authentication server   63
   Kerberos authentication service request   63
Key agreement   52, 58, 61
Key de-registration   48
Key distribution
   key distribution center   63
   manual key distribution   53
Key distribution center   63
Key encryption key   48, 60
Key establishment scheme   51
Key labeling   48, 52
Key management policy   51
Key refresh   57, 61, 82
Key registration   48
Key revocation   48
Key transport   51, 56
Key type   49, 52
Key wrapping   49, 53
Keying material   48, 51, 87
KINK   66
KRB_AS_REP   64
KRB_AS_REQ   63
KRB_TGS_REP   65
KRB_TGS_REQ   64

LAAA, *see* Local AAA server
Layer-2
   -device   114
   -function   114
   -handover   115
   -trigger   114 *passim*
LCP, *see* Link control protocol
LEAP   136, 144, 241, 260
Liberty alliance
   discovery service   277
   identity provider   276, 278, 279
   identity service   275, 276
   permission-based attribute sharing   278
Liberty-enabled sites   281
Liberty Identity-Federation Framework   275

Liberty Identity Services Interfaces
  Specifications 276
Liberty Identity Web Services Framework 275, 277, 281
Link control protocol 26
Linked identities 276
Local AAA server 12, 177, 189
Local proxy 14
Lock-step protocol 239

MAC, *see* Message authentication code
Man-in-the-middle attack 143, 271
Manual key transport 49
Master key 49, 63, 87, 95
Maximum transmission unit 239
M-bit 156
MS-CHAP 144, 249, 260
MD5 34, 74, 86, 164, 185, 198
Media access control
  MAC address 58, 222
Media independence 58
Message authentication code 34, 95, 258
Message Authenticator, *see* RADIUS
Message authenticator attribute 132, 139, 171
MITM, *see* Man-in-the-middle attack
MN, *see* Mobile node
Mobike 91
Mobile Foreign challenge extension 109, 184, 193
Mobile IP
  agent advertisement 102, 109
  agent advertisement challenge extension 109, 184
  agent discovery 102
  authentication extensions 107, 108, 179, 184, 199
  binding update 114
  bootstrapping 110, 113
  de-tunneling 100
  dynamic home address assignment 111
  dynamic home agent assignment 111
  key generation nonce reply extension 186, 195
  key generation nonce request extension 186, 193
  local home agent assignment 111
  redirected HA extension 113
  registration life time 100, 103, 105, 107
  registration reply 103, 107, 113, 183
  registration request 103, 107, 112, 167, 176, 181 *passim*
  requested ha extension 112

reverse tunneling 103, 106
security 101, 103, 106 *passim*
tunneling 103
Mobile IP-AAA signaling 176, 177, 179, 182, 186, 188, 196
Mobile IPv4 101, 105, 108, 111, 152, 167, 176, 184, 188
Mobile IPv6 102, 109
Mobile node 52, 99, 111, 167, 176, 181, 188, 190
Mobility binding 100
Mobility security association 179
MSA, *see* Mobility security association
MTU, *see* Maximum transmission unit

NAI, *see* Network access identifier
NAS, *see* Network access server
NASREQ 151, 160, 163, 164
NAT, *see* Network address translator
National Institute of Standards and Technology 47
NCP, *see* Network control protocol
Nemo, *see* Network mobility
Network access identifier 14, 111, 223, 271
Network access server 6, 21, 127, 152, 163
Network address translator 91
Network control protocol 26
Network discovery 118, 262
Network interface card 260, 261
Network mobility 52, 110, 125
Network selection 118
Network time protocol 64
NIC, *see* Network interface card
NIST, *see* National Institute of Standards and Technology
Nonce 49, 52, 61, 81, 82, 89–90, 93–4, 108, 179, 181–3, 186, 190, 193–5, 197, 220
  nonce reply extension 183, 186, 187, 195
  nonce request extension 183, 186, 193, 195, 197

Oakley 80, 85, 90
One-time password 37, 38, 40, 42, 165, 241, 242, 262
Online certificate status protocol 207, 210, 222
Open source implementation 97, 145, 172
OPENSSL 97
OSCP, *see* Online certificate status protocol
OTP, *see* One-time password

Packet cable 68
Packet Data Service Node 196

Pairwise key 48, 49, 50
PANA, *see* Protocol for carrying authentication for network access
PAP, *see* Password authentication protocol
Password
   one-time password 37, 38, 40, 42, 165, 241, 242, 262
   password authentication protocol 26, 135, 235
   password file 37, 38
   password sniffing 38, 249
Password authentication protocol 26, 135, 235
PDP, *see* Policy decision point
PDSN, *see* Packet Data Service Node
PEAP 145, 241, 242, 262, 284
Peer 43, 152, 158, 170
Peer-to-peer 52, 54, 66, 79, 81, 91, 163, 171, 243, 244, 245, 269, 281
Peer-to-peer security 161
Perfect forward secrecy 49, 62
PFS, *see* Perfect forward secrecy
PKCS#10 213, 214–15, 216, 218, 219
PKCS#7 213, 214, 216, 218, 230
PKI
   PKI management functions 178, 210, 219, 249
   PKI management protocols 210, 212, 213, 214, 219, 221, 222
   registration 206
PKI4IPsec 91, 228, 232
PKINIT 66
Point of presence 21, 30, 71, 114, 259
Point-to-point protocol 8, 26, 135, 244
Policy 12
   policy decision point 12
   policy framework 12, 13, 23
   policy repository 12, 20, 21
   policy server 12, 13
Policy decision point 12
Polling model 15–16
POP, *see* Point of presence
PPP, *see* Point-to-point protocol
   PPP extension working group 30, 45, 262
   PPP frame 26 *passim*, 237
Pre-paid (Diameter) 8, 9, 19, 151
Private key 32–3, 40–1, 50, 52, 57, 59, 60, 67, 68
   password based private key 32–3, 40–1
Proof of possession 206, 211–12, 219
Protocol for carrying authentication for network access 45
Proxy
   diameter proxy 149, 150, 168
   RADIUS proxy 134, 142

Proxy chaining 134, 142, 143, 170
Public key 32, 52, 67
Public key algorithms 50
Public key certificate 47, 50, 60, 61, 66, 68, 80, 203, 204
Public key cryptography for initial authentication in kerberos 66
Public key infrastructure, *see* PKI
Public seed 49
Pull sequence 12
Push sequence 12

Quality of service, QoS 8, 19, 64, 91, 271

RADIUS 7, 15, 19, 22, 127 *passim*, 176 *passim*, 237, 239, 249 *passim*
   access accept 128, 132, 136, 137, 183, 198–200
   access challenge 127, 129, 130, 131, 137
   access reject 128, 130, 132, 136, 137
   access request 127 *passim*, 198–200
   accounting 139
   accounting request 128, 132, 139, 141
   accounting response 128, 139, 141
   attributes 129, 130, 132, 134, 136, 143, 144
   client 127 *passim*
   EAP-Message attribute 139
   message-authenticator attribute 132, 139, 171
   request authenticator 131, 132, 133, 136, 137, 141
   response authenticator 131, 132, 141
   vendor-specific attributes 130, 144, 176, 197
   VSA, *see* Vendor-specific attributes
RADIUS++ 147, 148
Re-auth-answer (Diameter) 155, 164
Re-auth-request (Diameter) 155, 164
Re-keying 49, 82, 123
Reassembly 239
Registration reply 103 *passim*, 183, 186, 190, 193, 195, 200
Registration request 103, 104, 105, 107–9, 111–13, 167, 176 *passim*
Relay agent 142, 153, 156
Replay attack 36, 37, 38, 49, 60, 81, 108, 131
Resource management
   layer-2 resource management 114
Result-Code AVP (Diameter NAS) 155, 157, 160, 164 *passim*, 181, 190
Retransmission 43, 61, 123, 131, 141, 169, 239

Retransmission behavior 18
Reverse address translation 115, 116, 119
Reverse tunneling 103, 106
  direct delivery style 106
  encapsulated delivery style 106
Roaming agreements 15, 118, 142, 269
Roaming operations 141, 158, 284
Roaming relationship path 142, 143
ROAMOPS 141, 142
Robust secure network 58
RSN, *see* Robust secure network

SA, *see* Security association
SADB, *see* Security association database
Salt, salting 37, 38, 49, 134
SCEP, *see* Simple certificate enrollment protocol
SCTP 18, 123, 124, 148, 150, 157, 158, 169, 170, 239, 245
SDO, *see* Standard Development Organization
Seamless mobility 110, 116, 117, 118, 121
Seamoby 116, 118, 121
Secret seed 49
Secure association protocol 58
Secure hash standard 34–6, 75
Secure shell 39
Secure socket layer 39, 92, 213, 215
SecureID card 37, 38
Security association 10, 54, 57, 77, 78, 80, 81, 85, 117, 123, 161, 175, 178, 179, 180, 197
  AAA security association 54, 57, 161, 175 *passim*, 180, 251, 254
  mobility security association 179, 180
  pre-established security associations 251
  security association database 78
Security association database 78
Security gateway 72, 73, 76, 77
Security parameter index 77, 108, 179, 180
Security policy database 78, 110
Server-initiated messages 156, 169, 171
SHA1, *see* Secure hash standard
SIM, *see* Subscriber identity module
  SIM-based authentication 30, 31, 224, 257, 258
Simple certificate enrollment protocol 213, 221
Single sign-on 62, 63, 275, 276, 279, 281, 282, 283
Single sign out 281
SKEME 80, 90
Sniffing attack 37, 38
SNMP 114, 147, 148
SOAP 281
SPD, *see* Security policy database
  SPD selector 78, 79

SPI, *see* Security parameter index
Spoofing attack 60, 74
SSH, *see* Secure shell
SSL, *see* Secure socket layer
Standard Development Organization 196
Stream Control Transport Protocol, *see* SCTP
Subscriber identity module 30, 257
Supplicant 6, 8, 26, 43, 128, 237, 244, 260
Symmetric key algorithms 50, 75

Tamper-evident 32, 40, 225
Tamper-proof 205, 214, 225
Target access router 120
TCP (Diameter) 148, 150, 157, 158, 169, 170, 239
Third Generation Partnership Project 2 196
Threat model
  threat model analysis 44
Ticket granting server 63
Ticket granting ticket 63
TLS
  alert protocol 95
  certificate verify 93
  client certificate 94
  client hello 93
  client key exchange 93
  finished 93
  master secret 94
  pre-master secret 94
  server hello 93
  TLS handshake protocol 92, 95
  TLS record protocol 92, 95
Token 33, 36, 41, 125
Transient EAP key 56
Transient session key 57
Translation agent 153, 163, 171
Transport layer protocol 22
Transport layer security 44, 60, 73, 91, 96, 149
Trend analysis 13, 18
Triggers
  layer-2 trigger 114
Trust model 177
Trust relationship 10, 40
TTLS
  TTLS challenge 256
  TTLS server 250, 253, 254, 256
Tunneled TLS, *see* TTLS
Tunneling 100, 103 *passim*

Vendor-specific attributes 130, 197, 198
VPN
  VPN gateway 72, 91, 210, 231, 251, 266

W3C, *see* World Wide Web consortium
Web redirect   281, 286
WEP, *see* Wired equivalent privacy
Wi-Fi alliance   54
Windows NT key   261
Wired equivalent privacy   53, 260
Wireless Local Area Networks (WLAN)   29, 54
Wireless transport layer security   55
WLAN 802.11   119, 175
World Wide Web consortium   279, 281
WTLS   96

X.509 certificate   22, 197, 205, 207, 217, 228, 230, 279